Multisensor Fusion Estimation Theory and Application

Liping Yan · Lu Jiang · Yuanqing Xia

Multisensor Fusion Estimation Theory and Application

Springer

Liping Yan
School of Automation
Beijing Institute of Technology
Beijing, China

Lu Jiang
School of Artificial Intelligence
Beijing Technology and Business University
Beijing, China

Yuanqing Xia
School of Automation
Beijing Institute of Technology
Beijing, China

ISBN 978-981-15-9425-0 ISBN 978-981-15-9426-7 (eBook)
https://doi.org/10.1007/978-981-15-9426-7

This Springer imprint is published by the registered company Springer Nature Singapore Pte Ltd.
The registered company address is: 152 Beach Road, #21-01/04 Gateway East, Singapore 189721, Singapore

I keep the subject constantly before me and wait till the first dawnings open little by little into the full light.

—Sir Isaac Newton

This book is dedicated to our families,
for their endless love and support.

Preface

Multisensor data fusion refers to the combination of data from multiple sensors, either of the same or different types, to achieve more specific inferences than could be achieved by using a single independent sensor. It is a rapidly developing science and technology since 1970s. With the comprehensive and rapid development of science and technology, from the initial war as the background, it has developed into various aspects of civil applications, including industrial process monitoring, industrial robots, remote sensing, drug inspection, patient care systems, financial systems, and networked control systems. The theory of multisensor optimal estimation is an important part in the field of multisensor data fusion. In this field, a large number of valuable academic papers and monographs have been published at home and abroad.

In recent years, with the rapid development of computer and network technology, traditional data fusion methods face many new problems and challenges. The random packet loss, data asynchronism, multirate, non-Gaussian noise and other problems caused by various interferences and network transmission stimulate the research in this area. In the aspect of optimal estimation theory, most of the existing monographs and textbooks were based on Kalman filtering, mainly involving the fusion estimation of a single sensor or the fusion of multiple sensors with the same sampling rate of Gaussian noise systems. In the field of data fusion optimal estimation for asynchronous, multirate, non-Gaussian noise systems, there are few monographs involved. The authors have been engaged in the research on the theory and application of multisensor fusion estimation for a long time. In recent years, our research achievements have been recognized by experts of the same trade at home and abroad in the relevant fields, especially in data fusion of asynchronous and multirate sensors, fusion estimation of observation data under random and unreliable conditions, design of event-triggered filtering and fusion estimation algorithms, filtering and fusion estimation of non-Gaussian but heavy-tailed noise systems, etc. The authors sincerely felt that it is necessary to write a monograph that focuses on the theory and application of multisensor fusion estimation covering the above contents, that's why we have this book. This book is featured with a number of attractive and original research studies, we believe that the publication of this book will systematically bring new knowledge created by the authors to the multisensor data fusion, state estimation, target tracking, and integrated navigation, etc.

This book focuses on the basic theory and methods of multisensor data fusion state estimation and its application. The book consists of four parts with 12 chapters. In Part I, the basic framework and methods of multisensor optimal estimation (Chap. 1) and the basic concepts of Kalman filtering (Chap. 2), are briefly and systematically introduced. In Part II, the data fusion state estimation algorithms under networked environment are introduced, which includes five chapters, which are: optimal Kalman filtering fusion for linear dynamic systems with cross-correlated sensor noises (Chap. 3), distributed data fusion for multirate sensor networks (Chap. 4), optimal estimation for multirate systems with unreliable measurements and correlated noise (Chap. 5), multisensor distributed fusion estimation for systems with network delays, uncertainties and correlated noises (Chap. 6), and fusion estimation for asynchronous multirate multisensor systems with unreliable measurements and coupled noises (Chap. 7). Part III consists of three chapters, in which the fusion estimation algorithms under event-triggered mechanisms are introduced, including Chap. 8: event-triggered centralized fusion estimation for dynamic systems with correlated noises, Chap. 9: event-triggered distributed fusion estimation for WSN systems, and Chap. 10: event-triggered sequential fusion estimation for dynamic systems with correlated noises. Part IV consists of two chapters, in which fusion estimation for systems with heavy-tailed noises are introduced, including Chap. 11: distributed fusion estimation for multisensory systems with heavy-tailed noises and Chap. 12: sequential fusion estimation for multisensory systems with heavy-tailed noises.

The book is primarily intended for researchers and engineers in the field of data fusion and state estimation. It can also serve as complementary reading for both graduate and undergraduate students who are interested in target tracking, navigation, networked control, etc. The book will help to consolidate knowledge and extend skills and experiences of related researchers. This book will also bring fresh new ideas into education and will benefit students by exposing them to the very forefront of data fusion research; preparing them for possible future academic or industrial careers in target tracking and control applications.

Due to the limited level of the authors, it is inevitable that there are some shortcomings and mistakes in the book. We sincerely welcome the readers for their criticism and valuable comments.

Beijing, China
August 2020

Liping Yan
Lu Jiang
Yuanqing Xia

Acknowledgements

The authors would like to acknowledge support from National Natural Science Foundation of China (NSFC) under Grant 61773063, Beijing Natural Science Foundation of China under Grant 4202071, Beijing University Young Talents Program (YETP1212) and China scholarship Council (CSC). The authors would also like to thank all the co-operators of some chapters of the book, they are X. Rong Li from the University of New Orleans, USA, Q. M. Jonathan Wu from University of Windsor, Canada, and Mengyin Fu from Beijing Institute of Technology. The work of this book has also been helped by some students. We would like to thank Xiaodi Shi and Zirui Xing for the preparation of Chaps. 2 and 6, respectively, and Chenying Di for simulation and proofreading of Chaps. 11 and 12. We would like to thank students Hui Li, Yuqin Zhou, Yifan Chen, and Jiamin Li for the proofreading of this book. The responsible editor of this book has also paid hard work for the high-quality publication of this book, and hereby thanks.

The authors would also like to thank the generous and egoless loves from our families. Without their continuous support and forgiveness, this book would not have appeared in its current form. During the days of research, their encouragement, enquiries and enduring to us always accompanied with us, which make ourselves not feel lonely in those hard times.

Contents

Part I
Introduction to Optimal Fusion Estimation and Kalman Filtering: Preliminaries

Chapter 1
Introduction to Optimal Fusion Estimation

With the improvement of information technology, the performance of sensors has been greatly improved, and a large number of multi-sensor systems for various applications have emerged. The diversification and complexity of modern war threat also put forward higher requirements for traditional data or information processing systems. In addition, the diversity of information forms, the huge amount of information, the complexity of information relations, and the timeliness of information processing all require to put forward a new theory and technology that can effectively fuse multi-source information. In order to deal with this situation, information fusion came into being [3].

Multisensor data fusion is a rapidly developing science and technology since 1970s. It combines data from multiple sensors to achieve more specific inferences than could be achieved by using a single, independent sensor. It has drawn significant attention over the past decades for its widely applications in both military and nonmilitary fields. Multisensor data fusion is a new interdisciplinary technology, which involves signal processing, probability statistics, information theory, pattern recognition, artificial intelligence, fuzzy mathematics and other fields.

In recent years, with the development of sensor technology, signal detection and processing, and computer application technology, information fusion technology is more widely used. In addition to the application on various weapon platforms, it has also been widely used in many civil fields, such as industrial process monitoring, industrial robots, remote sensing, drug inspection, patient care system, financial system, ship collision avoidance and air traffic control systems. All the developed countries in the world have listed it as one of the priority technologies. In fact, when the measured (or recognized) target is interfered by multiple attributes or uncertain factors, it is necessary to use multi-sensor coordination to complete the common detection task. Therefore, the research of multi-sensor information fusion has a wide range of theoretical significance and application value.

L. Yan et al., *Multisensor Fusion Estimation Theory and Application*,
https://doi.org/10.1007/978-981-15-9426-7_1

1.1 Definition of Multisensor Data Fusion

Data fusion is defined in [2] as a multilevel, multifaceted process dealing with the detection, association, correlation, estimation, and combination of data and information from multiple sources to achieve refined state and identity estimation, and complete and timely assessments of situation and threat. The basic objective of data fusion is to derive more information, through combining, than is present in any individual element of input data.

The concept of multisensor data fusion is hardly new. As humans and animals evolved, they developed the ability to use multiple senses to help them survive. For example, as illustrated in Fig. 1.1, assessing the quality of an edible substance may not be possible using only the sense of vision. The combination of sight, touch, smell, and taste is far more effective. Similarly, when vision is limited by structures and vegetation, the sense of hearing can provide advanced warning of impending dangers. Thus, multisensory data fusion is naturally performed by animals and humans to assess more accurately the surrounding environment and to identify threats, thereby improving their chances of survival.

Although the concept of data fusion is not new, the emergence of new sensors, advanced processing techniques, improved processing hardware, and wideband communications have made real-time fusion of data increasingly viable.

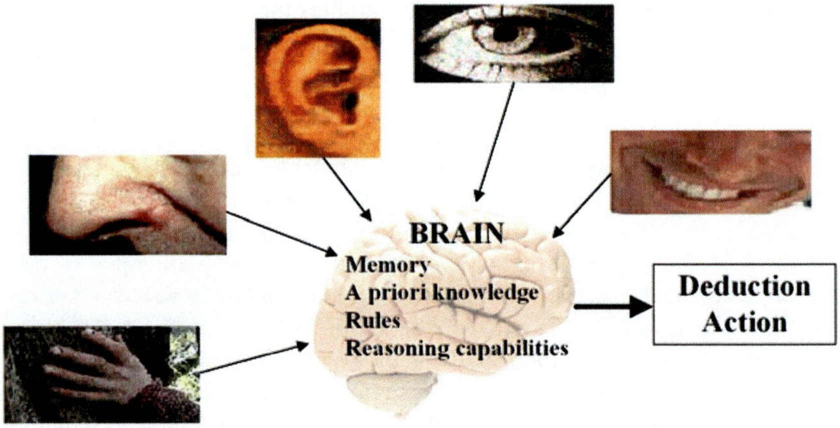

Fig. 1.1 The fusion process of the human brain

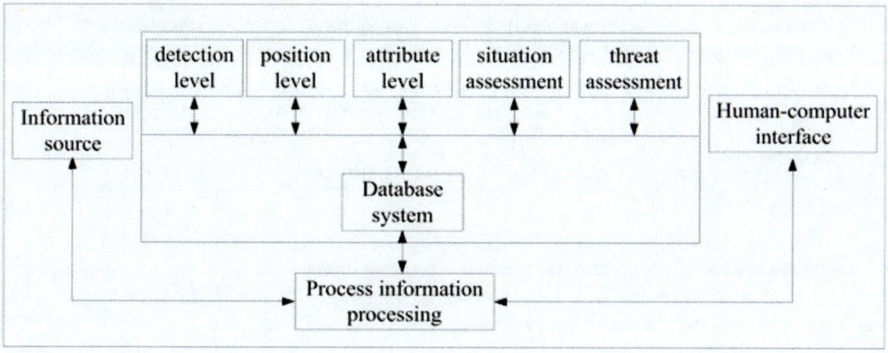

Fig. 1.2 Functional block diagram of information fusion systems

1.2 The Principle and Architecture of Multi-sensor Data Fusion

Multi–sensor information fusion is a basic phenomenon that commonly exists in human beings and other biological systems. It is actually a functional simulation of the complex processing of the human brain. According to the functional level of information abstraction, information fusion can be divided into five levels: detection level fusion, position level fusion (state estimation), attribute (target recognition) level fusion, situation assessment and threat estimation [4, 5]. The system flow diagram of each functional module of information fusion is shown in Fig. 1.2.

1.2.1 Detection Level Fusion

Detection level fusion is the fusion that performed directly on the detection decision and signal layer in a multi-sensor distributed detection system. It was initially used in military command, control and communication, and now its application has been extended to many fields such as weather forecasting, medical diagnosis and organizational management decision-making. There are four main structural models of detection level fusion, namely, parallel structure, decentralized structure, serial structure and tree structure, as shown in Fig. 1.3.

1.2.2 Position Level Fusion

Position level fusion is the fusion performed directly on the observation data of the sensors or the state estimation of the sensors, including the fusion in time and space.

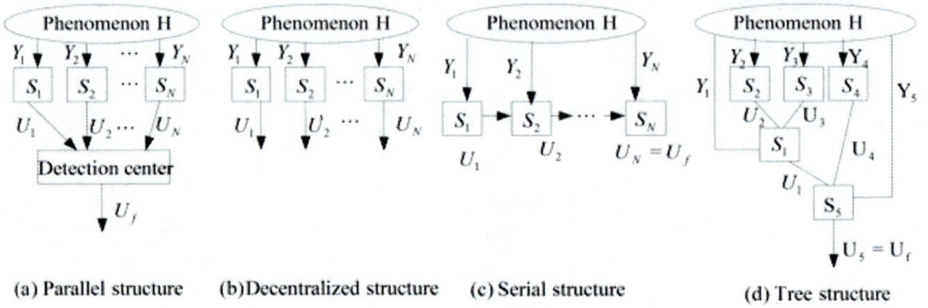

Fig. 1.3 Framework of detection level fusion

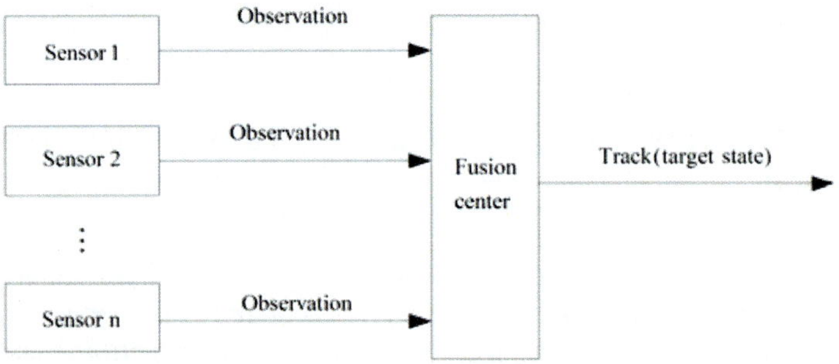

Fig. 1.4 Centralized fusion structure

It is the fusion of the tracking level, which belongs to the intermediate level. It is one of the most important fusion level. In recent years, there have been a lot of literatures focus on this level of fusion at home and abroad. Our research on the state fusion estimation algorithm is mainly carried out on this basis too [1].

For a single-sensor tracking system, it is mainly based on the sequence of the observations of the target, which is named as the detection report, to perform fusion. For example, trace with scan radar systems, infrared and sonar sensors and other multi-target tracking or estimation technologies all belong to this type of fusion. In the multi-sensor tracking system, there are several fusion structures such as centralized, distributed, compound and multi-level fusion structures.

For the centralized multi-sensor tracking system, firstly the measurements perform time fusion according to the time sequence of observations of the target. Then, multiple sensors perform spatial fusion on the observation of the same target at the same time. It includes the whole process of multi-sensor fusion tracking and state estimation. Some common systems of this type include multi-radar integrated tracking and multi-sensor marine monitoring and tracking systems. The schematic diagram of the centralized fusion structure is shown in Fig. 1.4.

In the distributed multi-sensor tracking systems, each sensor uses its own measurements to track the target separately, and send its estimation result to the fusion center (master station), and the fusion center then synthesizes the estimates of each

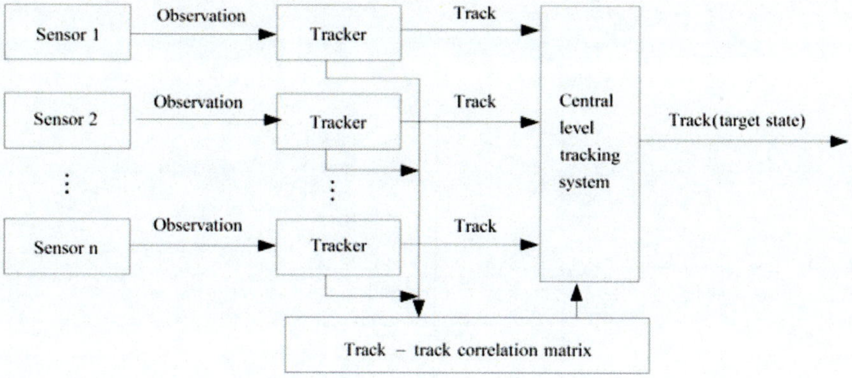

Fig. 1.5 Distributed fusion structure

sub-station into a joint target estimation. In general, the accuracy of distributed estimation is not as high as the centralized one, but because of its low demand for communication bandwidth, fast calculation speed, and good reliability and continuity, it has become a hot research topic in recent years. The schematic diagram of distributed fusion structure is shown in the Fig. 1.5.

Distributed systems can generally be divided into three types of fusion structures, such as distributed systems with feedback, distributed systems without feedback, and fully distributed systems [7].

(1) The hierarchical structure of distributed systems without feedback: each sensor node transmits its respective local estimation results to the central node to form a global estimation, which is the most commonly used distribution estimation structure.
(2) The hierarchical structure of distributed systems with feedback: the main difference between it and (1) is that the communication structure is different. That is, the global estimation of the central node can be fed back to each local node. This structure has the advantage of fault tolerance. When it is detected that the estimated result of a local node is poor, it is not necessary to exclude it from the system, but to use a better global result to modify the state of the local node, which not only improves the local node information, but also continues to use the information from the node.
(3) Fully distributed structure: in this generalized system structure, all nodes are connected by forms of network or chain to communicate. A node can enjoy the information from the nodes that connected to it. This also means that each local node can enjoy part of the global information to varying degrees, so that it is possible to obtain better estimates on many nodes. In extreme cases, each node can be used as a central node to obtain a global optimal estimation.

Hybrid position-level fusion is a hybrid structure combining centralized and distributed multi-sensor systems. The detection report of the sensor and the track infor-

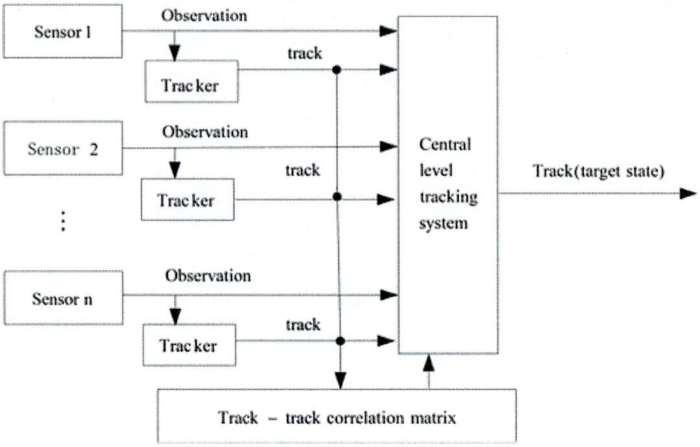

Fig. 1.6 Compound fusion structure

mation of the target state estimation are sent to the fusion center, where both time fusion and space fusion are performed. Because this kind of structure needs to give the detection report and the track estimation at the same time, and optimize the combination, it needs complicated processing logic. The hybrid structure can also be selected and transformed in a centralized and distributed structure according to the requirements of the problem. The communication and calculation of this structure are larger than other structures, because the sensors should send the detection report and track estimation information at the same time, and the communication link must be bidirectional. In addition, in the fusion center, besides processing the track information from local nodes, it is also necessary to give the detection report sent by the sensors, which doubles the calculation amount. However, it can meet the needs of many applications. Cruise missile control and active and passive radar combined guidance systems are typical hybrid structures. The schematic diagram of the hybrid fusion structure is shown in Fig. 1.6.

1.2.3 Attribute Level Fusion/Target Recognition Level Fusion

Target recognition is also called attribute classification or identity estimation (identification). According to the level of information abstraction, target recognition (identification) can be divided into three layers: decision-making layer, feature layer and data layer fusion. The three-layer fusion structure diagram and flow chart of identity recognition are shown in Figs. 1.7 and 1.8, respectively.

In the decision-level fusion method, each sensor is transformed to obtain an independent identity estimation, and then perform fusion on the classification of attributes from each sensor. In the feature-level fusion method, each sensor observes a target

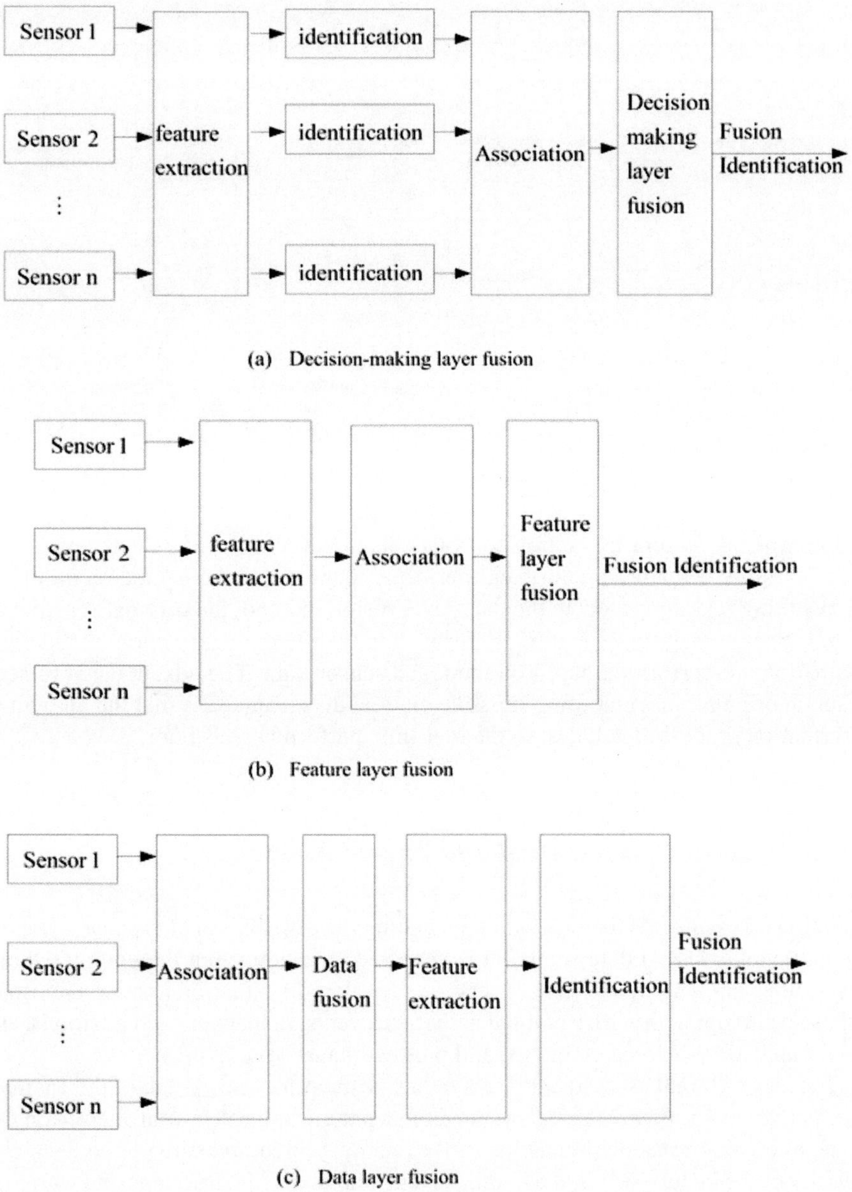

Fig. 1.7 Hierarchical structure of multi-sensor target recognition

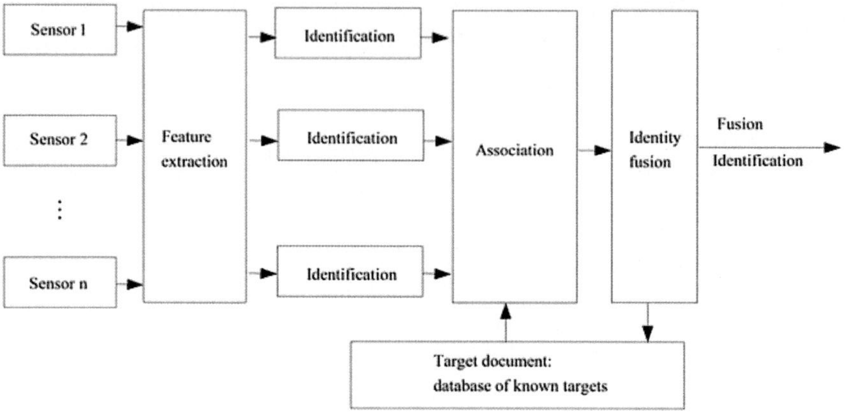

Fig. 1.8 Flow chart of identification

and completes feature extraction to obtain feature vectors from each sensor, and then fuses these feature vectors and performs identity determination according to the resulting feature vector. In the data-level fusion method, the original sensor data from the same magnitude is directly fused, and then feature extraction and identity estimation are performed based on the fused sensor data. The advantage is to keep as much original information as possible, but the disadvantage is that the amount of information processed is large, so the real-time performance is poor.

1.2.4 Situation Assessment and Threat Assessment

Situation assessment is the process of evaluating the distribution of combat power on the battlefield. Threat determination is achieved by quantifying the enemy's threat capabilities and enemy attempts. It can be seen that the situation assessment establishes an attempt to organize combat activities, events, maneuvers and positions, and the organization of force elements, and thus estimates what happened.

The task of threat assessment is to synthesize the prior knowledge of the enemy's destructive ability, maneuverability, movement pattern and behavioral attempt on this basis, to obtain the tactical meaning of the enemy's force, to estimate the degree or severity of the occurrence of the combat event, and to give instructions and warnings of combat intention. Its focus is to quantify the enemy's combat capabilities and estimate the enemy's attempts.

1.3 Advantages and Disadvantages for Multisensor Data Fusion

The ultimate goal of the multi-sensor information fusion system is to give an accurate assessment of the situation of the observed object in order to take appropriate countermeasures. Compared with single-sensor systems, multi-sensor systems have great advantages in quantifiable state estimation performance.

First, if several identical sensors are used (e.g., identical radars tracking a moving object), combining the observations would result in an improved estimation of the target position and velocity. A statistical advantage is gained by adding the N independent observations.

The second advantage gained using multiple sensors is the improved observability [6, 8, 9]. As illustrated in Fig. 1.9 is an example of image fusion. It can be clearly observed that in image (c), the two clocks can both be clearly seen by fusion of image (a) that focused on left and image (b) that focused on the right. Similarly, the edges and the content of the human brain can both be clearly observed in the fused image (f), which is the fusion of images MRI and CT.

The third advantage is that by using the relative placement or motion of multiple sensors, the observation process can be improved. For example, two sensors that measure angular directions to an object can be coordinated to determine the position of the object by triangulation. This technique is used in surveying and for commercial navigation. Similarly, sensors, one moving in a known way with respect to another, can be used to measure instantaneously an object's position and velocity with respect to the observing sensors. As illustrated in Fig. 1.10 is an example of target tracking,

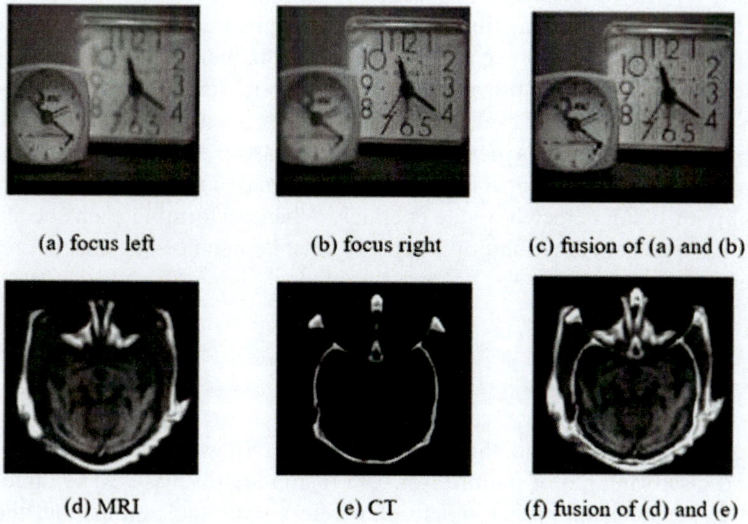

Fig. 1.9 Illustration of image fusion

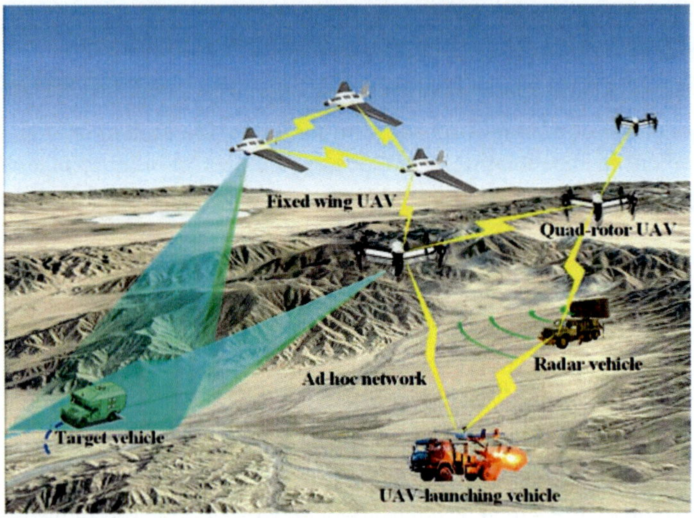

Fig. 1.10 Illustration of target tracking by multisensor systems

where there are several unmanned aircraft vehicles (UAVs) and ground vehicles connected by Ad hoc network, with the intention of tracking a ground vehicle. By fusion of the sensors' information loaded in the UAVs and the radar vehicle, the success rate of target tracking will be greatly improved.

Briefly, data fusion provides an efficient approach for the exploration of unknown environments. It has the following advantages: such as, it can improve the system reliability and stability, expand space coverage, expand time coverage, increase credibility, and shorten the responding time, reduce the ambiguity of information, improve the detection performance, improve the spatial resolution, increase the dimension of the measurement space, and improve the scale. However, compared with single-sensor systems, the complexity of multi-sensor systems is greatly increased, resulting in some disadvantages, such as increasing costs, increasing device size, weight, power consumption and other physical factors, and the increased radiation may increase the probability of being detected by the enemies. When performing each specific task, the performance benefits of multiple sensors must be balanced against the resulting disadvantages.

1.4 Conclusion

In this chapter, the definition, the architecture and framework, as well as the advantages and disadvantages of multisensor data fusion are discussed. The rapid development of sensors, the global spread of wireless communications, and the rapid improvements in computer processing and data storage enable new applications and methods for multisensor data fusion to be developed.

References

1. Bar-Shalom, Yaakov, X. Rong Li, and Thiagalingam Kirubarajan. 2001. *Estimation with Application to Tracking and Navigation*. Newyork: Wiley.
2. Edward, W., and L. James. 1990. *Multisensor Data Fusion*. Boston: Artech House.
3. He, You, Guohong Wang, Dajin Lu, and Yingning Peng. 2000. *Multisensor Information Fusion and Its Application*. Beijing: Electronic Industry Press.
4. Kang, Yaohong. 1997. *Data Fusion Theory and Application*. Xi'an: Xidian University Press.
5. Liggins, Martin E, David L Hall, and James Llinas. 2008. *Handbook of Multisensor Data Fusion: Theory and Practice*. CRC Press.
6. Piella, G. 2009. Image fusion for enhanced visualization: a variational approach. *International Journal of Computer Vision* 83 (1): 1–11.
7. Wen, Chenglin, and Donghua Zhou. 2002. *Multiscale Estimation Theory and Application*. Beijing: Tsinghua University Press.
8. Zhou, Zhiqiang, Sun Li, and Bo Wang. 2014. Multi-scale weighted gradient-based fusion for multi-focus images. *Information Fusion* 20: 60–72.
9. Zhou, Zhiqiang, Bo Wang, Sun Li, and Mingjie Dong. 2016. Perceptual fusion of infrared and visible images through a hybrid multi-scale decomposition with gaussian and bilateral filters. *Information Fusion* 30: 15–26.

Chapter 2
Kalman Filtering of Discrete Dynamic Systems

R.E. Kalman firstly presents the Kalman filter (KF), which treats the target state as a random variable based on Bayesian formulation [5]. By using the dynamic model and the measurement model, KF describes the motion of a target and how the measurement depends on the target state under the following assumptions:

(1) Both the dynamic model and the measurement model are linear functions of the state with additive Gaussian noises.
(2) The prior distribution of the target state follows the Gaussian distribution.

For linear systems with Gaussian noise, KF is an optimal filtering algorithm based on the minimum mean square error (MMSE) estimation criterion [1, 2, 4]. Many articles have described the theoretical derivation of KF [1–4, 6, 7]. The KF processes measurements sequentially to calculate the conditional mean and conditional covariance and its calculated covariance also represents the posterior Cramér-Rao lower bound (PCRLB) [8–10]. Therefore, given the assumptions, the KF is absolutely optimal.

In this chapter, considering a discrete linear dynamic system, the KF can be derived according to the following steps.

(1) Describe the dynamic systems that require state estimation, and use mathematical equations to represent the transfer of state mean and covariance.
(2) These mathematical equations form the basis for the derivation of the Kalman filter, where the mean of the state is the Kalman filter estimate of the state and the covariance of the state is the covariance of the Kalman filter state estimate.
(3) When a new measurement is obtained, we update the mean and covariance of the state. That is, we use measurements to recursively update the state estimation.

L. Yan et al., *Multisensor Fusion Estimation Theory and Application*,
https://doi.org/10.1007/978-981-15-9426-7_2

2.1 Overview of the Discrete-Time Kalman Filter

Let x_k denote the n-dimensional state of the target at time k. The discrete-time dynamic model of the target is described by

$$x_k = F_{k-1}x_{k-1} + G_{k-1}u_{k-1} + w_{k-1}, \quad k = 1, 2, \ldots \tag{2.1}$$

where $F_{k-1}, G_{k-1}, u_{k-1}$, and w_{k-1} are the state transition matrix, control gain, control input, and process noise at time $k-1$, respectively. Let z_k denote the m-dimensional measurement of the target by a sensor at time k. The measurement model is described by

$$z_k = H_k x_k + v_k, \quad k = 0, 1, \ldots \tag{2.2}$$

where H_k and v_k are the measurement matrix and measurement noise at time k, respectively.

The noise processes w_k and v_k are white, zero-mean, uncorrelated, and have known covariance matrices Q_k and R_k, respectively:

$$\begin{cases} E\{w_k\} = 0 \\ E\{v_k\} = 0 \\ E\{w_k w_j^{\mathrm{T}}\} = Q_k \delta_{kj} \\ E\{v_k v_j^{\mathrm{T}}\} = R_k \delta_{kj} \\ E\{v_k w_j^{\mathrm{T}}\} = 0 \end{cases} \tag{2.3}$$

where δ_{kj} is the Kronecker delta function, i.e.,

$$\delta_{kj} = \begin{cases} 1, & \text{if } k = j \\ 0, & \text{otherwise} \end{cases} \tag{2.4}$$

Assume that the initial state x_0 is Gaussian distributed with mean $\bar{x}_0$ and covariance P_0, i.e.,

$$x_0 \sim \mathcal{N}(x_0; \bar{x}_0, P_0) \tag{2.5}$$

Assume that the initial state, the process noise, and the measurement noise are mutually independent, i.e.,

$$\begin{cases} E\{x_0 w_k^{\mathrm{T}}\} = 0 \\ E\{x_0 v_k^{\mathrm{T}}\} = 0 \end{cases} \tag{2.6}$$

To estimate the state by processing noisy measurements sequentially, define Z_k to represent all the measurements up to time k as follows

$$Z_k = \{z_0, z_1, \ldots, z_k\}$$

Then, the conditional mean $\hat{x}_{k|k}$ and conditional covariance $P_{k|k}$ [2] are

$$\hat{x}_{k|k} = E\{x_k|Z_k\} \tag{2.7}$$
$$P_{k|k} = E\{(x_k - \hat{x}_{k|k})(x_k - \hat{x}_{k|k})^{\mathrm{T}}|Z_k\} \tag{2.8}$$

Suppose we have the initial state estimate $\hat{x}_{0|0}$ and associated covariance $P_{0|0}$. Our goal is to obtain the state estimate $\hat{x}_{k|k}$ and associated covariance $P_{k|k}$ at time k by processing the cumulative measurement Z_k. To obtain $\hat{x}_{k|k}$, $P_{k|k}$ from $\hat{x}_{k-1|k-1}$, $P_{k-1|k-1}$ involves a prediction step and measurement update step [1–4, 6, 7]. Next we describe these two steps.

2.1.1 Prediction

The prediction step is also called a time update step. Suppose an optimal linear estimate $\hat{x}_{k-1|k-1}$ of the state x_{k-1} at time $k-1$ is found. That is, $\hat{x}_{k-1|k-1}$ is the estimated value with the smallest mean square error in linear function of Z_{k-1}.

$$\hat{x}_{k-1|k-1} = E\{x_{k-1}|Z_{k-1}\} \tag{2.9}$$
$$P_{k-1|k-1} = E\{(x_{k-1} - \hat{x}_{k-1|k-1})(x_{k-1} - \hat{x}_{k-1|k-1})^{\mathrm{T}}|Z_{k-1}\} \tag{2.10}$$

The predicted state and associated covariance are defined, respectively by [2]

$$\hat{x}_{k|k-1} = E\{x_k|Z_{k-1}\} \tag{2.11}$$
$$P_{k|k-1} = E\{(x_k - \hat{x}_{k|k-1})(x_k - \hat{x}_{k|k-1})^{\mathrm{T}}|Z_{k-1}\} \tag{2.12}$$

Given Z_{k-1}, the predicted measurement $\hat{z}_{k|k-1}$ at time k is defined by [2]

$$\hat{z}_{k|k-1} = E\{z_k|Z_{k-1}\} \tag{2.13}$$

Due to the measurement residual $\tilde{z}_{k|k-1} = z_k - \hat{z}_{k|k-1}$, the measurement prediction covariance at time k is defined by

$$S_k = E\{\tilde{z}_{k|k-1}\tilde{z}_{k|k-1}^{\mathrm{T}}|Z_{k-1}\} \tag{2.14}$$

Therefore, a complete time update step, which uses the optimal estimate $\hat{x}_{k-1|k-1}$ of x_{k-1} to predict the state x_k, is described as follows.

$$\hat{x}_{k|k-1} = F_{k-1}\hat{x}_{k-1|k-1} + G_{k-1}u_{k-1} \tag{2.15}$$
$$P_{k|k-1} = F_{k-1}P_{k-1|k-1}F_{k-1}^{\mathrm{T}} + Q_{k-1} \tag{2.16}$$
$$\hat{z}_{k|k-1} = H_k\hat{x}_{k|k-1} \tag{2.17}$$
$$\tilde{z}_{k|k-1} = H_k\tilde{x}_{k|k-1} + v_k \tag{2.18}$$
$$S_k = H_kP_{k|k-1}H_k^{\mathrm{T}} + R_k \tag{2.19}$$

2.1.2 Update

The conditional mean $\hat{x}_{k|k}$ and associated covariance $P_{k|k}$ at time k are defined by [2]

$$\hat{x}_{k|k} = E\{x_k | Z_k\} \tag{2.20}$$

$$P_{k|k} = E\{(x_k - \hat{x}_{k|k})(x_k - \hat{x}_{k|k})^{\mathrm{T}} | Z_k\} \tag{2.21}$$

The updated state estimate and the updated covariance at time k is given by

$$\hat{x}_{k|k} = \hat{x}_{k|k-1} + K_k(z_k - \hat{z}_{k|k-1}) \tag{2.22}$$

$$P_{k|k} = P_{k|k-1} - K_k S_k K_k^{\mathrm{T}} \tag{2.23}$$

where the Kalman gain is expressed as

$$K_k = P_{k|k-1} H_k^{\mathrm{T}} S_k^{-1} \tag{2.24}$$

2.1.3 Alternate Forms of Updated Covariance and Kalman Gain

An alternate form of $P_{k|k}$ in (2.23) is

$$P_{k|k} = (I - K_k H_k) P_{k|k-1} \tag{2.25}$$

The Joseph form of the updated covariance is

$$P_{k|k} = (I - K_k H_k) P_{k|k-1} (I - K_k H_k)^{\mathrm{T}} + K_k R_k K_k^{\mathrm{T}} \tag{2.26}$$

The inverse of the updated covariance using the information filter [2] method is

$$P_{k|k}^{-1} = P_{k|k-1}^{-1} + H_k^{\mathrm{T}} R_k^{-1} H_k \tag{2.27}$$

Then, an alternate form of the Kalman gain is

$$K_k = P_{k|k} H_k^{\mathrm{T}} R_k^{-1} \tag{2.28}$$

2.2 Properties of the Kalman Filter

In this section, some important properties of the Kalman filter are introduced. Given the linear system of Eqs. (2.1)–(2.3), the error between the true state x_k and the estimated state $\hat{x}_k$ is denoted as $\tilde{x}_k$:

$$\tilde{x}_k = x_k - \hat{x}_k \tag{2.29}$$

Suppose we want to find the estimator that minimizes (at each time step) a weighted two-norm of the expected value of the estimation error $\tilde{x}_k$:

$$\min E\{\tilde{x}_k^{\mathrm{T}} S_k \tilde{x}_k\} \tag{2.30}$$

where S_k is a positive definite matrix. If S_k is diagonal with elements $S_k(l), \ldots, S_k(n)$, then the weighted sum is $S_k(l)E\{\tilde{x}_k^2(l)\} + \cdots + S_k(n)E\{\tilde{x}_k^2(n)\}$.

• If w_k and v_k are Gaussian, white, zero-mean, and uncorrelated, then the Kalman filter is the way to solve the above problem.

• If w_k and v_k are white, zero-mean, uncorrelated but not necessarily Gaussian, then the Kalman filter is the best linear solution to the above problem.

• If w_k and v_k are correlated or colored, then the Kalman filter can be modified to solve the above problem.

• For nonlinear systems, various nonlinear Kalman filters approximate the solution to the above problem.

Recall the measurement update equation from Eq. (2.22):

$$\hat{x}_{k|k} = \hat{x}_{k|k-1} + K_k(z_k - H_k\hat{x}_{k|k-1}) \tag{2.31}$$

The quantity $(z_k - H_k\hat{x}_{k|k-1})$ is called the innovations. This is the part of the measurement that contains new information about the state.

2.3 Alternate Propagation of Covariance

The propagation of the estimation-error covariance P can be used to find the closed equation of a scalar Kalman filter and to find a fast solution to the steady-state estimated error covariance. In this section, its alternate equation is derived.

2.3.1 Multiple State Systems

Recall the update equations by (2.16) and (2.27):

$$P_{k|k-1} = F_{k-1} P_{k-1|k-1} F_{k-1}^{\mathrm{T}} + Q_{k-1} \tag{2.32}$$

$$P_{k|k} = P_{k|k-1} - P_{k|k-1} H_k^{\mathrm{T}} (H_k P_{k|k-1} H_k^{\mathrm{T}} + R_k)^{-1} H_k P_{k|k-1} \tag{2.33}$$

If the $n \times n$ matrix $P_{k|k-1}$ can be decomposed into

$$P_{k|k-1} = A_k B_k^{-1} \tag{2.34}$$

where A_k and B_k are $n \times n$ matrices to be determined, then $P_{k+1|k}$ satisfies

$$P_{k+1|k} = A_{k+1} B_{k+1}^{-1} \tag{2.35}$$

where A_{k+1} and B_{k+1} are propagated as follows:

$$\begin{bmatrix} A_{k+1} \\ B_{k+1} \end{bmatrix} = \begin{bmatrix} F_k + Q_k F_k^{-\mathrm{T}} H_k^{\mathrm{T}} R_k^{-1} H_k & Q_k F_k^{-\mathrm{T}} \\ F_k^{-\mathrm{T}} H_k^{\mathrm{T}} R_k^{-1} H_k & F_k^{-\mathrm{T}} \end{bmatrix} \begin{bmatrix} A_k \\ B_k \end{bmatrix} \tag{2.36}$$

This can be seen by noting from Eq. (2.36) that

$$\begin{aligned} B_{k+1}^{-1} &= [F_k^{-1} H_k^{\mathrm{T}} R_k^{-1} H_k A_k + F_k^{-\mathrm{T}} B_k]^{-1} \\ &= [F_k^{-\mathrm{T}} (H_k^{\mathrm{T}} R_k^{-1} H_k A_k B_k^{-1} + I) B_k]^{-1} \\ &= B_k^{-1} [H_k^{\mathrm{T}} R_k^{-1} H_k A_k B_k^{-1} + I]^{-1} F_k^{\mathrm{T}} \end{aligned} \tag{2.37}$$

From Eq. (2.36),

$$A_{k+1} B_{k+1}^{-1} = [(F_k + Q_k F_k^{-\mathrm{T}} H_k^{\mathrm{T}} R_k^{-1} H_k) A_k + Q_k F_k^{-\mathrm{T}} B_k] B_{k+1}^{-1} \tag{2.38}$$

Substituting the expression for B_{k+1}^{-1} into this equation gives

$$\begin{aligned} A_{k+1} B_{k+1}^{-1} &= [(F_k + Q_k F_k^{-\mathrm{T}} H_k^{\mathrm{T}} R_k^{-1} H_k) A_k + Q_k F_k^{-\mathrm{T}} B_k] \\ &\quad \times B_k^{-1} (H_k^{\mathrm{T}} R_k^{-1} H_k A_k B_k^{-1} + I)^{-1} F_k^{\mathrm{T}} \\ &= [(F_k + Q_k F_k^{-\mathrm{T}} H_k^{\mathrm{T}} R_k^{-1} H_k) A_k B_k^{-1} + Q_k F_k^{-\mathrm{T}}] \\ &\quad \times (H_k^{\mathrm{T}} R_k^{-1} H_k A_k B_k^{-1} + I)^{-1} F_k^{\mathrm{T}} \end{aligned} \tag{2.39}$$

Substituting $P_{k|k-1}$ for $A_k B_k^{-1}$ in the above equation gives

$$\begin{aligned}
A_{k+1}B_{k+1}^{-1} &= [(F_k + Q_k F_k^{-\mathrm{T}} H_k^{\mathrm{T}} R_k^{-1} H_k)P_{k|k-1} + Q_k F_k^{-\mathrm{T}}] \\
&\quad \times [H_k^{\mathrm{T}} R_k^{-1} H_k P_{k|k-1} + I]^{-1} F_k^{\mathrm{T}} \\
&= [F_k P_{k|k-1} + Q_k F_k^{-\mathrm{T}} (H_k^{\mathrm{T}} R_k^{-1} H_k P_{k|k-1} + I)] \\
&\quad \times [H_k^{\mathrm{T}} R_k^{-1} H_k P_{k|k-1} + I]^{-1} F_k^{\mathrm{T}} \\
&= F_k P_{k|k-1} [H_k^{\mathrm{T}} R_k^{-1} H_k P_{k|k-1} + I]^{-1} F_k^{\mathrm{T}} + Q_k F_k^{-\mathrm{T}} F_k^{\mathrm{T}}
\end{aligned} \tag{2.40}$$

Applying the matrix inversion lemma to the term in Eq. (2.40)

$$\begin{aligned}
A_{k+1}B_{k+1}^{-1} &= F_k P_{k|k-1} [I - H_k^{\mathrm{T}} (H_k P_{k|k-1} H_k^{\mathrm{T}} + R_k)^{-1} H_k P_{k|k-1}] F_k^{\mathrm{T}} + Q_k \\
&= F_k [P_{k|k-1} - P_{k|k-1} H_k^{\mathrm{T}} (H_k P_{k|k-1} H_k^{\mathrm{T}} + R_k)^{-1} H_k P_{k|k-1}] F_k^{\mathrm{T}} + Q_k \\
&= F_k P_{k|k} F_k^{\mathrm{T}} + Q_k \\
&= P_{k+1|k}
\end{aligned} \tag{2.41}$$

According to the above proof, we have $A_{k+1}B_{k+1}^{-1} = P_{k+1|k}$.

Suppose that F, Q, H, and R are constant matrices. Using Eq. (2.36) to obtain a fast solution for the steady-state covariance of multidimensional systems (although not a closed-form solution):

$$\begin{bmatrix} A_{k+1} \\ B_{k+1} \end{bmatrix} = \begin{bmatrix} F + QF^{-\mathrm{T}} H^{\mathrm{T}} R^{-1} H & QF^{-\mathrm{T}} \\ F^{-\mathrm{T}} H^{\mathrm{T}} R^{-1} H & F^{-\mathrm{T}} \end{bmatrix} \begin{bmatrix} A_k \\ B_k \end{bmatrix} = \psi \begin{bmatrix} A_k \\ B_k \end{bmatrix} \tag{2.42}$$

We can easily get

$$\begin{bmatrix} A_k \\ B_k \end{bmatrix} = \psi^{k-1} \begin{bmatrix} P_1^- \\ I \end{bmatrix} \tag{2.43}$$

where $A_1 = P_{1|0}$ and $B_1 = I$ satisfy the original factoring of Eq. (2.36). Now we can successively square ψ a total of p times to obtain ψ^2, ψ^4, ψ^8 and so on, until ψ^{2^p} converges to a steady-state value:

$$\begin{bmatrix} A_\infty \\ B_\infty \end{bmatrix} \approx \psi^{2^p} \begin{bmatrix} P_{1|0} \\ I \end{bmatrix} \text{ for large } p. \tag{2.44}$$

$P_\infty^- = A_\infty B_\infty^{-1}$ is the steady-state covariance. We can also find the steady-state Kalman gain by simply iterating the filter equations from Eqs. (2.16), (2.22), (2.23) and (2.24), but the method in this section could be a much quicker way to find the steady-state gain. After finding P_∞^- as shown above, calculating $K_\infty = P_\infty^- H^{\mathrm{T}} (H P_\infty^- H^{\mathrm{T}} + R)^{-I}$ as the steady-state Kalman filter gain.

2.3.2 Divergence Issues

The theory introduced in this chapter makes the Kalman filter the optimal choice for state estimation. However, even if the theory is correct, it may not work very well when the Kalman filter is used to the real systems. The two main reasons for the failure of Kalman filter are finite precision and modeling errors.

In this chapter, it is assumed that the Kalman filter algorithm is infinitely accurate. In digital microprocessors, this arithmetic is of limited precision, which may cause divergence or even instability in the implementation of the Kalman filter.

The proposed theory also assumes that the system model is accurately known. It is assumed that F, Q, H and R are previously known matrices, and the noise sequences w_k and v_k are zero-mean uncorrelated white noise. If any of these assumptions are violated (as in actual implementation), that is, the assumption of the Kalman filter is not followed, the theory may not be entirely feasible.

In order to improve the performance of the filter in practical applications, the following strategies are proposed:

1. Improve arithmetic accuracy;
2. Use some form of square root filter;
3. Symmetrize P at each time step: $P = (P + P^{\mathrm{T}})/2$;
4. Properly initialize P to avoid large changes in P;
5. Use a fading-memory filter;
6. Use virtual process noise (especially for estimating "constants").

The adoption of the above strategies usually depends on the specific problem, and it also needs to be explored through simulation or experiment to obtain good results. A detailed explanation of the above strategies is given below.

Item 1 above, to improve arithmetic accuracy, it is only necessary to make the digital realization of the filter closer to the analog theory. In a PC-based implementation, it takes little effort to change all variables to double precision to improve arithmetic accuracy. Such subtle changes may cause differences between divergence and convergence. However, in microcontroller implementations, improving arithmetic accuracy may not be feasible.

Item 2 above, square root filter is a method to reconstruct the filter equations. Even if the physical accuracy achieved is unchanged, the square root filtering effectively improves the arithmetic accuracy.

Items 3 and 4 above involve forcing P to be symmetrical and initializing P appropriately. These two solutions usually do not significantly improve the convergence of the filter. However, since these steps are simple and prevent numerical problems, these steps should always be performed. Note from Eqs. (2.16), (2.23) and (2.24) that the P expression is already symmetric, so there is no need to force $P_{k|k-1}$. Depending on the equation used, $P_{k|k-1}$ can be symmetric or asymmetric. In Eqs. (2.16), (2.23) and (2.24), $P_{k|k}$ is mathematically equivalent, but not numerically equivalent. One of them has built-in symmetry, while the others do not. Rather than forcing asymmetric equations to be symmetric, it is obviously easier to use equations with inherent symmetry. There are some different ways to solve the problem. One method is as described in item 3 above, that is, after calculating P, set $P = (P + P^{\mathrm{T}})/2$. Other ways include forcing items below the diagonal to be equal to items above the diagonal, or forcing the eigenvalue of P to be positive.

Item 5 above is a simple method to make the filter to "forget" measurements made a long time ago and put more emphasis on the most recent measurements, which causes the filter to respond more sensitively to measurements. In theory, this will cause the Kalman filter to lose its optimality, but may restore convergence and stability. But in fact, a theoretically suboptimal filter will perform better than a theoretically optimal filter that cannot work due to modeling errors. The fading-memory filter is more sensitive to the latest measurements, which makes the filter less sensitive to the modeling errors and thereby become more robust.

Item 6 above, using virtual process noise, is also easy to implement. In fact, it can be achieved mathematically equivalent to the fading-memory filter of item 5. Adding virtual process noise is a way to express insufficient confidence in the system model. This causes the filter to pay more attention to measurement rather than process models.

2.4 Sequential Kalman Filtering

In this section, sequential Kalman filtering is derived. This is a method for implementing a Kalman filter without matrix inversion. Therefore, this method has great advantages in embedded systems that may not have matrix routines. However, it only makes sense to use sequential Kalman filtering under certain limited conditions, which will also be discussed in this section.

Recall the Kalman filter measurement update formulas by Eqs. (2.2), (2.19), (2.22),(2.24) and (2.25):

$$\begin{cases} z_k = H_k x_k + v_k \\ K_k = P_{k|k-1} H_k^{\mathrm{T}} (H_k P_{k|k-1} H_k^{\mathrm{T}} + R_k)^{-1} \\ \hat{x}_{k|k} = \hat{x}_{k|k-1} + K_k (z_k - H_k \hat{x}_{k|k-1}) \\ P_{k|k} = (I - K_k H_k) P_{k|k-1} \end{cases} \tag{2.45}$$

The calculation of K_k requires the matrix inversion of an $r \times r$ matrix, where r is the dimension of the measurements. Instead of measuring z_k at time k, we obtain r separate measurements at time k. That is, we sequentially measure z_1, then $z_2, \ldots,$ and finally z_r.

Define the shorthand notation $z_{i,k}$ for the ith element of the measurement vector z_k. Given a diagonal covariance of measured values R_k as

$$R_k = \begin{bmatrix} R_{1,k} & \cdots & 0 \\ \vdots & \ddots & \vdots \\ 0 & \cdots & R_{r,k} \end{bmatrix} \tag{2.46}$$

Define the shorthand notation $H_{i,k}$ to represent the ith row of H_k, and define $v_{i,k}$ to represent the ith element of v_k. Then we obtain

$$z_{i,k} = H_{i,k}x_k + v_{i,k}, \quad v_{i,k} \sim (0, R_{i,k}) \tag{2.47}$$

Instead of processing the measurements as a vector at time k, we will implement the measurement update equation of the Kalman filter one by one at a time.

Note that $K_{i,k}$ is the Kalman gain, which is used to process the ith measurement at time k. $\hat{x}_{i,k|k}$ is the optimal estimate after the ith measurement has been processed at time k, and $P_{i,k|k}$ is the estimation-error covariance after the ith measurement at time k has been processed. Based on the above definition, we have

$$\begin{cases} \hat{x}_{0,k|k} = \hat{x}_{k|k-1} \\ P_{0,k|k} = P_{k|k-1} \end{cases} \tag{2.48}$$

where $\hat{x}_{0,k|k}$ is the estimate after completing zero measurements, also known as a priori estimate. Similarly, $P_{0,k|k}$ is the estimated error covariance after processing zero measurements, so it is equivalent to the priori estimated error covariance.

The gain $K_{i,k}$ and covariance $P_{i,k|k}$ are obtained from the normal Kalman filter measurement-update equations, but they are suitable for scalar measurements $z_{i,k}$. For $i = 1, \ldots, r$, we have

$$\begin{cases} K_{i,k} = P_{i-1,k|k}H_{i,k}^{\mathrm{T}}(H_{i,k}P_{i-1,k|k}H_{i,k}^{\mathrm{T}} + R_{i,k})^{-1} \\ \hat{x}_{i,k|k} = \hat{x}_{i-1,k|k} + K_{i,k}(z_{i,k} - H_{i,k}\hat{x}_{i-1,k|k}) \\ P_{i,k|k} = (I - K_{i,k}H_{i,k})P_{i-1,k|k} \end{cases} \tag{2.49}$$

After processing all r measurements, set $\hat{x}_{k|k} = \hat{x}_{r,k|k}$, and $P_{k|k} = P_{r,k|k}$. A posteriori estimate and error covariance at time k can be obtained.

The process of sequential Kalman filter is as follows.

1. The system and measurement equations are given by

$$\begin{cases} x_k = F_{k-1}x_{k-1} + G_{k-1}u_{k-1} + w_{k-1}, \quad w_k \sim (0, Q_k) \\ z_k = H_k x_k + v_k, \quad v_k \sim (0, R_k) \end{cases} \tag{2.50}$$

where w_k and v_k are zero-mean uncorrelated white noise. The measurement covariance R_k is a diagonal matrix given as

$$R_k = \mathrm{diag}(R_{1,k}, \ldots, R_{r,k}) \tag{2.51}$$

2. The filter is initialized as

$$\begin{cases} \hat{x}_{0|0} = E\{x_0\} \\ P_{0|0} = E\{(x_0 - \hat{x}_{0|0})(x_0 - \hat{x}_{0|0})^{\mathrm{T}}\} \end{cases} \tag{2.52}$$

3. At each time k, the time-update process is given as

$$\begin{cases} \hat{x}_{k|k-1} = F_{k-1}\hat{x}_{k-1|k-1} + G_{k-1}u_{k-1} \\ P_{k|k-1} = F_{k-1}P_{k-1|k-1}F_{k-1}^{\mathrm{T}} + Q_{k-1} \end{cases} \tag{2.53}$$

which is the same as the standard Kalman filter.

4. At each time k, the measurement-update process are given as follows.
   (a) Initialize the posterior estimate and covariance as

$$\begin{cases} \hat{x}_{0,k|k} = \hat{x}_{k|k-1} \\ P_{0,k|k} = P_{k|k-1} \end{cases} \tag{2.54}$$

   (b) For $i = 1, \ldots, r$, we have

$$\begin{cases} K_{i,k} = P_{i-1,k|k}H_{i,k}^{\mathrm{T}}(H_{i,k}P_{i-1,k|k}H_{i,k}^{\mathrm{T}} + R_{i,k})^{-1} = \frac{P_{i,k|k}H_{i,k}^{\mathrm{T}}}{R_{i,k}} \\ \hat{x}_{i,k|k} = \hat{x}_{i-1,k|k} + K_{i,k}(z_{i,k} - H_{i,k}\hat{x}_{i-1,k|k}) \\ P_{i,k|k} = (I - K_{i,k}H_{i,k})P_{i-1,k|k}(I - K_{i,k}H_{i,k})^{\mathrm{T}} + K_{i,k}R_{i,k}K_{i,k}^{\mathrm{T}} \\ \quad = (I - K_{i,k}H_{i,k})P_{i-1,k|k} \end{cases} \tag{2.55}$$

   (c) Assign the posterior estimate and covariance as

$$\begin{cases} \hat{x}_{k|k} = \hat{x}_{r,k|k} \\ P_{k|k} = P_{r,k|k} \end{cases} \tag{2.56}$$

The above assumes that the measurement noise covariance R_k is diagonal. What if R_k is not diagonal? Suppose that $R_k = R$ is not diagonal, but a constant matrix. Perform Jordanian decomposition of R by finding the matrix S such that

$$R = S\hat{R}S^{-1} \tag{2.57}$$

where $\hat{R}$ is a diagonal matrix containing the eigenvalues of R, and S is an orthogonal matrix (i.e., $S^{-1} = S^{\mathrm{T}}$) containing the eigenvectors of R.

As stated in most linear system books, if R is symmetric positive definite, then this decomposition is always possible. Define a new measurement $\tilde{z}_k$ as

$$\begin{aligned} \tilde{z}_k &= S^{-1}z_k = S^{-1}(H_k x_k + v_k) \\ &= \tilde{H}_k x_k + \tilde{v}_k \end{aligned} \tag{2.58}$$

The covariance of $\tilde{v}_k$ can be obtained as

$$\begin{aligned} E\{\tilde{v}_k\tilde{v}_k^{\mathrm{T}}\} &= E\{S^{-1}\tilde{v}_k\tilde{v}_k^{\mathrm{T}}S^{-\mathrm{T}}\} \\ &= E\{S^{-1}\tilde{v}_k\tilde{v}_k^{\mathrm{T}}S\} \\ &= S^{-1}E\{\tilde{v}_k\tilde{v}_k^{\mathrm{T}}\}S \\ &= S^{-1}RS \\ &= \hat{R} \end{aligned} \tag{2.59}$$

Therefore, we introduce the normalized metric $\tilde{z}_k$ with diagonal noise covariance. Using the measurement $\tilde{z}_k$ instead of z_k, the measurement matrix $\tilde{H}_k$ instead of H_k, and the measurement noise covariance $\hat{R}$, then the sequential Kalman filter can be achieved.

Note that if R changes with time, this process is meaningless, because in this case, we will have to perform Jordan decomposition at each step of the Kalman filter.

In short, the sequential Kalman filter is used only if one of the following two conditions are met:

1. The measurement noise covariance R_k is diagonal.
2. The measurement noise covariance R is a constant.

Finally, note that the term "sequential filtering" is sometimes used synonymously with Kalman filters. In other words, sequential is usually used as a synonym for recursion. Sometimes, a standard Kalman filter is called a batch Kalman filter to distinguish it from a sequential Kalman filter.

2.5 Information Filtering

Information filtering is another form of Kalman filter, which spreads the reciprocal of P instead of P. In other words, the information filtering propagates the information matrix of the system.

$$P = E\{(x - \hat{x})(x - \hat{x})^{\mathrm{T}}\} \tag{2.60}$$

where P indicates the degree of uncertainty in the state estimate. When $P \to 0$, we fully understand x. As $P \to \infty$, we have zero knowledge of x. The information matrix is defined as

$$\Lambda = P^{-1} \tag{2.61}$$

where Λ represents the certainty in the state estimate. As Λ increases, our confidence in state estimation also gradually increases. Similarly, when $\Lambda \to 0$, we have no idea about x. When $\Lambda \to \infty$, we have perfect knowledge of x.

From Eqs. (2.16), (2.22), (2.23) and (2.24), the measurement update equation for P can be written as

$$P_{k|k}^{-1} = P_{k|k-1}^{-1} + H_k^{\mathrm{T}} R_k^{-1} H_k \tag{2.62}$$

Substituting the definition of Λ into this equation gives

$$\Lambda_{k|k} = \Lambda_{k|k-1} + H_k^{\mathrm{T}} R_k^{-1} H_k \tag{2.63}$$

In similar way, from Eq. (2.16), the time-update equation for P is obtained:

$$P_{k|k-1} = F_{k-1} P_{k-1|k-1} F_{k-1}^{\mathrm{T}} + Q_{k-1} \tag{2.64}$$

This implies that

$$\Lambda_{k|k-1} = (F_{k-1} \Lambda_{k-1|k-1} F_{k-1}^{\mathrm{T}} + Q_{k-1})^{-1} \tag{2.65}$$

Then applying the matrix inversion lemma to Eq. (2.65), we obtain

$$\Lambda_{k|k-1} = Q_{k-1}^{-1} - Q_{k-1}^{-1} F_{k-1} (\Lambda_{k-1}^{+} + F_{k-1}^{\mathrm{T}} Q_{k-1}^{-1})^{-1} F_{k-1}^{\mathrm{T}} + Q_{k-1}^{-1} \tag{2.66}$$

The information filter can be summarized as follows.

1. The dynamic system is given by the following equations:

$$\begin{cases} x_k = F_{k-1} x_{k-1} + G_{k-1} u_{k-1} + w_{k-1}, & w_k \sim (0, Q_k) \\ z_k = H_k x_k + v_k, \quad v_k \sim (0, R_k) \end{cases} \tag{2.67}$$

2. The Kalman filter is initialized as follows:

$$\begin{cases} \hat{x}_{0|0} = E\{x_0\} \\ \Lambda_{0|0} = \{E[(x_0 - \hat{x}_{0|0})(x_0 - \hat{x}_{0|0})^{\mathrm{T}}]\}^{-1} \end{cases} \tag{2.68}$$

3. The information filter is given by the following equations, which are computed for each time step $k = 1, 2, \ldots$:

$$\begin{cases} \Lambda_{k|k-1} = Q_{k-1}^{-1} F_{k-1} (\Lambda_{k-1|k-1} + F_{k-1}^{\mathrm{T}} Q_{k-1}^{-1} F_{k-1})^{-1} F_{k-1}^{\mathrm{T}} Q_{k-1}^{-1} \\ \Lambda_{k|k} = \Lambda_{k|k-1} + H_k^{\mathrm{T}} R_k^{-1} H_k \\ K_k = \Lambda_{k|k}^{-1} H_k^{\mathrm{T}} R_k^{-1} \\ \hat{x}_{k|k-1} = F_{k-1} \hat{x}_{k-1|k-1} + G_{k-1} u_{k-1} \\ \hat{x}_{k|k} = \hat{x}_{k|k-1} + K_k (z_k - H_k \hat{x}_{k|k-1}) \end{cases} \tag{2.69}$$

The information filter equation requires at least $n \times n$ matrix inversion, where n is the number of states. But the standard Kalman filter equation requires the inverse of the $r \times r$ matrix. Therefore, if the number of measurements r is larger than the number of states n (i.e., $r > n$), the use of information filter may improve computational efficiency. It can be considered that since the Kalman gain is given

$$K_k = P_{k|k} H_k^{\mathrm{T}} R_k^{-1} \tag{2.70}$$

Regardless of whether a standard Kalman filter or an information filter is used, we must perform the inverse operation of $r \times r$ matrix on R_k. But if R_k is a constant, then we can invert it during the initialization process, so the Kalman gain equation may not require $r \times r$ matrix to invert at all. The same idea applies to the inversion of Q_{k-1}.

If the initial uncertainty is ∞, we cannot set $P_{0|0} = \infty$ numerically, but we can set $\Lambda_{0|0} = 0$. For the case where the initial certainty is 0, information filtering is mathematically more accurate. If the initial uncertainty is zero, we can numerically set $\Lambda_{0|0} = \infty$, but we cannot numerically set $P_{0|0} = 0$. That is, for the case of zero initial uncertainty, the standard Kalman filter is more mathematically accurate.

2.6 Summary

In this chapter, we introduced the essence of the discrete-time Kalman filter. In the past few decades, this estimation algorithm has been used in almost every field of engineering. We have seen that the Kalman filter equations can be written in several different ways, and these expressions are mathematically equivalent, although each method may look completely different from the others. When the noise is Gaussian noise, the Kalman filter is the optimal estimator, and when the noise is not Gaussian noise, the Kalman filter is the optimal linear estimator. Generally speaking, if the basic assumption is not established, the performance of the Kalman filter may be degraded, so we simply mentioned some methods to compensate for the violation of the assumption. At the same time, this chapter also introduces some theories of sequential Kalman filter and information filter, which lays the foundation for the improved algorithm later.

References

1. Anderson, B.D., and J.B. Moore. 1979. *Optimal Filtering*. Englewood Cliffs, New Jersey: Prentice-Hall.
2. Bar-Shalom, Yarkov, X. Rong Li, and T. Kirubarajan. 2001. *Estimation with Applications to Tracking and Navigation: Theory, Algorithms and Software*. New York: Wiley.
3. Gelb, Arthur. 1974. *Applied Optimal Estimation*. MIT Press.
4. Jazwinski, Andrew. 1970. *Stochastic Processes And Filtering Theory*. Academic Press.
5. Kalman, R.E. 1960. A new approach to linear filtering and prediction problems. *Transactions of the ASME Journal of Basic Engineering* 82 (1): 35–45.
6. Kay, S.M. 1993. *Fundamentals of Statistical Signal Processing: Estimation Theory*. Englewood Cliffs, New Jersey: Prentice-Hall.
7. Mendel, J.M. 1986. *Lessons in Digital Estimation Theory*. Prentice-Hall.
8. Ristic, Branko, Sanjeev Arulampalam, and Neil Gordon. 2004. *Beyond the Kalman Filter*. Artech House.

9. Tichavsky, P., C.H. Muravchik, and A. Nehorai. 1998. Posterior Cramér-Rao bounds for discrete-time nonlinear filtering. *IEEE Transactions on Signal Processing* 46 (5): 1386–1396.
10. Van Trees, H.L., K.L. Bell, and Z. Tian. 2013. *Mathematical Methods for Physicists, Part I*, 2nd ed. New York: Wiley.

Part II
State Fusion Estimation for Networked Systems

Chapter 3
Fusion Estimation for Linear Systems with Cross-Correlated Sensor Noises

In this part, Kalman filtering fusion for linear dynamic systems will be introduced in five chapters. They are studied mainly aimed to solve the problems caused by network or wireless communication.

3.1 Introduction

Estimation fusion, or data fusion for estimation, is the problem of how to best utilize useful information contained in multiple sets of data for the purpose of estimating a quantity, e.g. a parameter or process [1, 12]. It originated in the military field, and is now widely used in military and civilian fields, e.g., target tracking and localization, guidance and navigation, surveillance and monitoring, etc. due to its improved estimation accuracy, enhanced reliability and survivability, etc. [29].

Most literatures are concerned about the fusion estimation of the linear systems with independent sensor noises. In the practical applications, the dynamic process is observed in a common noisy environment, so the noises of different sensors are generally correlated [22]. In this case, the traditional centralized fusion is also applicable and is still optimal in the sense of linear minimum mean square error (LMMSE). But, the computation becomes rather complex. The distributed fusion in this case may not be optimal.

Some researchers are working on the optimal Kalman filtering fusion with coupled sensor noises recently. The distributed fusion algorithms both without feedback and with feedback are presented in [22]. While, augmentations of some system parameters are existed, which may cause the additional work in computing.

In [9], distributed sequential estimation of a nonrandom parameter over noisy communication links was studied, where the sensor noises are correlated spatially across the sensor field, and a recursive algorithm for updating the sequential estimator was deduced. By reconstructing the measurements, Ref. [6] derived the distributed Kalman filtering fusion with feedback and without feedback, where the noise is

L. Yan et al., *Multisensor Fusion Estimation Theory and Application*,
https://doi.org/10.1007/978-981-15-9426-7_3

decoupled in sequence by the method given in [24]. By these methods, the augmentation of measurements and measurement matrices etc. are needed. The fusion algorithm given in [8] is similar. These methods have more complex form compared to [22], but they have the advantages of capability of handling the correlation of measurement noise and system noise besides decoupling the measurement noises. [23] studied the distributed fusion when the measurement noises are correlated across sensors and with the system noise at the same time step. When the noise of different sensors are cross-correlated and also coupled with the system noise of the previous step, we derive the optimal sequential fusion and the optimal distributed fusion algorithm in [27], and generate this result to the fusion of multirate sensor cases [15, 28]. When there is correlation between the process noise and the measurement noise and among measurement noises, a distributed weighted robust Kalman filter fusion algorithm is deduced for uncertain systems with multiple sensors in [7].

For fusion of the sensors with correlated noises, there are some results using linear transforms. Li et al. first propose this idea in [12], where the optimal batch fusion and the distributed fusion have been obtained in a unified form. In [17, 18], correlated measurement fusion Kalman filtering algorithms are obtained based on orthogonal transformation. By using the Cholesky factorization, the correlated noises can be decoupled and the optimal state estimations were derived in [3–5]. In [14], the authors proved that the sufficient condition of the lossless transform for distributed estimation with cross-correlated measurement noises is the transformation matrix of full column rank.

Briefly, there are some literatures deal with optimal Kalman filtering fusion of the dynamic systems with correlated noises. When the sensor noises are coupled, the centralized fusion can still be used and it is optimal in the sense of LMMSE. However, the computation and power requirements are too huge to be practical. Many people intend to generate the distributed fusion algorithm. While, the distributed fusion algorithms are not given explicitly in [4, 5, 12]. The explicit distributed fusion algorithms are presented in [6, 22, 24], etc., but augmentations of some system parameters are existed, and therefore it is not optimal when the computation complexity performance is concerned.

In recent years, wireless sensor networks (WSNs) have become a hot research topic in various fields due to their important roles in practical applications. In networked systems, the measurements or the local estimations are transmitted to the fusion center through the communication channel. Constrained by the limited resources and communication bandwidth, the computation and communication efficiencies are problems to which we should pay much attention.

Literatures [2, 19, 21, 30], etc. discussed the compression of data in the hope of reducing the communication requirements. A sufficient condition and a necessary and sufficient condition were given in [11, 30], respectively, for lossless of performance for distributed fusion with compressed data. A suboptimal and an optimal compression rules were derived in [2, 21], respectively, for state fusion estimation. In WSNs, the energy–saving related estimation algorithms are studied in recent years. The literatures include data compression methods [10, 19], quantization based methods [16, 25, 32], and the methods to slow down the information transmission rate of the sensors [13, 26, 31]. For distributed Kalman filtering fusion of measurements with cross–correlated noises, the simple form in considering of both optimality in accuracy as well as in computation and communication complexity, is still an open problem, and is the motivation of this chapter.

This chapter is organized as follows. Section 3.2 describes the problem formulation. In Sect. 3.3, the Cholesky decomposition and the linear transformation are introduced. Section 3.4 provides the optimal state fusion estimation algorithms, including the centralized fusion based on raw data, the centralized and the distributed fusion estimation algorithms based on the transformed measurement models. Performances, including the global optimization and the computation complexity and the communication requirements of each algorithm, are also analyzed in this section. Section 3.5 is the simulation and Sect. 3.6 reaches the conclusion.

3.2 Problem Formulation

Consider the following linear dynamic system,

$$x_{k+1} = A_k x_k + w_k, k = 0, 1, \ldots \quad (3.1)$$

$$z_{i,k} = C_{i,k} x_k + v_{i,k}, i = 1, 2, \ldots, N \quad (3.2)$$

where $x_k \in R^n$ is the system state, $A_k \in R^{n \times n}$ is the state transition matrix. w_k is the system noise and is assumed to be Gaussian distributed with

$$\begin{cases} E\{w_k\} = 0 \\ E\{w_k w_l^T\} = Q_k \delta_{kl} \end{cases} \quad (3.3)$$

where $Q_k \geq 0$, and δ_{kl} is the Kronecker delta function.

$z_{i,k} \in R^{m_i}$ is the observation of sensor i at time instant k. $C_{i,k} \in R^{m_i \times n}$ is the measurement matrix. $v_{i,k}$ is the measurement noise, and is assumed to be white Gaussian distributed, and is independent of w_k, i.e., for $k, l = 1, 2, \ldots; i, j = 1, 2, \ldots, N$, we have

$$\begin{cases} E\{v_{i,k}\} = 0 \\ E\{v_{i,k} v_{j,l}^T\} = R_{ij,k} \delta_{kl} \end{cases} \quad (3.4)$$

From the formula above one can see that the measurement noises by different sensors are correlated, i.e., $v_{i,k}$ and $v_{j,k}$ are cross-correlated for $i \neq j$ at time k with $E\{v_{i,k} v_{j,k}^T\} = R_{ij,k} \neq 0$. For simplicity, it is denoted that $R_{i,k} \underline{\underline{\triangle}} R_{ii,k} > 0$ for $i = 1, 2, \ldots, N$.

The initial state x_0 is independent of w_k and $v_{i,k}$ for $k = 1, 2, \ldots$ and $i = 1, 2, \ldots, N$, and is assumed to be Gaussian distributed with

$$\begin{cases} E\{x_0\} &= \bar{x}_0 \\ \text{cov}\{x_0\} &= E\{(x_0 - x_0)(x_0 - x_0)^T\} = \bar{P}_0 \end{cases} \tag{3.5}$$

where $\text{cov}\{x_0\}$ means the covariance of x_0.

The objective of this chapter is to generate the optimal estimation of state x_k by using the measurements $z_{i,k}$ based on the above descriptions.

3.3 Linear Transformation

For the system given in (3.1) and (3.2), let

$$z_k = [z_{1,k}^T, z_{2,k}^T, \ldots, z_{N,k}^T]^T \tag{3.6}$$

$$C_k = [C_{1,k}^T, C_{2,k}^T, \ldots, C_{N,k}^T]^T \tag{3.7}$$

$$v_k = [v_{1,k}^T, v_{2,k}^T, \ldots, v_{N,k}^T]^T \tag{3.8}$$

then we have

$$\begin{cases} E\{v_k\} = 0_{m\times 1}, m = \sum_{i=1}^{N} m_i \\ E\{v_k v_l^T\} = R_k \delta_{kl} \\ E\{w_k v_l^T\} = 0 \end{cases} \tag{3.9}$$

where

$$R_k = \text{cov}\{v_k\} = \begin{bmatrix} R_{11,k} & R_{12,k} & \cdots & R_{1N,k} \\ R_{21,k} & R_{22,k} & \cdots & R_{2N,k} \\ \vdots & \vdots & & \vdots \\ R_{N1,k} & R_{N2,k} & \cdots & R_{NN,k} \end{bmatrix} \tag{3.10}$$

is a symmetric positive definite matrix.

To decouple the measurement noises and to decrease the computation complexity, the cholesky decomposition is used to decompose R_k. Namely,

$$R_k = L_k^T L_k \tag{3.11}$$

where, L_k is a lower triangular matrix with strictly positive diagonal entries.

Let

$$T_k = L_k^{-T} \tag{3.12}$$

$$\bar{z}_k = T_k z_k \tag{3.13}$$

$$\bar{C}_k = T_k C_k \tag{3.14}$$

$$\bar{v}_k = T_k v_k \tag{3.15}$$

It follows that

$$\bar{z}_k = \bar{C}_k x_k + \bar{v}_k \tag{3.16}$$

where

$$\begin{cases} E\{\bar{v}_k\} = 0_{m\times 1}, m = \sum_{i=1}^{N} m_i \\ \bar{R}_k = \text{cov}\{\bar{v}_k\} = T_k R_k T_k^T = I_{m\times m} \\ E\{w_k \bar{v}_j^T\} = 0 \end{cases} \tag{3.17}$$

Hence, the coupled sensor noises are decoupled, whose covariance is identity matrix now.

3.4 The Optimal State Fusion Estimation Algorithms

3.4.1 The Centralized State Fusion Estimation with Raw Data

Based on system (3.1)–(3.2), the optimal state estimation can be generated by the centralized fusion.

Theorem 3.1 *Given the centralized fusion estimation $\hat{x}_{c,k-1|k-1}$ and $P_{c,k-1|k-1}$ at time $k-1$, the optimal state fusion estimation at time k can be computed by the following formula:*

$$\begin{cases} \hat{x}_{c,k|k-1} = A_{k-1}\hat{x}_{c,k-1|k-1} \\ P_{c,k|k-1} = A_{k-1}P_{c,k-1|k-1}A_{k-1}^T + Q_{k-1} \\ \hat{x}_{c,k|k} = \hat{x}_{c,k|k-1} + K_{c,k}(z_k - C_k\hat{x}_{c,k|k-1}) \\ K_{c,k} = P_{c,k|k-1}C_k^T(C_kP_{c,k|k-1}C_k^T + R_k)^{-1} \\ P_{c,k|k} = (I - K_{c,k}C_k)P_{c,k|k-1} \end{cases} \tag{3.18}$$

where subscript "c" denotes the centralized fusion, and where z_k, C_k and R_k are computed by Eqs. (3.6), (3.7) and (3.10), respectively.

Proof From the problem formulation and Eq. (3.9), it is obvious that v_k is white Gaussian and is uncorrelated with system noise w_k. Therefore, to generate the optimal state estimation, the traditional centralized fusion can be used.

3.4.2 The Centralized Fusion with Transformed Data

From Sect. 3.3, we can rewrite system (3.1)–(3.2) as,

$$\begin{cases} x_{k+1} = A_k x_k + w_k \\ \bar{z}_k = \bar{C}_k x_k + \bar{v}_k \end{cases} \tag{3.19}$$

where $x_k \in R^n$ is the system state, $A_k \in R^{n\times n}$ is the state transition matrix, w_k is the zero-mean white Gaussian process noises with covariance being $Q_k \geq 0$, and the initial state satisfies $E\{x_0\} = x_0$ and $\mathrm{cov}\{x_0\} = P_0$. $\bar{z}_k \in R^m$ is the measurement at time k, and $\bar{C}_k \in R^{m\times n}$ is the measurement matrix, where $m = \sum_{i=1}^{N} m_i$. Measurement noise $\bar{v}_k$ is zero-mean white Gaussian distributed and is uncorrelated with w_k and x_0, and $\mathrm{cov}\{\bar{v}_k\} = \bar{R}_k = I$.

The optimal state estimation of x_k could be generated by use of the following theorem.

Theorem 3.2 *Based on system (3.19), given the centralized fusion estimation $\hat{x}_{ct,k-1|k-1}$ and the estimation error covariance $P_{ct,k-1|k-1}$ at time $k-1$, the optimal estimation at time k can be computed by*

$$\begin{cases} \hat{x}_{ct,k|k-1} = A_{k-1}\hat{x}_{ct,k-1|k-1} \\ P_{ct,k|k-1} = A_{k-1}P_{ct,k-1|k-1}A_{k-1}^T + Q_{k-1} \\ \hat{x}_{ct,k|k} = \hat{x}_{ct,k|k-1} + K_{ct,k}(\bar{z}_k - \bar{C}_k\hat{x}_{ct,k|k-1}) \\ K_{ct,k} = P_{ct,k|k-1}\bar{C}_k^T(\bar{C}_k P_{ct,k|k-1}\bar{C}_k^T + \bar{R}_k)^{-1} \\ P_{ct,k|k} = (I - K_{ct,k}\bar{C}_k)P_{ct,k|k-1} \end{cases} \tag{3.20}$$

where subscript "ct" denotes the centralized fusion by use of the transformed data, and where $\bar{z}_k$, $\bar{C}_k$ and $\bar{R}_k$ are computed by Eqs. (3.13), (3.14) and (3.17), respectively.

Proof For the linear system (3.19), it is obvious that the system noise w_k is zero-mean white Gaussian distributed. The measurement noise $\bar{v}_k$ is also zero-mean white Gaussian distributed, and is uncorrelated with w_k and x_0. So, the optimal state fusion estimation can be generated by use of the centralized fusion algorithm.

Theorem 3.3 *The state estimation of the centralized fusion given in Theorem 3.2 with transformed data is equivalent to the centralized fusion given in Theorem 3.1 with the original observations in the sense of LMMSE.*

Proof From Eq. (3.18), $P_{c,k|k}$ and $P_{c,k|k-1}$ can be rewritten by using the information form of the Kalman filter as follows:

$$P_{c,k|k-1} = A_{k-1}P_{c,k-1|k-1}A_{k-1}^T + Q_{k-1} \tag{3.21}$$

$$(P_{c,k|k})^{-1} = (P_{c,k|k-1})^{-1} + C_k^T R_k^{-1} C_k \tag{3.22}$$

Similarly, from Eq. (3.20), we derive

$$P_{ct,k|k-1} = A_{k-1} P_{ct,k-1|k-1} A_{k-1}^T + Q_{k-1} \tag{3.23}$$
$$P_{ct,k|k}^{-1} = P_{ct,k|k-1}^{-1} + \bar{C}_k^T \bar{R}_k^{-1} \bar{C}_k \tag{3.24}$$

It is obviously that $P_{c,0|0} = P_{ct,0|0} = P_0$. Suppose $P_{c,k-1|k-1} = P_{ct,k-1|k-1}$, then, due to Eqs. (3.21) and (3.23), we have

$$P_{ct,k|k-1} = P_{c,k|k-1} \tag{3.25}$$

So, due to Eqs. (3.22) and (3.24), in order to prove $P_{ct,k|k} = P_{c,k|k}$, we just need to prove $\bar{C}_k^T \bar{R}_k^{-1} \bar{C}_k = C_k^T R_k^{-1} C_k$.

In fact,

$$\begin{aligned} \bar{C}_k^T \bar{R}_k \bar{C}_k &= (T_k C_k)^T (T_k R_k T_k^T)^{-1} (T_k C_k) \\ &= C_k^T T_k^T T_k^{-T} R_k^{-1} T_k^{-1} T_k C_k \\ &= C_k^T R_k^{-1} C_k \end{aligned} \tag{3.26}$$

Hence, $P_{ct,k|k} = P_{c,k|k}$.

From Theorem 3.3, it is obvious that the state estimation based on system (3.19) and system (3.1)–(3.2) are equivalent when the centralized fusion is concerned.

3.4.3 The Optimal State Estimation by Distributed Fusion

Rewrite $\bar{z}_k$, $\bar{C}_k$ and $\bar{v}_k$ in (3.16) to block form, we have

$$\bar{z}_k = [\bar{z}_{1,k}^T, \bar{z}_{2,k}^T, \ldots, \bar{z}_{N,k}^T]^T \tag{3.27}$$
$$\bar{C}_k = [\bar{C}_{1,k}^T, \bar{C}_{2,k}^T, \ldots, \bar{C}_{N,k}^T]^T \tag{3.28}$$
$$\bar{v}_k = [\bar{v}_{1,k}^T, \bar{v}_{2,k}^T, \ldots, \bar{v}_{N,k}^T]^T \tag{3.29}$$

where $\bar{z}_{i,k} \in R^{m_i \times 1}$, $\bar{C}_{i,k} \in R^{m_i \times n}$ and $\bar{v}_{i,k} \in R^{m_i \times 1}$. Then from Eq. (3.16), we have

$$\bar{z}_{i,k} = \bar{C}_{i,k} x_k + \bar{v}_{i,k}, \quad i = 1, 2, \ldots, N \tag{3.30}$$

From Eq. (3.17), it can be easily obtained that $\bar{v}_{i,k}$ is uncorrelated across different sensors, which is zero mean Gaussian distributed and is independent of system noise w_k and the initial state x_0. Therefore, based on Eqs. (3.1) and (3.30), the optimal state estimation could be generated by using the distributed fusion algorithm.

Theorem 3.4 *By using the linear system (3.1) and (3.30), the optimal state estimation can be derived by use of the distributed fusion algorithm. For simplicity, only the distributed fusion without feedback is given here.*

$$P_{d,k|k}^{-1} = P_{d,k|k-1}^{-1} + \sum_{i=1}^{N}(P_{d,i,k|k}^{-1} - P_{d,i,k|k-1}^{-1}) \tag{3.31}$$

$$\begin{aligned} P_{d,k|k}^{-1}\hat{x}_{d,k|k} = &\ P_{d,k|k-1}^{-1}\hat{x}_{d,k|k-1} + \sum_{i=1}^{N}(P_{d,i,k|k}^{-1}\hat{x}_{d,i,k|k} \\ &- P_{d,i,k|k-1}^{-1}\hat{x}_{d,i,k|k-1}) \end{aligned} \tag{3.32}$$

where the fused state prediction at time k is

$$\hat{x}_{d,k|k-1} = A_{k-1}\hat{x}_{d,k-1|k-1} \tag{3.33}$$

$$P_{d,k|k-1} = A_{k-1}P_{d,k-1|k-1}A_{k-1}^T + Q_{k-1} \tag{3.34}$$

and the local state estimations $\hat{x}_{d,i,k|k}$ *and* $P_{d,i,k|k}$ *are computed by Kalman filtering by using*

$$\begin{cases} \hat{x}_{d,i,k|k} = \hat{x}_{d,i,k|k-1} + K_{d,i,k}(\bar{z}_{i,k} - \bar{C}_{i,k}\hat{x}_{d,i,k|k-1}) \\ P_{d,i,k|k} = (I - K_{d,i,k}\bar{C}_{i,k})P_{d,i,k|k-1} \\ \hat{x}_{d,i,k|k-1} = A_{k-1}\hat{x}_{d,i,k-1|k-1} \\ P_{d,i,k|k-1} = A_{k-1}P_{d,i,k-1|k-1}A_{k-1}^T + Q_{k-1} \\ K_{d,i,k} = P_{d,i,k|k-1}\bar{C}_{i,k}^T(\bar{C}_{i,k}P_{d,i,k|k-1}\bar{C}_{i,k}^T + \bar{R}_{i,k})^{-1} \\ \bar{R}_{i,k} = I \end{cases} \tag{3.35}$$

where subscript "d" denotes the distributed fusion.

Proof From Eq. (3.35), by use of the information form of Kalman filtering, the update equations of Sensor i at time k can be rewritten as [20]

$$\begin{cases} \hat{x}_{d,i,k|k} = \hat{x}_{d,i,k|k-1} + P_{d,i,k|k}\bar{C}_{i,k}^T\bar{R}_{i,k}^{-1}(\bar{z}_{i,k} - \bar{C}_{i,k}\hat{x}_{d,i,k|k-1}) \\ P_{d,i,k|k}^{-1} = P_{d,i,k|k-1}^{-1} + \bar{C}_{i,k}^T\bar{R}_{i,k}^{-1}\bar{C}_{i,k} \end{cases} \tag{3.36}$$

Multiply the first equation with the second equation in Eq. (3.36), we obtain

$$\begin{aligned} P_{d,i,k|k}^{-1}\hat{x}_{d,i,k|k} &= (P_{d,i,k|k-1}^{-1} + \bar{C}_{i,k}^T\bar{R}_{i,k}^{-1}\bar{C}_{i,k})\hat{x}_{d,i,k|k-1} \\ &\quad + P_{d,i,k|k}^{-1}P_{d,i,k|k}\bar{C}_{i,k}^T\bar{R}_{i,k}^{-1}(\bar{z}_{i,k} - \bar{C}_{i,k}\hat{x}_{d,i,k|k-1}) \\ &= P_{d,i,k|k-1}^{-1}\hat{x}_{d,i,k|k-1} + \bar{C}_{i,k}^T\bar{R}_{i,k}^{-1}\bar{z}_{i,k} \end{aligned} \tag{3.37}$$

Then

$$\bar{C}_{i,k}^T\bar{R}_{i,k}^{-1}\bar{z}_{i,k} = P_{d,i,k|k}^{-1}\hat{x}_{d,i,k|k} - P_{d,i,k|k-1}^{-1}\hat{x}_{d,i,k|k-1} \tag{3.38}$$

On the other hand, from Eq. (3.20), we obtain

$$\begin{cases} \hat{x}_{ct,k|k} = \hat{x}_{ct,k|k-1} + P_{ct,k|k}\bar{C}_k^T\bar{R}_k^{-1}(\bar{z}_k - \bar{C}_k\hat{x}_{ct,k|k-1}) \\ P_{ct,k|k}^{-1} = P_{ct,k|k-1}^{-1} + \bar{C}_k^T\bar{R}_k^{-1}\bar{C}_k \end{cases} \tag{3.39}$$

Because $\bar{R}_k$ is a diagonal matrix, we can rewrite it as

$$\bar{R}_k = \mathrm{diag}\{\bar{R}_{1,k}, \bar{R}_{2,k}, \ldots, \bar{R}_{N,k}\} \tag{3.40}$$

where $\bar{R}_{i,k} = I_{m_i \times m_i}$.

Substitute Eqs. (3.27), (3.28) and (3.40) into Eq. (3.39), we obtain

$$\begin{cases} \hat{x}_{ct,k|k} = \hat{x}_{ct,k|k-1} + P_{ct,k|k}\sum_{i=1}^{N}[\bar{C}_{i,k}^T\bar{R}_{i,k}^{-1}(\bar{z}_{i,k} - \bar{C}_{i,k}\hat{x}_{ct,k|k-1})] \\ P_{ct,k|k}^{-1} = P_{ct,k|k-1}^{-1} + \sum_{i=1}^{N}(\bar{C}_{i,k}^T\bar{R}_{i,k}^{-1}\bar{C}_{i,k}) \end{cases} \tag{3.41}$$

Multiply the first equation by the second of Eq. (3.41), after reorganization, yield

$$P_{ct,k|k}^{-1}\hat{x}_{ct,k|k} = P_{ct,k|k-1}^{-1}\hat{x}_{ct,k|k-1} + \sum_{i=1}^{N}(\bar{C}_{i,k}^T\bar{R}_{i,k}^{-1}\bar{z}_{i,k}) \tag{3.42}$$

Substitute Eq. (3.38) to Eq. (3.42), we have

$$\begin{aligned} P_{ct,k|k}^{-1}\hat{x}_{ct,k|k} &= P_{ct,k|k-1}^{-1}\hat{x}_{ct,k|k-1} \\ &\quad + \sum_{i=1}^{N}(P_{d,i,k|k}^{-1}\hat{x}_{d,i,k|k} - P_{d,i,k|k-1}^{-1}\hat{x}_{d,i,k|k-1}) \end{aligned}$$

From the second equation of Eq. (3.36), we obtain

$$\bar{C}_{i,k}^T\bar{R}_{i,k}^{-1}\bar{C}_{i,k} = P_{d,i,k|k}^{-1} - P_{d,i,k|k-1}^{-1} \tag{3.43}$$

Substitute Eq. (3.43) to the second equation of Eq. (3.41), we have

$$P_{ct,k|k}^{-1} = P_{ct,k|k-1}^{-1} + \sum_{i=1}^{N}(P_{d,i,k|k}^{-1} - P_{d,i,k|k-1}^{-1}) \tag{3.44}$$

Combine Eq. (3.43) with Eq. (3.44), we have

$$\begin{cases} P_{ct,k|k}^{-1}\hat{x}_{ct,k|k} = \sum_{i=1}^{N}(P_{d,i,k|k}^{-1}\hat{x}_{d,i,k|k} - P_{d,i,k|k-1}^{-1}\hat{x}_{d,i,k|k-1}) \\ \qquad\qquad + P_{ct,k|k-1}^{-1}\hat{x}_{ct,k|k-1} \\ P_{ct,k|k}^{-1} = P_{ct,k|k-1}^{-1} + \sum_{i=1}^{N}(P_{d,i,k|k}^{-1} - P_{d,i,k|k-1}^{-1}) \end{cases} \tag{3.45}$$

Compare Eqs. (3.31)–(3.34) with Eqs. (3.20) and (3.45), deductively, we have

$$\begin{cases} \hat{x}_{d,k|k} = \hat{x}_{ct,k|k} \\ P_{d,k|k} = P_{ct,k|k} \\ \hat{x}_{d,k|k-1} = \hat{x}_{ct,k|k-1} \\ P_{d,k|k-1} = P_{ct,k|k-1} \end{cases} \tag{3.46}$$

According to the derivation process of Theorem 3.4, the distributed fusion algorithm is derived from the variation of the centralized fusion. So, we have the following corollary directly.

Corollary 3.1 *The distributed fusion given in Theorem 3.4 is equivalent to the centralized fusion given by Theorem 3.2, namely, they have the same state estimation for the same system given in Sect. 3.2.*

Theorem 3.5 *The state estimation by use of the distributed fusion algorithm of Theorem 3.4 with the transformed measurements of (3.30) is equivalent to the estimation by use of the centralized fusion algorithm with the original measurements of (3.2) given in Theorem 3.1.*

Proof From Theorem 3.3 and Corollary 3.1, we reach the conclusion.

From Theorem 3.5, one can see that by applying the linear transformation to the measurement equations and then by use of the distributed fusion algorithm given in Theorem 3.4, the optimal state estimation can be derived, which is equivalent to the centralized fusion estimation based on the original observations in estimation precision. While, the distributed algorithm with transformed data is more flexible and applicable, and has much less computation complexity compared to the centralized fusion algorithm with raw data. We will show this in detail in the next subsection.

Remark 3.1 In Theorem 3.4, to generate the state estimation, the distributed fusion without feedback is introduced. In fact, we should only change Eq. (3.35) to get the distributed fusion with feedback by use of the following formula.

$$\begin{cases} \hat{x}_{d,i,k|k} = \hat{x}_{d,i,k|k-1} + K_{d,i,k}(\bar{z}_{i,k} - \bar{C}_{i,k}\hat{x}_{d,i,k|k-1}) \\ P_{d,i,k|k} = (I - K_{d,i,k}\bar{C}_{i,k})P_{d,i,k|k-1} \\ \hat{x}_{d,i,k|k-1} = A_{k-1}\hat{x}_{d,k-1|k-1} \\ P_{d,i,k|k-1} = A_{k-1}P_{d,k-1|k-1}A_{k-1}^T + Q_{k-1} \\ K_{d,i,k} = P_{d,i,k|k-1}\bar{C}_{i,k}^T(\bar{C}_{i,k}P_{d,i,k|k-1}\bar{C}_{i,k}^T + \bar{R}_{i,k})^{-1} \\ \bar{R}_{i,k} = I \end{cases} \tag{3.47}$$

For the distributed fusion algorithm with feedback, it can be shown that it is equivalent to the distributed fusion without feedback in estimation precision, and is therefore globally optimal in the sense of LMMSE, whose advantage is that it has better reliability and robustness [7, 22].

3.4.4 The Complexity Analysis

The architectures of the centralized Kalman filtering fusion, the centralized fusion with the transformed measurements, and the optimal distributed fusion algorithms are shown in Figs. 3.1, 3.2 and 3.3, respectively. For the centralized fusion, the system parameters A_k, Q_k, $C_{i,k}$, $R_{ij,k}$ and the measurements $z_{i,k}$ for sensors $i = 1, 2, \ldots, N$ are sent to the remote fusion center directly. For the centralized fusion with transform, the system parameters and the measurements are sent to the local processor first, and after proper process (linear transform), the transformed data are sent to the remote fusion center. For the distributed fusion, the transformed data are sent to the local estimators, and the local estimations are sent to the remote fusion center. In the sequel, the computation and communication complexity of the three algorithms will be compared.

First, let's compare the centralized fusion algorithms by use of the original system and the system with transformed measurement equations.

It is known that to use the centralized fusion algorithm, besides the observations, it is necessary to send the state transition matrix A_k, the measurement matrix C_k, the covariance of the system noise Q_k and the covariance of measurement noise R_k to the fusion center. Nothing changes of A_k and Q_k after the transformation, neither does the dimension of C_k. Let's check R_k. Before transformation, the communication requirement to send the data of R_k to the fusion center is $m(m+1)/2$, where $m = \sum_{i=1}^{N} m_i$. However, as has been proven in Eq. (3.17), after the linear transformation, the measurement noise covariance becomes an identity matrix. So transmission of R_k is not necessary. This nice property certainly help to reduce the transmission burden to the fusion center. Particularly, when m is very large, the property will greatly reduce the communication demands and the computation complexity.

Next, let's compare the distributed fusion with the centralized fusion algorithm based on the system with transformed form. The centralized fusion algorithm should

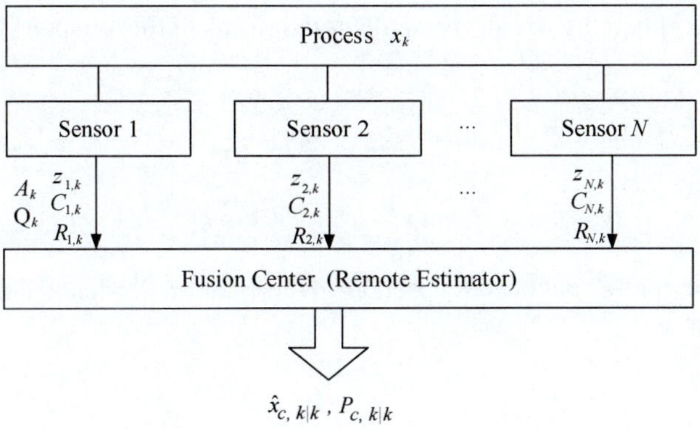

Fig. 3.1 Architecture of the centralized fusion by use of the raw measurements

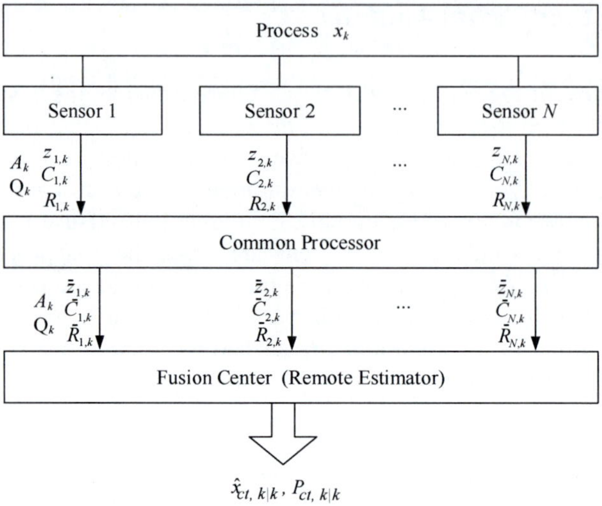

Fig. 3.2 Architecture of the centralized fusion by use of the transformed measurements

transmit $A_k, \bar{C}_k, Q_k, \bar{z}_k$ to the fusion center at each moment k, and the communication requirement is $\frac{3}{2}n^2 + \frac{1}{2}n + mn + m$, where $m = \sum_{i=1}^{N} m_i$. While, for the distributed fusion, the system model parameters and the local state estimations should be transmitted to the fusion center, and the communication requirement is $N(n^2 + 3n)$. It can be seen that when the state and the measurements dimensions are high, the centralized fusion will occupy larger transmission bandwidth. As far as the computation complexity is concerned, it is well known that the distributed fusion is more efficient as compared to the centralized fusion algorithm, because augmentation of measurements, measurement matrix, and measurement noises covariances are avoided.

Finally, let's compare the presented distributed algorithm with the distributed algorithm presented in [22]. To make it clear, we list the distributed fusion Eq. (21) of literature [22] here by use of the same notations as in this chapter:

$$
\begin{aligned}
P_{d,k|k}^{-1}\hat{x}_{d,k|k} = P_{d,k|k-1}^{-1}\hat{x}_{d,k|k-1} + C_k^T \sum_{i=1}^{N} R_{k,i*}^{-1}(R_{i,k})^{-1}C_{i,k}^{+} \\
\cdot(P_{d,i,k|k}^{-1}\hat{x}_{d,i,k|k} - P_{d,i,k|k-1}^{-1}\hat{x}_{d,i,k|k-1})
\end{aligned}
\tag{3.48}
$$

where, superscript "+" denotes the pseudo-inverse, and $R_{k,i*}^{-1}$ denotes the ith submatrix column of R_k^{-1}.

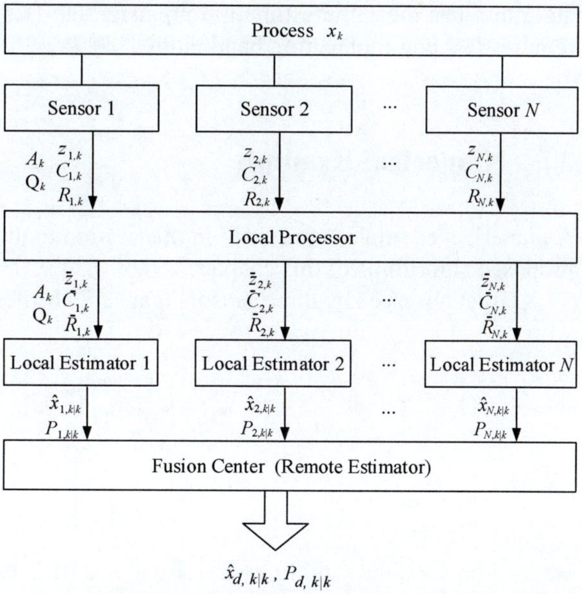

Fig. 3.3 Architecture of the distributed fusion

While, in the current presented distributed fusion algorithm, from Eq. (3.32), we obtain

$$P_{d,k|k}^{-1}\hat{x}_{d,k|k} = P_{d,k|k-1}^{-1}\hat{x}_{d,k|k-1} + \sum_{i=1}^{N}(P_{d,i,k|k}^{-1}\hat{x}_{d,i,k|k} - P_{d,i,k|k-1}^{-1}\hat{x}_{d,i,k|k-1})$$

The two algorithms are both shown to be equivalent to the centralized fusion in estimation precision, and therefore are both optimal in the sense of LMMSE. While, from Eq. (3.48), it is obvious that to generate the state fusion estimation, besides the necessary quantities of the presented distributed algorithm, $C_{i,k}$ and R_k are also required by the fusion center, which means it requires $\frac{1}{2}m(m+1)+mn$ more data transmission at each moment compared to the present distributed fusion algorithm. Moreover, compare Eq. (3.49) with Eq. (3.48), it can be seen that the computation complexity of the current presented algorithm is much less than [22], since there is an augmentation of $C_{i,k}$, i.e., C_k, and the inverse of R_k needed to compute and to multiply in Eq. (3.48).

To summarize, the centralized fusion with the transformed data reduced the communication burden compared to the centralized fusion with the raw data, and the distributed algorithm has even less computation complexity compared to the centralized fusion algorithm when the transformed system is concerned. As has been proven, the three algorithms are all globally optimal in the sense of LMMSE. Compared to the distributed fusion algorithm presented in [22], the presented distributed algorithm has the same estimation precision but less computation complexity and requires less communication bandwidth.

3.5 Numerical Example

A numerical example is provided in this section to illustrate the effectiveness of the proposed algorithms of this chapter.

A target observed by three sensors that could be described by Eqs. (3.1) and (3.2) with

$$A_k = \begin{bmatrix} 1 & T \\ 0 & 1 \end{bmatrix} \cdot 0.95 \tag{3.49}$$

$$Q_k = \begin{bmatrix} \frac{T^3}{3} & \frac{T^2}{2} \\ \frac{T^2}{2} & T \end{bmatrix} \cdot q \tag{3.50}$$

where $T = 1s$ is the sampling rate, and $q = 0.01$ is the disturbance parameter. Sensors 1 and 2 observe the first dimension of position, and Sensor 3 observes the second dimension of velocity, i.e., the measurement matrices are

$$C_{1,k} = [1 \quad 0] \tag{3.51}$$

$$C_{2,k} = [1 \quad 0] \tag{3.52}$$

$$C_{3,k} = [0 \quad 1] \tag{3.53}$$

The measurement noises covariances is given by

$$R_k = \begin{bmatrix} 1.0 & 0.5 & 0.005 \\ 0.5 & 1.0 & 0.005 \\ 0.005 & 0.005 & 0.01 \end{bmatrix} \tag{3.54}$$

The initial conditions are given by

$$\bar{x}_0 = \begin{bmatrix} 10 \\ 0.5 \end{bmatrix}, \quad \bar{P}_0 = \begin{bmatrix} 1 & \frac{1}{T} \\ \frac{1}{T} & \frac{2}{T^2} \end{bmatrix} \tag{3.55}$$

The Monte Carlo simulation results are shown in Figs. 3.4, 3.5, 3.6 and 3.7, where "CF" denotes the centralized fusion algorithm by using the original measurements as shown in Theorem 3.1, "CTF" denotes centralized fusion by using the transformed data as shown in Theorem 3.2, "DF" denotes the distributed fusion presented by [22] and "DTF" denotes the presented distributed fusion by using the transformed data given in Theorem 3.4.

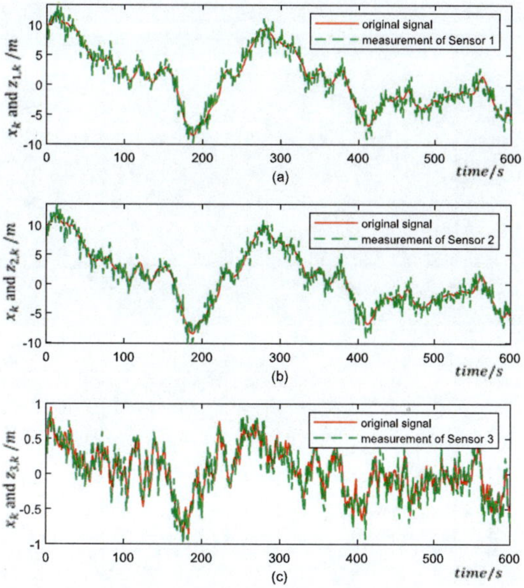

Fig. 3.4 First dimension of the original signal and the measurements

In Fig. 3.4, from (a) through (c), the measurements of Sensors 1 to 3 are shown in green dashed line, where Sensor 1 and 2 observe the position dimension, and Sensor 3 observes the velocity. For comparison, the original signals of the corresponding dimensions are shown in red solid line. It can be seen that the measurements are disturbed by noises.

Figure 3.5 shows the state estimations of the position dimension. From Fig. 3.5a–d are the estimations generated by use of the centralized fusion (CF), the centralized fusion with transformed data (CTF), the distributed fusion algorithm presented in [22] (DF) and the distributed fusion algorithm presented in this chapter by use of the linear transformation (DTF). For comparison, the estimations are shown in green dashed line, while the original signal is shown in red solid line. From this figure, it can be seen that all algorithms generate good estimations.

Figure 3.6 shows the state estimation errors by different algorithms. The green dashed lines in Fig. 3.6a–c are the estimation errors by algorithm CTF, DT and DTF, respectively, and the estimation error of CT is shown in each subfigure in read solid line for comparison. It can be seen that the estimation errors by CTF, DT, and DTF are exactly the same as that of CT, which means the four algorithms are equivalent in state estimation.

Figure 3.7 shows the trace of the covariances of the state estimation errors. It is obvious that all the lines are coincided, which further verify the conclusion drawn from Fig. 3.6.

The simulation in this section shows that if the state estimation precision is concerned, the presented distributed fusion algorithm based on linear transform is equiv-

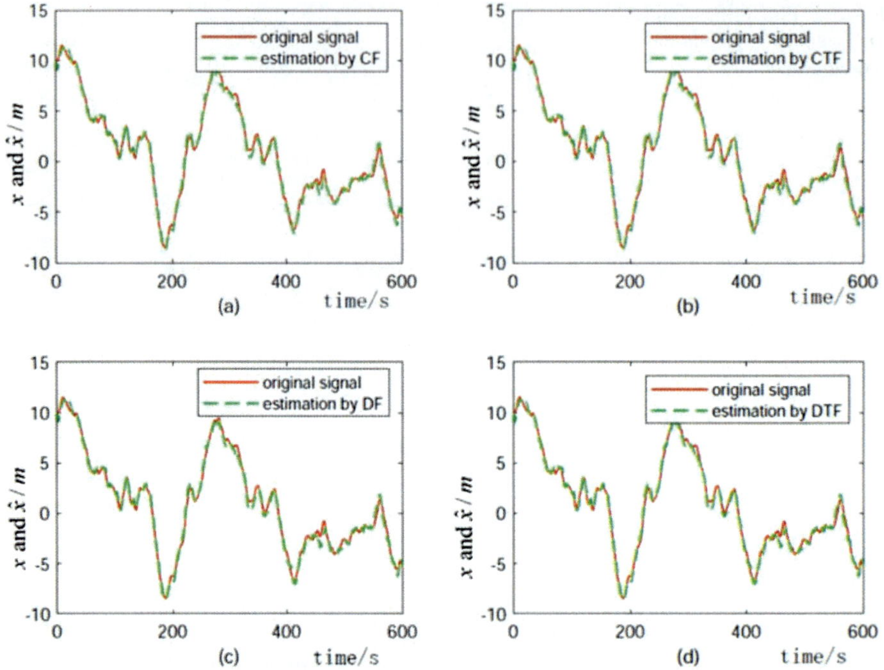

Fig. 3.5 First dimension of the state estimations

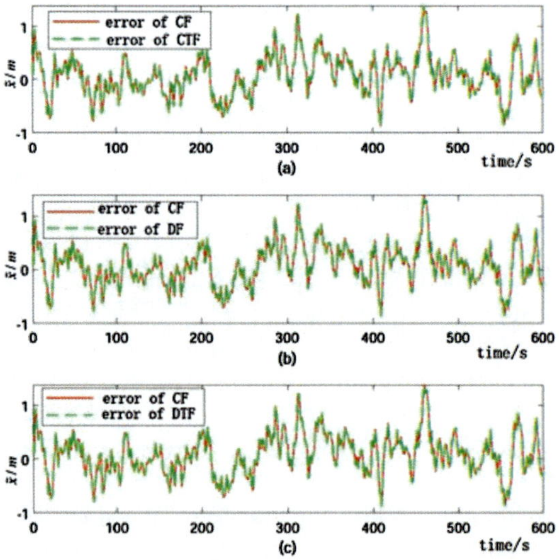

Fig. 3.6 State estimation errors

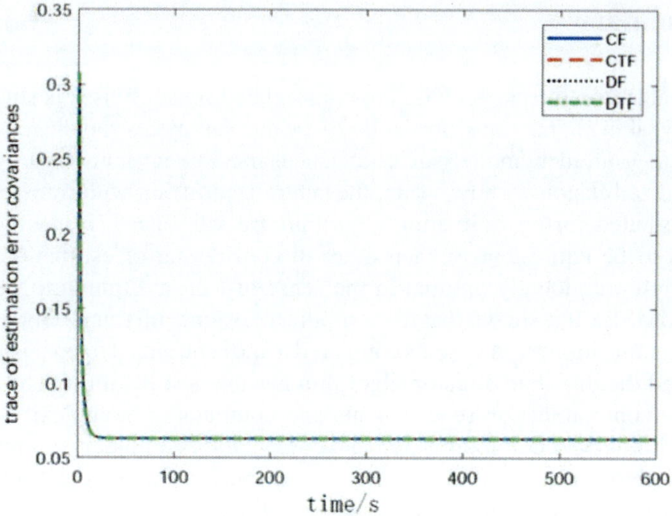

Fig. 3.7 Trace of the covariances of the state estimation errors

alent to the centralized fusion with and without linear transformation, which is also equivalent to the distributed fusion algorithm presented by [22].

In considering of the running time, the average time for 100 runs for each algorithms, namely, CF, CTF, DF and DTF are 0.0468, 0.0156, 0.0780 and 0.0624, respectively. Therefore, the centralized fusion after transformation is faster than the centralized fusion with raw data. The distributed fusion by using the transformed data is more efficient than the distributed fusion presented in [22]. While, the running time of the distributed fusion algorithms are longer than the centralized fusion algorithms. The above conclusions may rely on the following reasons: (1) The dimensions of states and measurements in this simulation are not large enough, the advantages of the distributed fusion algorithms are not obvious. (2) The computing of the inverse of the state estimation error covariances of the local estimations and the state prediction error covariances are time cost. By comparing the distributed or the centralized fusion with and without linear transform, respectively, it can be seen that linear transformation is really an efficient way to reduce computation complexity, let alone the communication or network transmission of data is concerned, which will show even more advantages of linear transformation.

Briefly, the simulation results in this section illustrate the effectiveness of the presented algorithms. When the dimensions of the measurements and the states are not high, the centralized fusion is fine. While, when the dimensions of the measurements are high, the distributed fusion with transformed data have potential advantages.

3.6 Summary

The optimal fusion of sensors with cross–correlated sensor noises is studied in this chapter. By taking linear transformations to the measurements and the related parameters, new measurement models are established, and the sensor noises are decoupled. The centralized fusion with raw data, the centralized fusion with transformed data, and a distributed fusion estimation algorithm are introduced, respectively, which are shown to be equivalent to each other in considering of estimation precision, and therefore are globally optimal in the sense of Linear Minimum Mean Square Errors (LMMSE). It is shown that the centralized fusion with transformed data needs lower communication requirements compared to the centralized fusion using raw data directly, and the distributed fusion algorithm has the best flexibility and robustness and proper communication requirements and computation complexity among the three algorithms (less communication and computation complexity compared to the existed distributed Kalman filtering fusion algorithms). A numerical example is used to show the effectiveness of the proposed algorithms.

References

1. Bar-Shalom, Yarkov, X. Rong Li, and T. Kirubarajan. 2001. *Estimation with Applications to Tracking and Navigation: Theory, Algorithms and Software*. New York: Wiley.
2. Chiuso, Alessandro, and Luca Schenato. 2011. Information fusion strategies and performance bounds in packet-drop networks. *Automatica* 47 (7): 1304–1316.
3. Duan, Zhansheng, Chongzhao Han, and Tangfei Tao. 2004. Optimal multi-sensor fusion target tracking with correlated measurement noises. *IEEE International Conference on Systems*, vol. 2, 1272–1278., Man and Cybernetics Netherlands: Hague.
4. Duan, Zhansheng, X. Rong Li. 2008. The optimality of a class of distributed estimation fusion algorithm. In *11th International Conference on Information Fusion*, 1–6, Cologne, Germany, 30 June–3 July.
5. Duan, Zhansheng, and X. Rong Li. 2011. Lossless linear transformation of sensor data for distributed estimation fusion. *IEEE Transactions on Signal Processing* 59 (1): 362–372.
6. Feng, Jianxin, and Ming Zeng. 2012. Optimal distributed Kalman filtering fusion for a linear dynamic system with cross-correlated noises. *International Journal of Systems Science* 43 (2): 385–398.
7. Feng, Jianxin, Zidong Wang, and Ming Zeng. 2013. Distributed weighted robust Kalman filter fusion for uncertain systems with autocorrelated and cross-correlated noises. *Information Fusion* 14 (1): 78–86.
8. Feng, Xiaoliang. 2009. *Fusion estimation for sensor network systems with correlated measurements and OOSMs*. Dissertation to Hangzhou Dianzi University for the Degree of Master.
9. Jayaweera, Sudharman K., and Carlos Mosquera. 2008. Distributed sequential estimation with noisy, correlated observations. *IEEE Signal Processing Letters* 15 (12): 741–744.
10. Li, J., and G. AlRegib. 2009. Distributed estimation in energy-constrained wireless sensor networks. *IEEE Transactions on Signal Processing* 57 (10): 3746–3758.
11. Li, X. Rong, and Keshu Zhang. 2001. Optimal linear estimation fusion — part iv : Optimality and efficiency of distributed fusion. In *4th International Conference Information Fusion*, 19–268, New Orleans, LA.
12. Li, X. Rong, Yunmin Zhu, Jie Wang, and Chongzhao Han. 2003. Optimal linear estimation fusion i: Unified fusion rules. *IEEE Transactions on Information Theory* 49 (9): 2192–2208.

13. Liang, Yan, Tongwen Chen, and Quan Pan. 2009. Multi-rate optimal state estimation. *International Journal of Control* 82 (11): 2059–2076.
14. Liu, Xiangli, Zan Li, Xiangyang Liu, and Binzhe Wang. 2013. The sufficient condition for lossless linear transformation for distributed estimation with cross-correlated measurement noises. *Journal of Process Control* 23 (10): 1344–1349.
15. Liu, Yulei, Liping Yan, Yuanqing Xia, Mengyin Fu, and Bo Xiao. 2013. Multirate multisensor distributed data fusion algorithm for state estimation with cross-correlated noises. In *Proceedings of the 32th Chinese Control Conference*, 4682–4687, Xi'an, China, July 26–28.
16. Msechu, J.J, S.I. Roumeliotis, A. Ribeiro, and G.B. Giannakis. 2008. Decentralized quantized Kalman filtering with scalable communication cost. *IEEE Transactions on Signal Processing* 56 (8): 3727–3741.
17. Ran, Chenjian, and Zili Deng. 2009. Correlated measurement fusion Kalman filters based on orthogonal transformation. In *Chinese Control and Decision Conference (CCDC 2009)*, 1138–1143, Guilin, China, June 17–19.
18. Ran, Chenjian, and Zili Deng. 2009. Two correlated measurement fusion Kalman filtering algorithms based on orthogonal transformation and their functional equivalence. In *Joint 48th IEEE Conference on Decision and Control and 28th Chinese Control Conference*, 2351–2356, Shanghai, P.R. China, December 16–18.
19. Schizas, I.D, G.B. Giannakis, and Zhiquan Luo. 2007. Distributed estimation using reduced-dimensionality sensor observations. *IEEE Transactions on Signal Processing* 55 (8): 4284–4299.
20. Simon, D. 2006. *Optimal State Estimation: Kalman, H_∞, and Nonlinear Approaches*. New York: John Wiley and Sons, Inc., Publication.
21. Song, Enbin, Yunmin Zhu, and Jie Zhou. 2005. Sensors' optimal dimensionality compression matrix in estimation fusion. *Automatica* 41 (12): 2131–2139.
22. Song, Enbin, Yunmin Zhu, Jie Zhou, and Zhisheng You. 2007. Optimal Kalman filtering fusion with cross-correlated sensor noises. *Automatica* 43 (8): 1450–1456.
23. Sun, Shuli, and Zili Deng. 2004. Multi-sensor optimal information fusion Kalman filter. *Automatica* 40 (6): 1017–1023.
24. Wen, Chenglin, Chuanbo Wen, and Yuan Li. 2008. An optimal sequential decentralized filter of discrete-time systems with cross-correlated noises. In *Proceedings of the 17th World Congress The International Federation of Automatic Control*, 7560–7565, Seoul, Korea.
25. Xu, Jian, Jianxun Li, and Sheng Xu. 2012. Data fusion for target tracking in wireless sensor networks using quantized innovations and Kalman filtering. *Science China Information Sciences* 55 (3): 530–544.
26. Yan, Liping, Lu Jiang, Yuanqing Xia, Mengyin Fu. 2016. State estimation and data fusion for multirate sensor networks. *International Journal of Adaptive Control and Signal Processing* 30 (1): 3–15.
27. Yan, Liping, X. Rong Li, Yuanqing Xia, and Mengyin Fu. 2013. Optimal sequential and distributed fusion for state estimation in cross-correlated noise. *Automatica* 49 (12): 3607–3612.
28. Yan, Liping, Jun Liu, Lu Jiang, Yuanqing Xia, Bo Xiao, Yang Liu, and Mengyin Fu. 2016. Optimal sequential estimation for multirate dynamic systems with unreliable measurements and correlated noise. In *2016 35th Chinese Control Conference (CCC)*, 4900–4905.
29. Yan, Liping, Yuanqing Xia, and Mengyin Fu. 2017. Optimal fusion estimation for stochastic systems with cross–correlated sensor noises. *Science China Information Sciences* 60 (12): 120205:1–120205:14.
30. Zhang, Keshu, X. Rong Li, Peng Zhang, and Haifeng Li. 2003. Optimal linear estimation fusion - part vi: sensor data compression. *Sixth International Conference of Information Fusion* 1: 221–228, 8-11 July.
31. Zhang, Wen-An, Steven Liu, and Li Yu. 2014. Fusion estimation for sensor networks with nonuniform estimation rates. *IEEE Transactions on Circuits and Systems I Regular Papers* 61 (5): 1485–1498.
32. Zhang, Zhi, Jianxun Li, and Liu Liu. 2015. Distributed state estimation and data fusion in wireless sensor networks using multi-level quantized innovation. *Science China Information Sciences* 59: 1–15.

Chapter 4
Distributed Data Fusion for Multirate Sensor Networks

This chapter introduces the state estimation and data fusion of linear dynamic systems observed by multirate sensors in the environment of wireless sensor networks (WSNs). The sampling, estimation, transmission rates in the WSNs are different. In this chapter, the centralized form, the sequential form and the distributed form of optimal linear estimation are derived. Simulation results illustrate the feasibility and effectiveness of the proposed approaches.

4.1 Introduction

The estimation of the state of a dynamic system is called filtering– the reason for the use of the word "filter" is that the process is "filtering out" the noise to obtaining the "best estimate" from noisy data [1]. Therefore, filtering is often used to eliminate undesired parts of the signal, which is the noise in this case. In the control system, the states of the (noisy) dynamic system required by the controller are obtained by filtering the signal. There is a method that is very commonly used in signal processing–in the frequency domain as well as in the spatial domain, which is Filtering of signals. The latter is done, for example, to select signals coming from a certain direction [1].

We can complete the state estimation of the Wireless Sensor Networks (WSNs) under the centralized or the distributed fusion architecture, and distinguish this by whether using the raw data for data fusion directly. This results in a large amount of energy consumption and the potential for a critical failure point at the central collector node, through the use of centralized fusion in the WSNs, which makes it not an effective solution. An alternative solution is to use each sensor in the WSNs as not only a sensor, but also as an estimator with both sensing and computation capabilities, i.e., to estimate in the network [3, 4]. In this case, the estimates are generated only by collecting measurements from its neighbors, which may reduce the communication burden and increase survivability, flexibility, and reliability.

L. Yan et al., *Multisensor Fusion Estimation Theory and Application*,
https://doi.org/10.1007/978-981-15-9426-7_4

In WSNs, there are many algorithms for saving energy because they are important. For example, they mainly include data-compression methods [6, 9] and quantization methods [2, 10, 17]. Both approaches reduce energy consumption in transmitting and receiving packets by reducing the size of data packet. For the considered above estimation problem, another straight-forward and useful way to save energy is to slow down the sensor's information transmission rate. Accordingly, a great number of multi-rate fusion methods are proposed, such as [13, 14]. A transmission rate method that allows the sensor to measure and transmit measurements with a period that is several times of the sampling period is proposed in [16]. This method has strong applicability because it is not only applicable to both cases where the measurement noises are mutually correlated and are uncorrelated, but also applicable to the case where the sensors are not time–synchronized. The sensors may measure and transmit measurements with a period that is several times of the sampling period, which is intuitively called as transmission rate method in [7]. In recent years, many methods combining the transmission rate method and the distributed fusion method are studied, such as [5, 8, 11]. Nevertheless, these articles did not consider energy issues. The nature of the WSNs is a collection of multiple distributed sensors, which means that the traditional single sensor fusion algorithm is not applicable to this network.

This chapter derives the optimal state estimation in the centralized form, the sequential form and the distributed form, respectively. We propose a multi-rate approach where the sensors estimate states at a faster time scale and exchange information with neighbors at a slower time scale to reduce communication costs. Literature [7] used a lifting technique to improve the applicability of the multi-rate approach, modeling the multi-rate estimation system as a single-rate discrete-time system with multiple stochastic parameters. Based on the obtained system model, centralized fusion estimation, sequential fusion estimation, and distributed fusion estimation are studied, respectively. Among them, there are two stages of the distributed fusion estimation. At the first stage, local estimations of every sensor in WSNs are generated by collecting its own measured values, and then a fusion estimation is generated by collecting the local estimations from the neighboring sensors and the neighbour's neighbor sensors. It can be seen from the fusion process that the two-stage fusion estimation method obtains more information from different sensors than the single-stage fusion estimation method to generate the fusion estimation. Therefore, the proposed two-stage estimation method helps steer each local estimation closer to the global optimal one.

This chapter is organized as follows. Section 4.2 formulates the problem. In Sect. 4.3, the optimal estimation algorithms are presented. Simulation is shown in Sect. 4.4 and conclusions are drawn in Sect. 4.5.

4.2 Problem Formulation

The discrete-time linear dynamic system with N sensors can be described by

$$x_{k_{i+1}} = A_p x_{k_i} + B_p w_{p,k_i}, \quad i = 0, 1, 2, \ldots \tag{4.1}$$

$$y_{l,k_i} = C_{pl} x_{k_i} + v_{pl,k_i}, \quad l = 1, 2, \ldots, N \tag{4.2}$$

where $x_{k_i} \in R^n$ is the system state at time k_i. $A_p \in R^{n\times n}$ is the system transition matrix. $B_p \in R^n$ is a constant matrix. w_{p,k_i} is the system noise and is assumed to be Gaussian distributed with zero mean and variance being Q_p, $Q_p \geq 0$. $y_{l,k_i} \in R^{m_l}$ is the k_ith measurement of sensor l. $C_{pl} \in R^{m_l\times n}$ is the measurement matrix. v_{pl,k_i} is the measurement noise and is assumed to be Gaussian distributed with zero mean and variance being R_{pl}. The initial state x_0 is independent of w_{p,k_i} and v_{pl,k_i}, whose mean is $\bar{x}_0$, and variance is $\bar{P}_0$.

In Fig. 4.1, the WSNs consist of 12 sensors, where every sensor only communicates with its neighbors and can communicate with each other within one hop. Notice that a sensor is always connected to itself. There are two steps of the distributed algorithm. In the first step, a local estimate of every sensor $l, l \in \{1, 2, \ldots, N\}$ is generated by use of information of its own. The second step is that the fusion estimation of the target sensor l is generated by use of the local estimations of its neighbors and the neighbour's neighbor being called the neighborhood and is denoted by $\mathcal{N}_l$. Note that only measurements of the sensors in the neighborhood are used for fusion to generate the distributed fusion estimate. For example, the information of sensors $1, 2, 5$ (neighbors of sensor 1) and sensors $3, 4, 6, 7, 9$ (neighbors of sensors $2, 5$) are used to generate the fusion estimation of sensor 1, $\mathcal{N}_1 = \{1, 2, 3, 4, 5, 6, 7, 9\}$. Similarly, the centralized and the sequential fusion estimation of sensors 1 are generated by fusing the information of sensors in $\mathcal{N}_1$, respectively.

It is assumed that the dynamic change of the stochastic process (4.1) is not too rapidly, then it is a waste of energy to brutal forcibly collect every measurement at sampling instants k_i. Figure 4.2 shows an example of the multi-rate state fusion estimation. To reduce the energy waste, the period of measurements sent by every sensor l to its neighbors h_m is assumed to be larger than the sampling period h_p. Denote $t_i, i = 0, 1, 2, \ldots$ as the measurement transmission instants, then

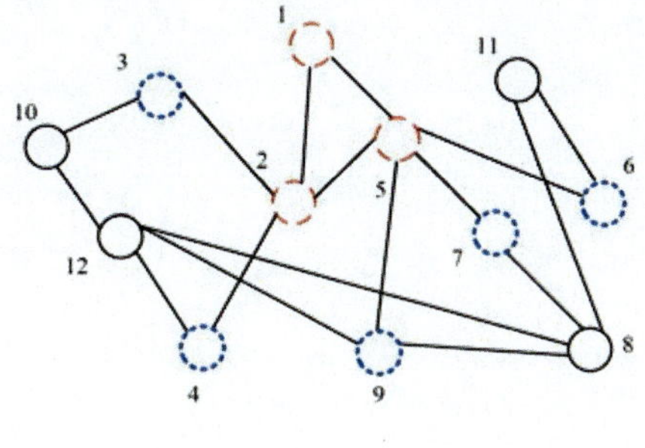

Fig. 4.1 The WSNs with $N = 12$ sensor nodes

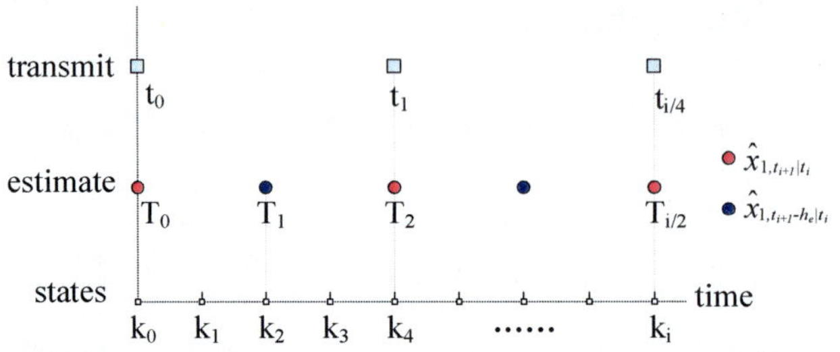

Fig. 4.2 Multi-rate sampling, estimation and transmission for a = 2, b = 2

$h_m = t_i - t_{i-1}, i = 0, 1, 2, \ldots$. Thus, for every sensor in the WSNs, the measurements are collected, and the Kalman filter is used. The local estimations are calculated and outputted with a period of h_m. In practice, one may expect to obtain estimates not only at the instances t_i but also at instances over the interval $(t_{i-1}, t_i]$. This means that updating the estimates at a rate higher than the estimates output rate is an expectation for this study. Suppose that we can update estimates at instances T_i, and $T_i - T_{i-1} = h_e, i = 0, 1, 2, \ldots$. In this generic case, there are three rates in the estimated system, namely, the measurement sampling rate (also the system state updating rate), the measurement transmitting rate (also the estimate outputting rate) and the estimate updating rate. In what follows, we will transform the multirate estimation system model into a single-rate system model for further development by using the lifting technique.

For simplicity and without loss of generality, the relationship between the measurement transmission period h_m and the estimated update period h_e and the measurement sampling period h_p is assumed to be integer multiples. Specifically, let $h_e = ah_p$ and $h_m = bh_e$, where a and b are positive integers and chosen as small as possible in practice under energy constraints of the sensor networks. Then, by applying the difference equation in Eq. (4.1), the following state equation with a state updating period of h_e is obtained:

$$x_{T_{i+1}} = A_e x_{T_i} + B_e w_{e,T_i}, \quad i = 0, 1, 2, \ldots \tag{4.3}$$

where $A_e = A_p^a, \quad B_e = [A_p^{a-1}B_p \quad \cdots \quad A_pB_p \quad B_p]$ and

$$w_{e,T_i} = [w_{p,T_i}^T \quad w_{p,T_i+h_p}^T \quad \cdots \quad w_{p,T_i+(a-1)h_p}^T]^T$$

Similarly, the difference equation in Eq. (4.3) is used to recursively obtain the following state equation with a state updating period of h_m:

$$x_{t_{i+1}} = A_m x_{t_i} + B_m w_{m,t_i}, \quad i = 0, 1, 2, \ldots \tag{4.4}$$

where $A_m = A_e^b, \quad B_m = [A_e^{b-1}B_e \quad \cdots \quad A_eB_e \quad B_e]$ and

$$w_{m,t_i} = [w_{e,t_i}^T \quad w_{e,t_i+h_e}^T \quad \cdots \quad w_{e,t_i+(b-1)h_e}^T]^T$$

We can obtain the corresponding observation models as follows:

$$y_{l,t_i} = C_{pl}x_{t_i} + v_{pl,t_i}, \quad l = 1, 2, \ldots N \tag{4.5}$$

By following the similar procedures for obtaining Eq. (4.4), one has for $j = 1, \ldots, b-1$ that

$$x_{t_{i+1}-jh_e} = A_{mj}x_{t_i} + B_{mj}w_{m,t_i} \tag{4.6}$$

where $A_{mj} = A_e^{b-j}$, and

$$\begin{aligned} B_{m1} &= [A_e^{b-2}B_e \; \cdots \; A_eB_e \; B_e \; 0] \\ B_{m2} &= [A_e^{b-3}B_e \; \cdots \; B_e \; 0 \; 0] \\ &\vdots \\ B_{m(b-1)} &= [B_e \; 0 \cdots 0] \end{aligned}$$

Define

$$x_{o,t_i} = [x_{t_i}^T \, x_{t_i-h_e}^T \cdots x_{t_i-(b-1)h_e}^T]^T \tag{4.7}$$

then the following augmented single-rate estimation system model is obtained from Eqs. (4.4)–(4.6):

$$\begin{cases} x_{o,t_{i+1}} = Ax_{o,t_i} + Bw_{m,t_i} \\ y_{l,t_i} = C_l x_{o,t_i} + v_{pl,t_i} \end{cases} \tag{4.8}$$

where $l = 1, 2, \ldots N$, $i = 0, 1, 2, \ldots$, $C_l = [C_{pl} \; 0 \; \cdots \; 0]$ and

$$A = \begin{bmatrix} A_m & 0 & \cdots & 0 \\ A_{m1} & 0 & \cdots & 0 \\ \vdots & \vdots & \vdots & \vdots \\ A_{m(b-1)} & 0 & \cdots & 0 \end{bmatrix}, \quad B = \begin{bmatrix} B_m \\ B_{m1} \\ \vdots \\ B_{m(b-1)} \end{bmatrix}$$

Let $Q_m = E\{w_{m,t_i}w_{m,t_i}^T\}$. The initial state x_{o,t_0} is uncorrelated with w_{m,t_i} and $v_{pl,t_i}, l = 1, \ldots, N$. We use the transmission rate method and it results in a multi-rate estimation system to reduce energy consumption. By using the lifting technique, we finally model the multi-rate estimation system as a single-rate system with multiple stochastic parameters as in Eq. (4.8). Based on the system model (4.8), we will propose the centralized, the sequential and the distributed fusion algorithms, respectively.

Remark 4.1 We transmit the information of neighbors to the fusion node through the neighbors. Take Fig. 4.1 for an example, Sensor 1 receives the information of the Sensors 3 and 4 through Sensor 2 and receives the information of Sensors 6, 7, 9 through Sensor 5. This chapter assumes that the speed of information transmission is very quick and the time delay in information transmission of above case is ignored.

4.3 The Data Fusion Algorithms for State Estimation

For $r = 1, 2, \ldots, N_l$, denote

$$\mathbb{Y}_r^{t_i} = \{y_{r,t}, 0 < t \le t_i\} \tag{4.9}$$

$${}^r\mathbb{Y}_{t_i} = \{y_{j,t}, t_{i-1} < t \le t_i, j = 1, 2, \ldots, r\} \tag{4.10}$$

$$\begin{aligned} \mathbb{Y}^{t_i,r} &= \{y_{j,t}, 0 < t \le t_i, j = 1, 2, \ldots, r\} \\ &= \{\mathbb{Y}_j^{t_i}\}_{j=1}^r = \{{}^r\mathbb{Y}_k\}_{k=1}^{t_i} \end{aligned} \tag{4.11}$$

then $\mathbb{Y}_r^{t_i}$ denotes the measurements observed by sensor r during time $(0, t_i]$. ${}^r\mathbb{Y}_{t_i}$ stands for the measurements sampled by sensors $1, 2, \ldots, r$ during time $(t_{i-1}, t_i]$. $\mathbb{Y}^{t_i,r}$ stands for the measurements observed by sensors $1, 2, \ldots, r$ up to and including time t_i. $\mathbb{Y}^{t_i,N_l}$ then stands for the measurements observed by all the sensors in $\mathcal{N}_l$ up to and including time t_i, where $N_l = |\mathcal{N}_l|$ denotes the number of the sensors in $\mathcal{N}_l$.

In the sequel, the centralized fusion algorithm, the sequential fusion algorithm and the distributed fusion algorithm are derived to obtain the state estimation.

4.3.1 The Centralized Fusion

Theorem 4.1 (The Centralized Fusion) *For system (4.8), suppose that the state estimation $\hat{x}^c_{ol,t_{i-1}|t_{i-1}}$ and its estimation error covariance $P^c_{ol,t_{i-1}|t_{i-1}}$ are obtained based on data $\mathbb{Y}^{t_{i-1},N_l}$, then we have the state estimation of x_{o,t_i} with the fusion center of sensor l by*

$$\begin{cases} \hat{x}^c_{ol,t_i|t_i} &= \hat{x}^c_{ol,t_i|t_{i-1}} + K^c_{l,t_i}(y_{l,t_i} - \bar{C}_l\hat{x}^c_{ol,t_i|t_{i-1}}) \\ P^c_{ol,t_i|t_i} &= (I - K^c_{l,t_i}\bar{C}_l)P^c_{ol,t_i|t_{i-1}}(I - K^c_{l,t_i}\bar{C}_l)^T + K^c_{l,t_i}R_l K^{c,T}_{l,t_i} \\ K^c_{l,t_i} &= P^c_{ol,t_i|t_{i-1}}\bar{C}_l^T(\bar{C}_l P^c_{ol,t_i|t_{i-1}}\bar{C}_l^T + R_l)^{-1} \\ \hat{x}^c_{ol,t_i|t_{i-1}} &= A\hat{x}^c_{ol,t_{i-1}|t_{i-1}} \\ P^c_{ol,t_i|t_{i-1}} &= AP^c_{ol,t_{i-1}|t_{i-1}}A^T + BQ_mB^T \end{cases} \tag{4.12}$$

where $y_{l,t_i} = \mathrm{col}\{y_{r,t_i}\}_{r\in\mathcal{N}_l}$, $\bar{C}_l = \mathrm{col}\{C_r\}_{r\in\mathcal{N}_l}$, $R_l = \mathrm{diag}\{R_{p1}, R_{p2}, \ldots, R_{pN_l}\}$.

4.3.2 The Sequential Fusion

Theorem 4.2 (The Sequential Fusion) *For system Eq. (4.8), suppose that the state estimation $\hat{x}^s_{ol,t_{i-1}|t_{i-1}}$ and its estimation error covariance $P^s_{ol,t_{i-1}|t_{i-1}}$ are obtained based on data $\mathbb{Y}^{t_i-1,N_l}$. During time $(t_{i-1}, t_i]$, y_{r,t_i}, $r = 1, 2, \ldots, N_l$ are obtained, then we can get the measurements to calculate the state estimation of x_{o,t_i} with the* *fusion center of sensor l by*

$$\begin{cases} \hat{x}^s_{or,t_i|t_i} &= \hat{x}^s_{o(r-1),t_i|t_i} + K^s_{r,t_i}[y_{r,t_i} - C_r\hat{x}^s_{o(r-1),t_i|t_i}] \\ P^s_{or,t_i|t_i} &= P^s_{o(r-1),t_i|t_i} - K^s_{r,t_i}C_rP^s_{o(r-1),t_i|t_i} \\ K^s_{r,t_i} &= P^s_{o(r-1),t_i|t_i}C_r^T[C_rP^s_{o(r-1),t_i|t_i}C_r^T + R_{pr}]^{-1} \end{cases} \tag{4.13}$$

where $\hat{x}^s_{or,t_i|t_i}$ and $P^s_{or,t_i|t_i}$ are the state estimation of x_{o,t_i} and the corresponding estimation error covariance based on the observations of sensors from $1, 2, \dots$ to r. When $r = N_l$, the optimal sequential estimation $\hat{x}^s_{ol,t_i|t_i} = \hat{x}^s_{oN_l,t_i|t_i}$ and the corresponding estimation error covariance $\hat{P}^s_{ol,t_i|t_i} = \hat{P}^s_{oN_l,t_i|t_i}$ are obtained.

For $r = 0$, we have

$$\begin{cases} \hat{x}^s_{o0,t_i|t_i} &= \hat{x}^s_{ol,t_i|t_{i-1}} = A\hat{x}^s_{ol,t_{i-1}|t_{i-1}} \\ P^s_{o0,t_i|t_i} &= P^s_{ol,t_i|t_{i-1}} = AP^s_{ol,t_{i-1}|t_{i-1}}A^T + BQ_mB^T \end{cases} \tag{4.14}$$

Proof By the projection theorem, we have the state prediction of x_{o,t_i} with the fusion center of sensor l based on data $\mathbb{Y}^{t_{i-1},N_l}$ by

$$\begin{aligned} \hat{x}^s_{ol,t_i|t_{i-1}} &= E\{x_{o,t_i}|\mathbb{Y}^{t_{i-1},N_l}\} \\ &= E\{Ax_{o,t_{i-1}} + Bw_{m,t_{i-1}}|\mathbb{Y}^{t_{i-1},N_l}\} \\ &= A\hat{x}^s_{ol,t_{i-1}|t_{i-1}} \end{aligned} \tag{4.15}$$

Then, the state prediction error covariance is as follows

$$P^s_{ol,t_i|t_{i-1}} = E\{\tilde{x}^s_{ol,t_i|t_{i-1}}\tilde{x}^{s,T}_{ol,t_i|t_{i-1}}\} \tag{4.16}$$

From Eqs. (4.1) and (4.15), we have the state prediction error by

$$\begin{aligned} \tilde{x}^s_{ol,t_i|t_{i-1}} &= x_{o,t_i} - \hat{x}^s_{ol,t_i|t_{i-1}} \\ &= A\tilde{x}^s_{ol,t_{i-1}|t_{i-1}} + Bw_{m,t_{i-1}} \end{aligned} \tag{4.17}$$

Substituting Eq. (4.17) into Eq. (4.16) yields

$$P^s_{ol,t_i|t_{i-1}} = AP^s_{ol,t_{i-1}|t_{i-1}}A^T + BQ_mB^T \tag{4.18}$$

Denote $\hat{x}^s_{o0,t_i|t_i} = \hat{x}^s_{ol,t_i|t_{i-1}}$ and $P^s_{o0,t_i|t_i} = P^s_{ol,t_i|t_{i-1}}$ and as assumed in Theorem 4.2, we can obtain y_{r,t_i}, $r = 1, 2, \dots, N_l$ during $(t_{i-1}, t_i]$, then $\hat{x}^s_{or,t_i|t_i}$ is obtained by applying the projection theorem

$$\begin{aligned} \hat{x}^s_{or,t_i|t_i} &= E\{x_{o,t_i}|\mathbb{Y}^{t_{i-1},N_l}, {}^r\mathbb{Y}_{t_i}\} \\ &= E\{x_{o,t_i}|\mathbb{Y}^{t_{i-1},N_l}, {}^{r-1}\mathbb{Y}_{t_i}, y_{r,t_i}\} \\ &= \hat{x}^s_{o(r-1),t_i|t_i} + \mathrm{cov}\{\tilde{x}^s_{o(r-1),t_i|t_i}, \tilde{y}_{r,t_i}\}\mathrm{var}\{\tilde{y}_{r,t_i}\}^{-1}\tilde{y}_{r,t_i} \end{aligned} \tag{4.19}$$

where

$$\tilde{y}_{r,t_i} = y_{r,t_i} - \hat{y}_{r,t_i} \tag{4.20}$$

$$\text{var}\{\tilde{y}_{r,t_i}\} = E\{\tilde{y}_{r,t_i}\tilde{y}_{r,t_i}^T\} \tag{4.21}$$

and

$$\begin{aligned} \hat{y}_{r,t_i} &= E\{y_{r,t_i}|\mathbb{Y}^{t_{i-1},N_l}, {}^{r-1}\mathbb{Y}_{t_i}\} \\ &= E\{C_r x_{o,t_i} + v_{pr,t_i}|\mathbb{Y}^{t_{i-1},N_l}, {}^{r-1}\mathbb{Y}_{t_i}\} \\ &= C_r\hat{x}^s_{o(r-1),t_i|t_i} \end{aligned} \tag{4.22}$$

So,

$$\begin{aligned} \tilde{y}_{r,t_i} &= y_{r,t_i} - \hat{y}_{r,t_i} \\ &= y_{r,t_i} - C_r\hat{x}^s_{o(r-1),t_i|t_i} \\ &= C_r x_{o,t_i} + v_{pr,t_i} - C_r\hat{x}^s_{o(r-1),t_i|t_i} \\ &= C_r\tilde{x}^s_{o(r-1),t_i|t_i} + v_{pr,t_i} \end{aligned} \tag{4.23}$$

Therefore,

$$\begin{aligned} \text{cov}\{\tilde{x}^s_{o(r-1),t_i|t_i}, \tilde{y}_{r,t_i}\} &= E\{\tilde{x}^s_{o(r-1),t_i|t_i}\tilde{y}_{r,t_i}^T\} \\ &= E\{\tilde{x}^s_{o(r-1),t_i|t_i}[C_r\tilde{x}^s_{o(r-1),t_i|t_i} + v_{pr,t_i}]^T\} \\ &= P^s_{o(r-1),t_i|t_i}C_r^T \end{aligned} \tag{4.24}$$

and

$$\begin{aligned} \text{var}\{\tilde{y}_{r,t_i}\} &= E\{\tilde{y}_{r,t_i}\tilde{y}_{r,t_i}^T\} \\ &= E\{[C_r\tilde{x}^s_{o(r-1),t_i|t_i} + v_{pr,t_i}][C_r\tilde{x}^s_{o(r-1),t_i|t_i} + v_{pr,t_i}]^T\} \\ &= C_r P^s_{o(r-1),t_i|t_i}C_r^T + R_{pr} \end{aligned} \tag{4.25}$$

Substitute Eqs. (4.24), (4.25) and the second equation of Eq. (4.23) into Eq. (4.19) yields

$$\hat{x}^s_{or,t_i|t_i} = \hat{x}^s_{o(r-1),t_i|t_i} + K^s_{r,t_i}[y_{r,t_i} - C_r\hat{x}^s_{o(r-1),t_i|t_i}] \tag{4.26}$$

where

$$\begin{aligned} K^s_{r,t_i} &= \text{cov}\{\tilde{x}^s_{o(r-1),t_i|t_i}, \tilde{y}_{r,t_i}\}\text{var}\{\tilde{y}_{r,t_i}\}^{-1} \\ &= P^s_{o(r-1),t_i|t_i}C_r^T[C_r P^s_{o(r-1),t_i|t_i}C_r^T + R_{pr}]^{-1} \end{aligned} \tag{4.27}$$

Then, by using Eq. (4.27), we can obtain the estimation error covariance as follow

$$
\begin{aligned}
P^s_{or,t_i|t_i} &= E\{\tilde{x}^s_{or,t_i|t_i}\tilde{x}^{s,T}_{or,t_i|t_i}\} \\
&= E\{[x_{o,t_i} - \hat{x}^s_{or,t_i|t_i}][x_{o,t_i} - \hat{x}^s_{or,t_i|t_i}]^T\} \\
&= E\{[(I - K^s_{r,t_i}C_r)\tilde{x}^s_{o(r-1),t_i|t_i} - K^s_{r,t_i}v_{pr,t_i}][(I - K^s_{r,t_i}C_r)\tilde{x}^s_{o(r-1),t_i|t_i} \\
&\quad - K^s_{r,t_i}v_{pl,t_i}]^T\} \\
&= [I - K^s_{r,t_i}C_r]P^s_{o(r-1),t_i|t_i}[I - K^s_{r,t_i}C_r]^T + K^s_{r,t_i}R_{pr}K^{s,T}_{r,t_i} \\
&= P^s_{o(r-1),t_i|t_i} - K^s_{r,t_i}C_rP^s_{o(r-1),t_i|t_i}
\end{aligned}
\tag{4.28}
$$

Combining Eqs. (4.26)–(4.28), we obtain

$$
\begin{cases}
\hat{x}^s_{or,t_i|t_i} &= \hat{x}^s_{o(r-1),t_i|t_i} + K^s_{r,t_i}[y_{r,t_i} - C_r\hat{x}^s_{o(r-1),t_i|t_i}] \\
P^s_{or,t_i|t_i} &= P^s_{o(r-1),t_i|t_i} - K^s_{r,t_i}C_rP^s_{o(r-1),t_i|t_i} \\
K^s_{r,t_i} &= P^s_{o(r-1),t_i|t_i}C^T_r[C_rP^s_{o(r-1),t_i|t_i}C^T_r + R_{pr}]^{-1}
\end{cases}
\tag{4.29}
$$

If $\hat{x}^s_{ol,t_i|t_i} = \hat{x}^s_{oN_l,t_i|t_i}$ and $P^s_{ol,t_i|t_i} = P^s_{oN_l,t_i|t_i}$ are denoted, the optimal sequential estimation $\hat{x}^s_{ol,t_i|t_i}$ and its estimation error covariance $P^s_{ol,t_i|t_i}$ with sensor l as the fusion censor are obtained, and the proof is completed.

4.3.3 Two-Stage Distributed Fusion

To obtain the optimal distributed estimation, the measurements of each sensor are first used to generate he local Kalman estimators.

Theorem 4.3 (The Local Estimation) *For system (4.8), suppose that the optimal state estimation $\hat{x}_{ol,t_{i-1}|t_{i-1}}$ and its estimation error covariance $P_{ol,t_{i-1}|t_{i-1}}$ based on data $\mathbb{Y}^{t_{i-1}}_l$ are obtained, then we can get the state estimation of x_{o,t_i} based on data $\mathbb{Y}^{t_i}_l$ by:*

$$
\begin{cases}
\hat{x}_{ol,t_i|t_i} &= \hat{x}_{ol,t_i|t_{i-1}} + K_{l,t_i}(y_{l,t_i} - C_l\hat{x}_{ol,t_i|t_{i-1}}) \\
P_{ol,t_i|t_i} &= (I - K_{l,t_i}C_l)P_{ol,t_i|t_{i-1}}(I - K_{l,t_i}C_l)^T + K_{l,t_i}R_{pl}K^T_{l,t_i} \\
K_{l,t_i} &= P_{ol,t_i|t_{i-1}}C^T_l(C_lP_{ol,t_i|t_{i-1}}C^T_l + R_{pl})^{-1} \\
\hat{x}_{ol,t_i|t_{i-1}} &= A\hat{x}_{ol,t_{i-1}|t_{i-1}} \\
P_{ol,t_i|t_{i-1}} &= AP_{ol,t_{i-1}|t_{i-1}}A^T + BQ_mB^T
\end{cases}
\tag{4.30}
$$

where $\hat{x}_{ol,t_i|t_i}$ and $P_{ol,t_i|t_i}$ are the state estimation of x_{o,t_i} based on data $\mathbb{Y}^{t_i}_l$ and the corresponding estimation error covariance.

Next, we apply a fusion criterion weighted by matrices in the linear minimum variance sense to generate the fusion estimation for every sensor l, $l \in [1, N]$. When each sensor in WSNs has local estimates calculated based on the estimators in Theorem 4.3, the fusion estimation of each sensor l then is generated based on collecting these local estimations from its neighborhood $\mathcal{N}_l$ and the distributed fusion rules in [11]. After the two steps, we complete the distributed fusion estimation.

Theorem 4.4 (The Distributed Fusion) *Let* $\hat{x}_{or,t_i|t_i}, r \in \mathcal{N}_l$ *be the local estimations obtained by Theorem 4.3. Let the estimation errors be* $\tilde{x}_{or,t_i|t_i} = x_{or,t_i} - \hat{x}_{or,t_i|t_i}$. *Assume that* $\tilde{x}_{or,t_i|t_i}$ *and* $\tilde{x}_{oj,t_i|t_i}$ *are correlated, and the covariance and cross-covariance matrices are defined as* $P_{rr,t_i|t_i} = E\{\tilde{x}_{or,t_i|t_i}\tilde{x}^T_{or,t_i|t_i}\}$ *and* $P_{rj,t_i|t_i} = E\{\tilde{x}_{or,t_i|t_i}\tilde{x}^T_{oj,t_i|t_i}\}(j \in \mathcal{N}_l, r \neq j)$ *, respectively. Then, we can get the optimal fusion estimation of* $x_{o,t_i|t_i}$ *at sensor* $l, l \in [1, N]$ *by*

$$\hat{x}^d_{ol,t_i|t_i} = \sum_{r=1}^{N_l} \alpha_{r,t_i}\hat{x}_{or,t_i|t_i} \tag{4.31}$$

$$P^d_{ol,t_i|t_i} = (e^T \Sigma_{t_i}^{-1} e)^{-1} \tag{4.32}$$

where, we have the optimal matrix weights α_{r,t_i} *by*

$$\begin{aligned} \alpha_{t_i} &= (e^T \Sigma_{t_i}^{-1} e)^{-1} e^T \Sigma_{t_i}^{-1} \\ &= [\alpha_{1,t_i}\ \alpha_{2,t_i}\ \cdots\ \alpha_{N_l,t_i}] \end{aligned} \tag{4.33}$$

where $\Sigma_{t_i} = [P_{rj,t_i|t_i}], r, j \in \mathcal{N}_l$ *is a symmetric positive definite matrix with dimension* bnN_l *and* $e = [\underbrace{I_n \ \cdots \ I_n}_{N_l}]^T$, $I_n \in R^{bn \times bn}$ *is an identity matrix. We can compute* $P_{rj,t_i|t_i}, r, j = 1, 2, \ldots\ \mathcal{N}_l$ *by*

$$P_{rj,t_i|t_i} = [I - K_{r,t_i} C_r][A P_{rj,t_{i-1}|t_{i-1}} A^T + B Q_m B^T][I - K_{j,t_i} C_j]^T \tag{4.34}$$

We can prove that the covariance matrix of the fusion estimation error $P^d_{ol,t_i|t_i}$ *meets*

$$P^d_{ol,t_i|t_i} \leq P_{rr,t_i|t_i}, r \in \mathcal{N}_l. \tag{4.35}$$

Which means the fusion estimation is effective in the sense of minimizing the estimation error covariance.

Proof

$$\begin{aligned} P_{rj,t_i|t_i} &= E\{[x_{o,t_i} - \hat{x}_{or,t_i|t_i}] \cdot [x_{o,t_i} - \hat{x}_{oj,t_i|t_i}]^T\} \\ &= E\{[(I - K_{r,t_i} C_r)\tilde{x}_{or,t_i|t_{i-1}} - K_{r,t_i} v_{pr,t_i}] \\ &\quad \cdot [(I - K_{j,t_i} C_j)\tilde{x}_{oj,t_i|t_{i-1}} - K_{j,t_i} v_{pj,t_i}]^T\} \\ &= (I - K_{r,t_i} C_r) E\{\tilde{x}_{or,t_i|t_{i-1}}\tilde{x}^T_{oj,t_i|t_{i-1}}\}(I - K_{j,t_i} C_j)^T + K_{r,t_i} \\ &\quad \cdot E\{v_{pr,t_i} v^T_{pl,t_i}\} K^T_{j,t_i} - (I - K_{r,t_i} C_r) E\{\tilde{x}_{or,t_i|t_{i-1}} v^T_{pj,t_i}\} K^T_{j,t_i} \\ &\quad - K_{r,t_i} E\{v_{pr,t_i}\tilde{x}^T_{oj,t_i|t_{i-1}}\}(I - K_{j,t_i} C_j)^T \\ &= [I - K_{r,t_i} C_r][A P_{rj,t_{i-1}|t_{i-1}} A^T + B Q_m B^T][I - K_{j,t_i} C_j]^T \end{aligned} \tag{4.36}$$

Let

$$F(x) = (y_{t_i} - ex_{o,t_i})^T \Sigma_{t_i}^{-1} (y_{t_i} - ex_{o,t_i}) \tag{4.37}$$

where $y_{t_i} = [\hat{x}_{o1,t_i|t_i} \ \hat{x}_{o2,t_i|t_i} \ \cdots \ \hat{x}_{oN_l,t_i|t_i}]^T$, $e = [\underbrace{I_n \ \cdots \ I_n}_{N_l}]^T$, and $\Sigma = [P_{rj,t_i|t_i}]$ is a $bnN_l \times bnN_l$ symmetrical positive definite matrix.

For each $t_i = 1, 2, \ldots$, obviously, $F(x)$ is a quadratic form of x_{o,t_i} and thus a convex function of x_{o,t_i}. So we can find the minimum of $F(x)$ by $\frac{dF(x)}{dx} = 0$, from which we obtain

$$x_{o,t_i} = (e^T \Sigma_{t_i}^{-1} e)^{-1} e^T \Sigma_{t_i}^{-1} y_{t_i} \tag{4.38}$$

So, from the unbiasedness property, the optimal distributed estimation is given by

$$\hat{x}_{ol,t_i|t_i}^d = \alpha_{t_i} y_{t_i} = \sum_{r=1}^{N_l} \alpha_{r,t_i} \hat{x}_{or,t_i|t_i} \tag{4.39}$$

where $\alpha_{t_i} = [\alpha_{1,t_i} \ \alpha_{2,t_i} \ \cdots \ \alpha_{N_l,t_i}] = (e^T \Sigma_{t_i}^{-1} e)^{-1} e^T \Sigma_{t_i}^{-1}$ is a $bn \times bnN_l$ matrix.

So, we can obtain the error covariance of the optimal distributed estimation as

$$\begin{aligned} P_{ol,t_i|t_i}^d &= \sum_{r=1}^{N_l} \sum_{j=1}^{N_l} \alpha_{r,t_i} P_{rj,t_i|t_i} \alpha_{j,t_i}^T \\ &= \alpha_{t_i} \Sigma_{t_i} \alpha_{t_i}^T \\ &= (e^T \Sigma_{t_i}^{-1} e)^{-1} \end{aligned} \tag{4.40}$$

Notice that by setting $\alpha_{r,t_i} = I_n$ and $\alpha_{j,t_i} = 0$ $(j = 1, 2 \ldots, N_l, j \neq r)$ in Eq. (4.40), $P_{ol,t_i|t_i}^d$ is equivalent to $P_{rr,t_i|t_i}$. Therefore, $P_{ol,t_i|t_i}^d \leq P_{rr,t_i|t_i}$, $r \in \mathcal{N}_l$ is obtained.

Remark 4.2 At each fusion center, the using information in this chapter and [15] are the same. Due to the two distributed fusion algorithms have the same estimation accuracy, both are optimal in the sense of weighted least square. The presented algorithm does not reuse information, so its computational complexity is smaller. Take Fig. 4.1 for an example, Sensor 1 uses the information from Sensors 1, 2, 3, 4, 5, 6, 7, 9 to get the state fusion estimation. By using the presented algorithm, the state fusion estimation at sensor 1 is obtained based on the local state estimations of every sensor in $\mathcal{N}_l$. In [15], firstly, the local estimation of sensor 1 is generated by fusing the information of Sensors 1, 2 and 5; then the local estimation of Sensor 2 is generated by fusing the information of Sensors 1, 2, 3, 4, 5, and the local estimation of Sensor 5 is generated by fusing the information of Sensors 1, 2, 5, 6, 7, 9; finally, the state fusion estimation at Sensor 1 is generated by fusing the local estimation of Sensors 1, 2, and 5. Thus, in this fusion process, to get the state fusion estimation, the information of Sensors 1, 2 and 5 is repeatedly used. The next section is the simulation of an example. From the simulation results, one can see that the presented distributed fusion algorithm has similar estimation accuracy with [15] but less CPU time in the same hardware environment by use of matlab. Therefore, it is easy to notice that the presented distributed algorithm in this chapter has better performance.

Remark 4.3 The above three algorithms are estimation accuracy in different meanings, that is, the centralized fusion algorithm (also called the optimal batch fusion: OBF) is linear minimum mean square error (LMMSE) optimal, the optimal distributed fusion algorithm (ODF) is weighted least square (WLS) optimal, and the optimal sequential fusion (OSF) is LMMSE optimal among sequential use of measurements. They are not necessarily equivalent [12] gives a typical example. When computational complexity is concerned, the OSF and the OBF are comparable, but the OSF is more efficient. The reason is that OBF needs time to collect the information of all sensors in a given time, and OSF handles data sequentially and the estimation results have nothing to do with the order of the sensors to be fused, when there is stagnation in the collected data. Therefore, any available sensor information can be processed in the OSF fusion center. Comparing the OSF with the ODF, one can see that the computational complexity of the OSF is comparable with that of the ODF for the generation of the local estimates. However, we need a further fusion step to get the optimal distributed estimation. When parallel computation is allowed, the ODF could be better, along with better flexibility, robustness and survivability than the OBF and the OSF. The simulation in the next section is a proper example.

4.4 Numerical Example

In this section, an example is given to illustrate the effectiveness of the presented algorithms. The tracking system of two sensors can be described by Eqs. (4.1) and (4.2), where [15]

$$A_p = \begin{bmatrix} 1 & h_p \\ 0 & 1 \end{bmatrix}, \quad B_p = \begin{bmatrix} 1.81 \\ 1 \end{bmatrix} \tag{4.41}$$

where h_p is the sampling period, and we take $h_p = 0.5$ s in the simulation. We deploy a wireless sensor network with 12 sensor nodes to monitor the target in this example, and the topology of the WSNs is shown in Fig. 4.1.

$$\begin{cases} C_{p1} = [1 \quad 0], & C_{p2} = [0.8 \quad 0] \\ C_{p3} = [0.7 \quad 0], & C_{p4} = [0.6 \quad 0] \\ C_{p5} = [0.5 \quad 0], & C_{p6} = [0.4 \quad 0] \\ C_{p7} = [0.3 \quad 0], & C_{p8} = [0.2 \quad 0] \\ C_{p9} = [1 \quad 0], & C_{p10} = [0.8 \quad 0] \\ C_{p11} = [0.6 \quad 0], & C_{p12} = [0.7 \quad 0] \end{cases} \tag{4.42}$$

Table 4.1 Average CPU run time

	PDA	ZA	PCA	PSA
Time	12.1250	40.7344	4.6563	4.9219

Table 4.2 Time averaged RMSE for position and velocity

	PDA	ZA	PCA	PSA
Position	0.3403	0.3869	0.3261	0.3261
Velocity	0.1903	0.1986	0.1580	0.1580

We set the system noise covariance and the measurement noise covariances as:

$$Q_p = 0.1 \tag{4.43}$$

$$\begin{cases} R_{p1} = 0.4, & R_{p2} = 0.7, & R_{p3} = 0.4 \\ R_{p4} = 0.4, & R_{p5} = 0.3, & R_{p6} = 0.2 \\ R_{p7} = 0.3, & R_{p8} = 0.3, & R_{p9} = 0.5 \\ R_{p10} = 0.4, & R_{p11} = 0.3, & R_{p12} = 0.1 \end{cases} \tag{4.44}$$

First, we set $a = b = 2$, which means the generation period of every sensor to generate estimations is $h_m = 2s$, which is four times of the sampling period. The period of updating estimations is $h_e = 1s$, which is two times of the sampling period. The proposed method saves energy consumption in communication and calculation by slowing down the measurement transmission rate and the estimation updating rate. The initial time is $t_0 = 0$, and the initial state is given by $\bar{x}_o = [1 \ \ 0.5]^T$, $\bar{P}_o = \text{diag}\,\{0.25, 0.25\}$.

In the sequel, the fusion estimation is generated according to Theorems 4.1–4.4. Tables 4.1 and 4.2 and Figs. 4.3, 4.4 and 4.5 show the results from 100 Monte Carlo simulations, where Sensor 1 is the fusion center.

In Table 4.1, the results by use of model (4.8) and the presented distributed algorithm in Theorem 4.4 are denoted by "PDA", the results by use of [15] are denoted by "ZA", the results by use of the presented centralized algorithm in Theorem 4.1 are denoted by "PCA", the results by use of the presented sequential algorithm in Theorem 4.2 are denoted by "PSA".

The average CPU time of 100 runs is listed in Table 4.1. From Table 4.1, one can obviously see that the run time of the presented centralized algorithm (PCA) is the smallest, followed by the sequential algorithm (PSA) and the presented distributed algorithm (PDA), and [15] (ZA) is the largest. Therefore, the advantage of the presented three algorithms in this chapter is that the amount of calculation is small.

The time average of the root mean square error (RMSE) results is listed in Table 4.2. From Table 4.2, for the average RMSE of position and velocity, PCA

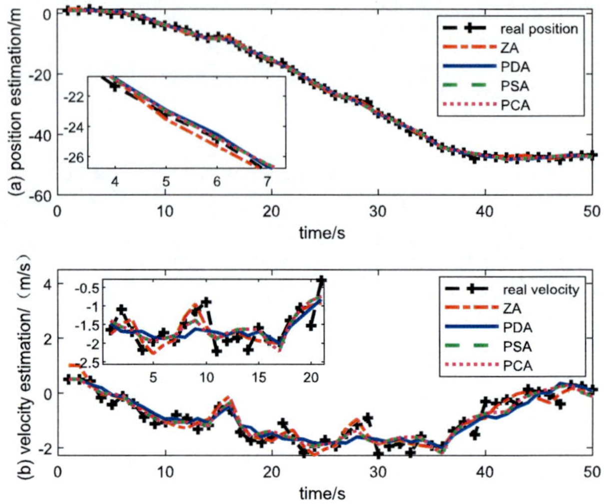

Fig. 4.3 The signal and the estimates

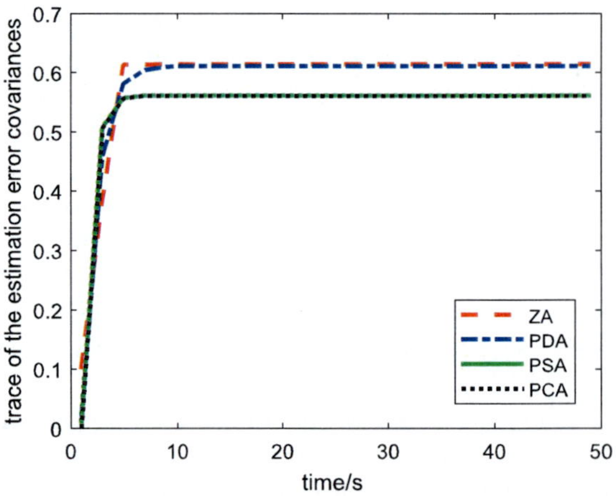

Fig. 4.4 The Trace of the estimation error covariances

and PSA are the optimal, followed by ZA, and PDA is the worst in the sense of minimizing RMSE. Therefore, from Table 4.2, PCA and PSA are superior to ZA and PDA.

From Fig. 4.3, the black dashed line with plus indicates the real position and real velocity; the green dash-dotted line represents the state estimates by PSA; the red dashed line is the state estimates by ZA; the blue dashed line denotes the state estimates by PDA; and the purple dotted line indicates the state estimates by PCA. It can be seen from Fig. 4.3 that the results of the four algorithms are similar.

In Fig. 4.4, the red dashed line indicates the trace of the state estimation error covariances by use of the algorithm ZA, blue dash-dotted line represents the PDA, the green solid line is the PSA and the black dotted line denotes the PCA. One can see that PCA and PSA are equivalent, and more effective than PDA and ZA.

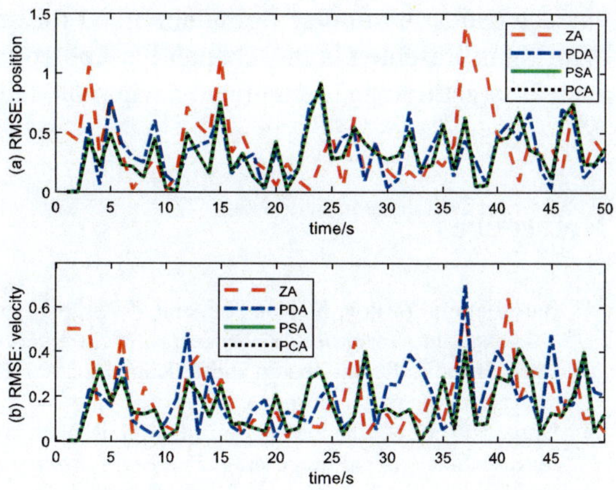

Fig. 4.5 Root mean square errors: position and velocity

In Fig. 4.5, the blue dashed line, the red dash-dotted line, the black solid line and the green dotted line represent the simulation curves of the RMSE (position and velocity) of the state estimation of the ZA, PDA, PSA and PCA algorithms, respectively. One can see that for the result of position RMSE, PSA and PCA are the optimal, followed by PDA, and ZA is the worst. For the results of velocity RMSE, the PSA and PCA are better than that of PDA and ZA.

Remark 4.4 In the simulation, we compared the centralized fusion, the sequential fusion, the distributed fusion algorithms and the algorithm in [15]. From the simulation results, one can see that the presented centralized, the sequential and the distributed algorithms all have good estimation effects. The results show that the centralized fusion and the sequential fusion are equivalent and the optimal in the sense of minimum RMSE and minimizing the trace of the estimation error covariances, and [15] is the worst. In addition, considering the CPU time in matlab simulation under the same hardware environment, the centralized fusion is the optimal, followed by the sequential fusion and the distributed fusion, and [15] is the worst.

4.5 Summary

In this chapter, we derived the centralized, the sequential and the distributed fusion estimation algorithms for estimating the states of a discrete time linear stochastic system in the WSNs environment. To reduce communication costs, we set the sampling, estimation updating and transmitting speed of sensors in the WSNs to be different. Through theoretical proof and simulation, the following conclusions are obtained: The presented centralized, sequential, and distributed fusion algorithms are effective; the presented centralized fusion algorithm and the sequential fusion algorithm are the optimal in the sense of minimizing the mean square error; compared with the existing results, the distributed fusion algorithm proposed in this chapter has the characteristics of simpler formulation and the smaller amount of calculation. The algorithms presented in this chapter have potential value in many application fields, such as target tracking, integrated navigation, fault detection and control, etc.

References

1. Bar-Shalom, Yarkov, X. Rong Li, and T. Kirubarajan. 2001. *Estimation with Applications to Tracking and Navigation: Theory, Algorithms and Software*. New York: Wiley.
2. Chen, Huimin, Keshu Zhang, and X. Rong Li. 2004. Optimal data compression for multisensor target tracking with communication constraints. In *IEEE Conference on Decision and Control.*
3. Dimakis, A.G, S. Kar, J.M.F. Moura, M.G. Rabbat, and A. Scaglione. 2010. Gossip algorithms for distributed signal processing. *Proceedings of the IEEE* 98 (11): 1847–1864.
4. Giridhar, A, and P.R. Kumar. 2006. Toward a theory of in-network computation in wireless sensor networks. *Communications Magazine of the IEEE* 44 (5): 98–107.
5. Jing, Ma, and Shuli Sun. 2012. Distributed fusion filter for multi-rate multi-sensor systems with packet dropouts. In *Intelligent Control and Automation.*
6. Li, J, and G. AlRegib. 2009. Distributed estimation in energy-constrained wireless sensor networks. *IEEE Transactions on Signal Processing* 57 (10): 3746–3758.
7. Liang, Yan, Tongwen Chen, and Quan Pan. 2009. Multi-rate optimal state estimation. *International Journal of Control* 82 (11): 2059–2076.
8. Ma, Jing, Honglei Lin, and Shuli Sun. 2012. Distributed fusion filter for asynchronous multirate multi-sensor non-uniform sampling systems. In *2012 15th International Conference on Information Fusion (FUSION).*
9. Msechu, J.J, S.I. Roumeliotis, A. Ribeiro, and G.B. Giannakis. 2008. Decentralized quantized Kalman filtering with scalable communication cost. *IEEE Transactions on Signal Processing* 56 (8): 3727–3741.
10. Schizas, I.D, G.B. Giannakis, and Zhiquan Luo. 2007. Distributed estimation using reduced-dimensionality sensor observations. *IEEE Transactions on Signal Processing* 55 (8): 4284–4299.
11. Sun, Shuli, and Zili Deng. 2004. Multi-sensor optimal information fusion Kalman filter. *Automatica* 40 (6): 1017–1023.
12. Yan, Liping, X. Rong Li, Yuanqing Xia, and Mengyin Fu. 2013. Optimal sequential and distributed fusion for state estimation in cross-correlated noise. *Automatica* 49 (12): 3607–3612.
13. Yan, Liping, Baosheng Liu, and Donghua Zhou. 2007. Asynchronous multirate multisensor information fusion algorithm. *IEEE Transactions on Aerospace and Electronic Systems* 43 (3): 1135–1146.
14. Yan, Liping, Donghua Zhou, Mengyin Fu, and Yuanqing Xia. 2010. State estimation for asynchronous multirate multisensor dynamic systems with missing measurements. *IET Signal Processing* 4 (6): 728–739.
15. Zhang, Wen-An, Gang Feng, and Li Yu. 2012. Multi-rate distributed fusion estimation for sensor networks with packet losses. *Automatica* 58 (9): 2016–2028.
16. Zhang, Wen-An, Steven Liu, and Li Yu. 2014. Fusion estimation for sensor networks with nonuniform estimation rates. *IEEE Transactions on Circuits and Systems I Regular Papers* 61 (5): 1485–1498.
17. Zhu, H, I.D. Schizas, and G.B. Giannakis. 2009. Power-efficient dimensionality reduction for distributed channel-aware Kalman tracking using WSNs. *IEEE Transactions on Signal Processing* 57 (8): 3193–3207.

Chapter 5
State Estimation for Multirate Systems with Unreliable Measurements

In this chapter, the sequential fusion estimation problem is studied when multi-rate multisensor data with randomly unreliable measurements are concerned in case of noise correlation.

5.1 Introduction

In the field of target tracking and navigation, multi-sensor data fusion has been widely applied. Most of the data fusion algorithms are built on the premise that the sensor observation information is reliable. However, in practical problems, due to the limitation of communication and sensor fault, etc., data missing or unreliable measurements will happen inevitably. In addition, a lot of literatures aimed to solve the estimation problem with the assumption that the measurement noise between various sensors is irrelevant, and also independent of the system noise. However, noise correlation is more practical. In this chapter, the multi-rate multi-sensor data fusion state estimation with unreliable observations under correlated noises is studied. A numerical example is given to show the feasibility and effectiveness of the given algorithm.

Nowadays, multi-sensor systems have been widely used in target tracking and state estimation problems. Most of the multi-sensor fusion state estimation algorithms assume the sensor noises, and the system noise are independent [1–7]. However, in the actual target tracking problems, the measurement noise is likely to be correlated because the measurements are usually observed in the same noise environment. The observation noise is usually coupled, and the observation noise and the process noise are usually correlated [2–10]. Sensor noise is correlated possibly due to common interference or noise propagation where common interferers are propagated through the network via member sensors and create spatial correlated noise among sensors [2]. When the distributed data fusion structure is used in state fusion estimation, the noise correlation among various sensors can not be fully utilized, and there is a certain degree of information loss [11].

L. Yan et al., *Multisensor Fusion Estimation Theory and Application*,
https://doi.org/10.1007/978-981-15-9426-7_5

In the literatures listed above, unreliability or missing measurements due to the limitation of communication and sensor faults are rarely concerned, which is inclined to encounter in application fields including communication and navigation. For a class of linear dynamic systems missing data with a Bernoulli distribution, the algorithms presented by the team of Wang are promising in that it has proper computation complexity and can generate nearly optimal state estimation [8]. If the missing of measurements is Markov distributed, based on a discrete-time linear system, Kalman filtering with intermittent observations was studied in [4], where the arrival of the observations is modeled as a random process, and the statistical convergence property of Kalman filter is given. By use of peak covariance as the estimation of filtering deterioration resulted from packet losses, the stability of Kalman filtering with Markovian packet losses was studied in [3] based on a linear time-invariant system. However, in these interesting studies on fusion of unreliable measurements, multiscale or multirate multisensor data fusion was not considered. In [5, 6], state estimation with multirate sensors was studied, but how to deal with unreliable measurements was not considered. In [9], optimal state estimation with multirate sensors and unreliable measurements was studied, but noise correlation was not considered.

Due to the introduction of network, the problems of data missing, time delay, asynchronous multi-rate data fusion have appeared. When the network load is too heavy, the data will inevitably encounter the network congestion in the transmission process, which will cause the loss of data. When the data loss is serious, the system will be unstable, so it is very important to study the multi-sensor data fusion state estimation in condition of multi-sensor sequential data fusion, where the data measured by each sensor carry out Kalman filter sequentially, the sensor that has the highest sampling rate is taken as the first, and then the other sensors are used according to the decrease of the sampling rates.

5.2 Problem Formulation

When the observed data is missing or unreliable in a certain probability, the discrete-time linear dynamic system of the multirate sensor with N sensors for the same target can be described by

$$x_{k+1} = A_k x_k + w_k, \tag{5.1}$$

$$z_{i,k_i} = \gamma_{i,k_i} C_{i,k_i} x_{i,k_i} + v_{i,k_i} \tag{5.2}$$

where, $i = 1, 2, \ldots, N$, $x_k \in \mathbb{R}^n$ is the system state at the highest sampling rate at time k. $A_k \in \mathbb{R}^{n\times n}$ is the system transition matrix. $i = 1, 2, \ldots, N$ denote sensors, whose sampling rates are B_i, respectively. w_k is the system noise and is assumed to be Gaussian distributed with zero mean and satisfies:

$$E\{w_k\} = 0 \tag{5.3}$$

$$E\{w_k w_l^T\} = Q_k \delta_{kl} \tag{5.4}$$

where $Q_k \geq 0$, δ_{kl} is Kronecker delta function; $z_{i,k_i} \in \mathbb{R}^{q_i \times 1}(q_i \leq n)$ is the k_ith measurement of sensor i with sampling rate B_i. The sampling rates of different sensors meet:

$$B_i = \frac{B_1}{l_i}, i = 1, 2, \ldots, N \tag{5.5}$$

where l_i are known positive integers and $l_1 = 1$; $C_{i,k_i} \in \mathbb{R}^{q_i \times n}$ is the measurement matrix. $v_{i,k_i} \in \mathbb{R}^{q_i \times 1}$ is the measurement noise and is assumed to be Gaussian distributed with zero mean and satisfies:

$$E\{v_{i,k_i} v_{j,k_i}^T\} = \begin{cases} R_{ij,k}, & \text{if } k_i l_i = k_j l_j \\ 0, & \text{otherwise} \end{cases} \tag{5.6}$$

$$E\{v_{i,k_i} w_{k-1}\} = \begin{cases} S_{i,k}, & \text{if } k = l_i k_i \\ 0, & \text{otherwise} \end{cases} \tag{5.7}$$

where, $i = 1, 2, \ldots, N$.

The initial value of the state vector x_0 is a random vector, whose mean and estimation error covariance are $\bar{x}_0$ and $\bar{P}_0$, respectively. It is assumed that x_0, w_k and v_{i,k_i} are mutually independent.

The variable $\gamma_{i,k_i} \in R$ is a stochastic sequence that takes values from 0 and 1 with Bernoulli distribution, whose expectation is $\bar{\gamma}_i$, which is used to describe the degree of measurements missing or faulty. It is assumed that γ_{i,k_i} is independent of w_k, v_{i,k_i} and x_0, $i = 1, 2, \ldots, N$. So, we have

$$\text{Prob}[\gamma_{i,k_i} = 1] = E[\gamma_{i,k_i}] = \bar{\gamma}_i \tag{5.8}$$

$$\text{Prob}[\gamma_{i,k_i} \neq 1] = 1 - \bar{\gamma}_i \tag{5.9}$$

$$\sigma_{\gamma_i}^2 = E[\gamma_{i,k_i} - \bar{\gamma}_i]^2 = (1 - \bar{\gamma}_i)\bar{\gamma}_i \tag{5.10}$$

In this chapter, linear system filter is studied on condition that multiple sensors observe a single target at different sampling rates synchronously. The multirate multisensor dynamic system is described at different scales. The scale index and the sensor index are both denoted by i. The bigger i, the coarser the scale is. For instance, $i = 1$ denotes the finest scale, in which the sensor samples fastest. $i = 2, 3, \ldots, N - 1$ denote coarser scales. While, $i = N$ denotes the coarsest scale, where the sensor samples slowest. It is assumed that each sensor should sample uniformly. k_i denotes time series. k denotes time and k_i denotes the k_ith measurement of Sensor i and its sampling time is $l_i k_i$, we have

$$x_{i,k_i} = x_{l_i k_i}, i = 1, 2, \ldots, N \tag{5.11}$$

where, $l_1 = 1, k_1 = k, x_{1,k_1} = x_k$.

Assuming there are h sensors (Sensor $i_1, i_2, \ldots i_h$) satisfying k mod $l_{i_p} = 0$ at the moment $k = 1, 2, \ldots$, where $1 \leq i_1, i_2, \ldots, i_h \leq N$. Then measurement equation (5.2) can be rewritten as:

$$z_{i_p,\frac{k}{l_{i_p}}} = \gamma_{i_p,\frac{k}{l_{i_p}}} C_{i_p,\frac{k}{l_{i_p}}} x_k + v_{i_p,\frac{k}{l_{i_p}}} \tag{5.12}$$

5.3 The Sequential Fusion Algorithm

Theorem 5.1 *Suppose there are h sensors (Sensor $i_1, i_2, \ldots, i_h$) satisfy k mod $l_{i_p} = 0$ at moment k ($k = 1, 2, \ldots$), where $1 \leq i_1, i_2, \ldots, i_h \leq N$, $p = 1, 2, \ldots, h$. System model (5.1) and (5.2) can be rewritten as:*

$$x_{k+1} = A_k x_k + w_k \tag{5.13}$$

$$z_{i_p,\frac{k}{l_{i_p}}} = \gamma_{i_p,\frac{k}{l_{i_p}}} C_{i_p,\frac{k}{l_{i_p}}} x_k + v_{i_p,\frac{k}{l_{i_p}}} \tag{5.14}$$

where,

$$z_{i_p,\frac{k}{l_{i_p}}} = z_{i_p,k_{i_p}} \tag{5.15}$$

$$C_{i_p,\frac{k}{l_{i_p}}} = C_{i_p,k_{i_p}} \tag{5.16}$$

$$v_{i_p,\frac{k}{l_{i_p}}} = v_{i_p,k_{i_p}} \tag{5.17}$$

$$\gamma_{i_p,\frac{k}{l_{i_p}}} = \gamma_{i_p,k_{i_p}} \tag{5.18}$$

We have

$$\begin{cases} \hat{x}_{i_p,k|k} = \hat{x}_{i_{p-1},k|k} + K_{i_p,\frac{k}{l_{i_p}}} [z_{i_p,\frac{k}{l_{i_p}}} - C_{i_p,\frac{k}{l_{i_p}}} \hat{x}_{i_{p-1},k|k}] \\ P_{i_p,k|k} = P_{i_{p-1},k|k} - K_{i_p,\frac{k}{l_{i_p}}} [\Delta^T_{i_{p-1},\frac{k}{l_{i_p}}} + C_{i_p,\frac{k}{l_{i_p}}} P_{i_{p-1},k|k}] \end{cases} \tag{5.19}$$

where,

$$K_{i_p,\frac{k}{l_{i_p}}} = \begin{cases} [P_{i_{p-1},k|k} C^T_{i_p,\frac{k}{l_{i_p}}} + \Delta_{i_{p-1},\frac{k}{l_{i_{p-1}}}}] \\ \quad \cdot [C_{i_p,\frac{k}{l_{i_p}}} P_{i_{p-1},k|k} C^T_{i_p,\frac{k}{l_{i_p}}} + R_{i_p,\frac{k}{l_{i_p}}} \\ \quad + C_{i_p,\frac{k}{l_{i_p}}} \Delta_{i_{p-1},\frac{k}{l_{i_{p-1}}}} + \Delta^T_{i_{p-1},\frac{k}{l_{i_{p-1}}}} C^T_{i_p,\frac{k}{l_{i_p}}}]^{-1}, \\ \qquad\qquad \text{if } |\rho_{i_p,\frac{k}{l_{i_p}}}| \leq \chi^2_\alpha(q_{i_p}) \\ 0, \qquad\qquad \text{if } |\rho_{i_p,\frac{k}{l_{i_p}}}| > \chi^2_\alpha(q_{i_p}) \end{cases} \tag{5.20}$$

$$\Delta_{i_{p-1},k} = \prod_{1}^{m=p-1} [I - K_{i_m,k} C_{i_m,\frac{k}{l_{im}}}] S_{i_p,k} - \sum_{q=2}^{p-1} \prod_{q}^{m=p-1} [I - K_{i_m,k} C_{i_m,\frac{k}{l_{im}}}]$$
$$\cdot K_{i_{q-1},k} R_{i_{q-1},i_p,k} - K_{i_{p-1},k} R_{i_{q-1},i_p,k} \tag{5.21}$$
$$\rho_{i_p,\frac{k}{l_{ip}}} = \tilde{z}^T_{i_p,\frac{k}{l_{ip}}} G^{-1}_{i_p,\frac{k}{l_{ip}}} \tilde{z}_{i_p,\frac{k}{l_{ip}}} \tag{5.22}$$

and where,

$$\tilde{z}_{i_p,\frac{k}{l_{ip}}} = z_{i_p,\frac{k}{l_{ip}}} - C_{i_p,\frac{k}{l_{ip}}} \hat{x}_{i_{p-1},k|k} \tag{5.23}$$

$$G_{i_p,\frac{k}{l_{ip}}} = C_{i_p,\frac{k}{l_{ip}}} P_{i_{p-1},k|k} C^T_{i_p,\frac{k}{l_{ip}}} + R_{i_p,\frac{k}{l_{ip}}} + C_{i_p,\frac{k}{l_{ip}}} S^T_{i_p,\frac{k}{l_{ip}}}$$
$$+ S_{i_p,\frac{k}{l_{ip}}} C^T_{i_p,\frac{k}{l_{ip}}} \tag{5.24}$$

When $p = h$, the optimal sequential data fusion estimation of x_k at moment k is:

$$\hat{x}_{k|k} = \hat{x}_{i_h,k|k} \tag{5.25}$$
$$P_{k|k} = P_{i_h,k|k} \tag{5.26}$$

In Eq. (5.20), q_{i_p} denotes the dimension of $z_{i_p,k_{i_p}}$.

Proof When $k \bmod l_{i_p} = 0$, $p = 1, 2, \ldots, h$, denote

$$z_{i_p,\frac{k}{l_{ip}}} = z_{i_p,k_{i_p}} \tag{5.27}$$
$$C_{i_p,\frac{k}{l_{ip}}} = C_{i_p,k_{i_p}} \tag{5.28}$$
$$v_{i_p,\frac{k}{l_{ip}}} = v_{i_p,k_{i_p}} \tag{5.29}$$
$$\gamma_{i_p,\frac{k}{l_{ip}}} = \gamma_{i_p,k_{i_p}} \tag{5.30}$$

and for any $p = 1, 2, \ldots, h$, define

$$Z_{i_p,k} = \{z_{i_p,1}, z_{i_p,2}, \ldots, z_{i_p,k}\} \tag{5.31}$$
$$Z^{i_p}_{1,k} = \{z_{i_1,k}, z_{i_2,k}, \ldots, z_{i_p,k}\} \tag{5.32}$$
$$\bar{Z}^{i_p}_{1,k} = \{Z^{i_p}_{1,l}\}^k_{l=1} \tag{5.33}$$

where, $Z_{i_p,k}$ denotes the measurements observed by sensor i_p before and including time k; $Z^{i_p}_{1,k}$ stands for the measurements sampled by sensors $i_1, i_2, \ldots, i_p$ at time k; $\bar{Z}^{i_p}_{1,k}$ stands for the measurements observed by sensor $i_1, i_2, \ldots, i_p$ before and including time k; $\bar{Z}^N_{1,k}$ denotes the measurements observed by all the sensors before and including time k.

From Eq. (5.11), we have

$$x_{i_p,k_{i_p}} = x_{l_{i_p}k_{i_p}} = x_k \tag{5.34}$$

Then, system model (5.1) and (5.2) can be rewritten as:

$$x_{k+1} = A_k x_k + w_k \tag{5.35}$$

$$z_{i_p,\frac{k}{l_{i_p}}} = \gamma_{i_p,\frac{k}{l_{i_p}}} C_{i_p,\frac{k}{l_{i_p}}} x_k + v_{i_p,\frac{k}{l_{i_p}}} \tag{5.36}$$

where, $p = 1, 2, \ldots, h$.

If all the data are reliable, by the projection theorem, based on data $Z^N_{1,k-1}$, the state prediction of x_k is calculated by the following formula:

$$\hat{x}_{k|k-1} = A_{k-1}\hat{x}_{k-1|k-1} \tag{5.37}$$

$$P_{k|k-1} = A_{k-1}P_{k-1|k-1}A^T_{k-1} + Q_{k-1} \tag{5.38}$$

Denote

$$\hat{x}_{0,k|k} = \hat{x}_{k|k-1} \tag{5.39}$$

$$P_{0,k|k} = P_{k|k-1} \tag{5.40}$$

$$\Delta_{0,k} = S_{1,k} \tag{5.41}$$

Without loss of generality, assume the processing sequence of observation data at each moment is $i_1, i_2, \ldots, i_h$, and denote

$$\hat{x}_{i_p,k|k} = E[x_k|\bar{Z}^N_{1,k-1}, Z^{i_p}_{1,k}] \tag{5.42}$$

Then, by using projection theorem, we have

$$\begin{aligned} \hat{x}_{i_p,k|k} &= E[x_k|\bar{Z}^N_{1,\frac{k}{l_{i_p}}-1}, Z^{i_{p-1}}_{1,\frac{k}{l_{i_p}}}, z_{i_p,\frac{k}{l_{i_p}}}] \\ &= \hat{x}_{i_{p-1},k|k} + K_{i_p,\frac{k}{l_{i_p}}}\tilde{z}_{i_p,\frac{k}{l_{i_p}}} \end{aligned} \tag{5.43}$$

where,

$$\begin{aligned} \tilde{z}_{i_p,\frac{k}{l_{i_p}}} &= z_{i_p,\frac{k}{l_{i_p}}} - \hat{z}_{i_p,\frac{k}{l_{i_p}}} \\ &= z_{i_p,\frac{k}{l_{i_p}}} - C_{i_p,\frac{k}{l_{i_p}}}\hat{x}_{i_{p-1},k|k} \\ &= C_{i_p,\frac{k}{l_{i_p}}}\tilde{x}_{i_{p-1},k|k} + v_{i_p,\frac{k}{l_{i_p}}} \end{aligned} \tag{5.44}$$

According to formula (5.43) and (5.44), we obtain

$$\hat{x}_{i_p,k|k} = \hat{x}_{i_{p-1},k|k} + K_{i_p,\frac{k}{l_{ip}}}[z_{i_p,\frac{k}{l_{ip}}} - C_{i_p,\frac{k}{l_{ip}}}\hat{x}_{i_{p-1},k|k}] \tag{5.45}$$

where,

$$\begin{aligned} K_{i_p,\frac{k}{l_{ip}}} &= \text{cov}[\tilde{x}_{i_{p-1},k|k}, \tilde{z}_{i_p,\frac{k}{l_{ip}}}]\text{var}[\tilde{z}_{i_p,\frac{k}{l_{ip}}}]^{-1} \\ &= E[\tilde{x}_{i_{p-1},k|k}\tilde{z}^T_{i_p,\frac{k}{l_{ip}}}]E[\tilde{z}_{i_p,\frac{k}{l_{ip}}}\tilde{z}^T_{i_p,\frac{k}{l_{ip}}}]^{-1} \\ &= [P_{i_{p-1},k|k}C^T_{i_p,\frac{k}{l_{ip}}} + \Delta_{i_{p-1},\frac{k}{l_{i_{p-1}}}}][C_{i_p,\frac{k}{l_{ip}}}P_{i_{p-1},k|k}C^T_{i_p,\frac{k}{l_{ip}}} \\ &+ R_{i_p,\frac{k}{l_{ip}}} + C_{i_p,\frac{k}{l_{ip}}}\Delta_{i_{p-1},\frac{k}{l_{ip}}} + \Delta^T_{i_{p-1},\frac{k}{l_{i_{p-1}}}}C^T_{i_p,\frac{k}{l_{ip}}}]^{-1} \end{aligned} \tag{5.46}$$

$$\begin{aligned} \Delta_{i_{p-1},k} &= E[\tilde{x}_{i_{p-1},k|k}v^T_{i_p,\frac{k}{l_{ip}}}] \\ &= E\{[x_k - \hat{x}_{i_{p-1},k|k}]v^T_{i_p,\frac{k}{l_{ip}}}\} \\ &= [I - K_{i_{p-1},\frac{k}{l_{i_{p-1}}}}C_{i_{p-1},\frac{k}{l_{i_{p-1}}}}]E[\tilde{x}_{i_{p-2},k|k}v^T_{i_p,\frac{k}{l_{ip}}}] - K_{i_{p-1},\frac{k}{l_{i_{p-1}}}} \\ &\quad E[v_{i_{p-1},\frac{k}{l_{i_{p-1}}}}v^T_{i_p,\frac{k}{l_{ip}}}] \\ &= \prod_{1}^{m=p-1}[I - K_{i_m,\frac{k}{l_{im}}}C_{i_m,\frac{k}{l_{im}}}]S_{i_p,\frac{k}{l_{ip}}} - \sum_{q=2}^{p-1}\prod_{q}^{m=p-1}[I - K_{i_m,\frac{k}{l_{im}}}C_{i_m,\frac{k}{l_{im}}}] \\ &\quad \cdot K_{i_{q-1},\frac{k}{l_{i_{q-1}}}}R_{i_{q-1},i_p,k} - K_{i_{p-1},\frac{k}{l_{i_{p-1}}}}R_{i_{q-1},i_p,k} \end{aligned} \tag{5.47}$$

From Eqs. (5.1), (5.44) and (5.45), we have

$$\begin{aligned} \tilde{x}_{i_p,k|k} &= x_k - \hat{x}_{i_p,k|k} \\ &= x_k - \hat{x}_{i_{p-1},k|k} - K_{i_p,\frac{k}{l_{ip}}}[z_{i_p,\frac{k}{l_{ip}}} - C_{i_p,\frac{k}{l_{ip}}}\hat{x}_{i_{p-1},k|k}] \\ &= (I - K_{i_p,\frac{k}{l_{ip}}}C_{i_p,\frac{k}{l_{ip}}})\tilde{x}_{i_{p-1},k|k} - K_{i_p,\frac{k}{l_{ip}}}v_{i_p,\frac{k}{l_{ip}}} \end{aligned} \tag{5.48}$$

So, the estimation error covariance is

$$\begin{aligned} P_{i_p,k|k} &= E[\tilde{x}_{i_p,k|k}\tilde{x}^T_{i_p,k|k}] \\ &= E\{[(I - K_{i_p,\frac{k}{l_{ip}}}C_{i_p,\frac{k}{l_{ip}}})\tilde{x}_{i_{p-1},k|k} - K_{i_p,\frac{k}{l_{ip}}}v_{i_p,\frac{k}{l_{ip}}}] \cdot [(I - K_{i_p,\frac{k}{l_{ip}}} \\ &\quad \cdot C_{i_p,\frac{k}{l_{ip}}})\tilde{x}_{i_{p-1},k|k} - K_{i_p,\frac{k}{l_{ip}}}v_{i_p,\frac{k}{l_{ip}}}]^T\} \\ &= [I - K_{i_p,\frac{k}{l_{ip}}}C_{i_p,\frac{k}{l_{ip}}}]P_{i_{p-1},k|k}[I - K_{i_p,\frac{k}{l_{ip}}}C_{i_p,\frac{k}{l_{ip}}}]^T + K_{i_p,\frac{k}{l_{ip}}} \\ &\quad \cdot R_{i_p,\frac{k}{l_{ip}}}K^T_{i_p,\frac{k}{l_{ip}}} - [I - K_{i_p,\frac{k}{l_{ip}}}C_{i_p,\frac{k}{l_{ip}}}]\Delta_{i_{p-1},\frac{k}{l_{i_{p-1}}}}K^T_{i_p,\frac{k}{l_{ip}}} - K_{i_p,\frac{k}{l_{ip}}} \\ &\quad \cdot \Delta^T_{i_{p-1},\frac{k}{l_{i_{p-1}}}}[I - K_{i_p,\frac{k}{l_{ip}}}C_{i_p,\frac{k}{l_{ip}}}]^{-1} \end{aligned}$$

$$= P_{i_{p-1},k|k} - K_{i_p,\frac{k}{l_{i_p}}}[C_{i_p,\frac{k}{l_{i_p}}} P_{i_{p-1},k|k} + \Delta^T_{i_{p-1},\frac{k}{l_{i_{p-1}}}}] \tag{5.49}$$

When $p = h$, the optimal sequential fusion estimation is $\hat{x}_{k|k} = \hat{x}_{i_h,k|k}$, $P_{k|k} = P_{i_h,k|k}$, $k = 1, 2, \ldots$.

If there exist unreliable measurements, we will use the statistical property of $\tilde{z}_{i_p,\frac{k}{l_{i_p}}}$ to judge whether measurement $z_{i_p,\frac{k}{l_{i_p}}}$ is normal or faulty.

From the problem formulation, one can notice that when $z_{i_p,\frac{k}{l_{i_p}}}$ is reliable, $\tilde{z}^T_{i_p,\frac{k}{l_{i_p}}}$ is Gaussian distributed with zero mean and covariance $G_{i_p,\frac{k}{l_{i_p}}}$, i.e., $\tilde{z}_{i_p,\frac{k}{l_{i_p}}} \sim \mathcal{N}(0, G_{i_p,\frac{k}{l_{i_p}}})$, where

$$\begin{aligned} G_{i_p,\frac{k}{l_{i_p}}} &= \text{cov}[\tilde{z}_{i_p,\frac{k}{l_{i_p}}}] \\ &= E\{[C_{i_p,\frac{k}{l_{i_p}}} \tilde{x}_{i_{p-1},k|k} + v_{i_p,\frac{k}{l_{i_p}}}][C_{i_p,\frac{k}{l_{i_p}}} \tilde{x}_{i_{p-1},k|k} + v_{i_p,\frac{k}{l_{i_p}}}]^T\} \\ &= C_{i_p,\frac{k}{l_{i_p}}} P_{i_{p-1},k|k} C^T_{i_p,\frac{k}{l_{i_p}}} + R_{i_p,\frac{k}{l_{i_p}}} + C_{i_p,\frac{k}{l_{i_p}}} S^T_{i_p,\frac{k}{l_{i_p}}} \\ &\quad + S_{i_p,\frac{k}{l_{i_p}}} C^T_{i_p,\frac{k}{l_{i_p}}} \end{aligned} \tag{5.50}$$

Denote

$$\rho_{i_p,\frac{k}{l_{i_p}}} = \tilde{z}^T_{i_p,\frac{k}{l_{i_p}}} G^{-1}_{i_p,\frac{k}{l_{i_p}}} \tilde{z}_{i_p,\frac{k}{l_{i_p}}} \tag{5.51}$$

Then, $\rho_{i_p,\frac{k}{l_{i_p}}} \sim \chi^2(q_{i_p})$ obeys chi-square distribution with q_{i_p} degrees of freedom, whose mean and variance are q_{i_p} and $2q_{i_p}$, respectively, where q_{i_p} is the dimension of $z_{i_p,\frac{k}{l_{i_p}}}$.

Hence, $\rho_{i_p,\frac{k}{l_{i_p}}}$ can be used as a measurement evaluation index to judge whether $z_{i_p,\frac{k}{l_{i_p}}}$ is normal or faulty. In this hypothesis test problem, the null hypothesis is: $H_0 : \gamma_{i_p,\frac{k}{l_{i_p}}} = 1$; the alternative hypothesis is: $H_1 : \gamma_{i_p,\frac{k}{l_{i_p}}} \neq 1$. The rejection region is $(\chi^2_\alpha(q_{i_p}), +\infty)$, where $\chi^2_\alpha(q_{i_p})$ is the boundary value of a unilateral chi-square distribution with confidence α, $1 \le i_p \le N$. That is to say, if $|\rho_{i_p,\frac{k}{l_{i_p}}}| > \chi^2_\alpha(q_{i_p})$, $z_{i_p,\frac{k}{l_{i_p}}}$ is taken as faulty and will not be used in the fusion estimation. If $|\rho_{i_p,\frac{k}{l_{i_p}}}| \le \chi^2_\alpha(q_{i_p})$, $z_{i_p,\frac{k}{l_{i_p}}}$ is deemed reliable and will be used in state estimation. Therefore, the innovation matrices should be modified, and we have Eq. (5.20).

5.4 Numerical Example

We use an example to illustrate the performance of the proposed algorithm in this section.

A tracking system with three sensors is described by the following formula:

$$x_{k+1} = A_k x_k + \Gamma_k w_k \tag{5.52}$$

$$z_{i,k_i} = \gamma_{i,k_i} C_{i,k_i} x_{i,k_i} + v_{i,k_i} \tag{5.53}$$

where, $i = 1, 2, 3$.

$$A_k = \begin{bmatrix} 1 & T_s & \frac{T_s^2}{2} \\ 0 & 1 & T_s \\ 0 & 1 & 1 \end{bmatrix} \tag{5.54}$$

$$Q_k := Q = q\Gamma\Gamma^T \tag{5.55}$$

$$\Gamma_k = [\frac{T_s^2}{2} \ \ T_s \ \ 1] \tag{5.56}$$

where $T_s = 0.01$s denotes the sampling period of the sensor with the highest sampling rate. $q = 0.01$. w_k denotes system noise and is assumed to be Gaussian distributed with zero mean and its variance is σ_w^2. Γ_k is the gain of the process noise. v_{i,k_i} denotes the observation noise of the k_i-th measurement of Sensor i, which could be sampled from $v_{i,k}$ with the ratio of l_i, where $l_1 = 1, l_2 = 3, l_3 = 4$, and $v_{i,k}$ is generated from the following equation,

$$v_{i,k} = \beta_i w_{k-1} + \eta_{i,k} \tag{5.57}$$

From Eq. (5.57), the observation noise of different sensors are correlated, and they are also coupled with the system noise w_{k-1}, and the correlation intensity depends on the value of β_i. $\eta_{i,k}$ is Gaussian white noise with zero mean and variance being $\sigma_{\eta_i}^2$. Sensors 1, 2, 3 observe the same target with different sampling rates and the ratio of the sample rates is $1 : \frac{1}{3} : \frac{1}{4}$. The measurement matrices are as follows

$$C_{1,k_1} := C_1 = [1 \ \ 0 \ \ 0] \tag{5.58}$$

$$C_{2,k_2} := C_2 = [1 \ \ 0 \ \ 0] \tag{5.59}$$

$$C_{3,k_3} := C_3 = [0 \ \ 1 \ \ 0] \tag{5.60}$$

where, Sensor 1 and Sensor 2 observe position and Sensor 3 observes velocity.

The covariance matrix of observation noise is given by:

$$R_k = \begin{bmatrix} \beta_1^2\sigma_w^2 + \sigma_{\eta_1}^2 & \beta_1\beta_2\sigma_w^2 & \beta_1\beta_3\sigma_w^2 \\ \beta_2\beta_1\sigma_w^2 & \beta_2^2\sigma_w^2 + \sigma_{\eta_2}^2 & \beta_2\beta_3\sigma_w^2 \\ \beta_3\beta_1\sigma_w^2 & \beta_3\beta_2\sigma_w^2 & \beta_3^2\sigma_w^2 + \sigma_{\eta_3}^2 \end{bmatrix} \tag{5.61}$$

The variance between observation noise v_{i,k_i} and system noise w_{k-1} satisfy

$$S_k = [\beta_1\sigma_w^2\Gamma_{k-1} \;\; \beta_2\sigma_w^2\Gamma_{k-1} \;\; \beta_3\sigma_w^2\Gamma_{k-1}] \tag{5.62}$$

The initial condition is:

$$\bar{x}_0 = \begin{bmatrix} 10 \\ 0.5 \\ 0 \end{bmatrix}, \bar{P}_0 = \begin{bmatrix} 1 & 0 & 0 \\ 0 & 1 & 0 \\ 0 & 0 & 1 \end{bmatrix} \tag{5.63}$$

In this section, we will fuse the information of three sensors to estimate the target state, and compare the difference between the estimations obtained by different algorithms in case of noise correlation.

In this example, we set $\sigma_w^2 = 0.01$, $\sigma_{\eta_1}^2 = 0.25$, $\sigma_{\eta_2}^2 = 0.09$, $\sigma_{\eta_1}^3 = 0.01$. Each sensor with a probability of 0.05 to produce unreliable measurements. When evaluating the reliability of the observations, we set confidence level $\alpha = 0.05$. For $\beta_i (i = 1, 2, 3)$, we set $\beta_1 = 5$, $\beta_2 = 5$, $\beta_3 = 0.1$, which means there are certain degrees of correlation between sensor noises and the system noise.

We select 300 sampling points for 100 Monte Carlo simulations, and observe the effectiveness of the present algorithm. The simulation results are shown in Figs. 5.1 and 5.2, as well as in Tables 5.1 and 5.2.

In Fig. 5.1, (a) and (b) denote simulation curves of position RMSEs and velocity RMSEs, respectively, in which the blue solid lines and the red dashed lines denote the estimation RMSEs by Kalman filtering of Sensor 1 (KF) and the RMSEs by the presented optimal sequential fusion algorithm (OSF), respectively. One can find that position RMSEs and velocity RMSEs by the optimal sequential fusion algorithm are both smaller than those by Kalman filtering of Sensor 1, which shows that the sequential fusion algorithm is effective. Table 5.1 shows the time-averaged RMSEs for position and velocity of the OSF and the KF, from which one can draw the same conclusion as from Fig. 5.1.

Figure 5.2 shows the difference between RMSEs of sequential fusion algorithm with reliability test and without test, where the blue solid lines denote RMSEs of sequential fusion algorithm with test and red dashed lines denote the RMSEs of sequential fusion algorithm without evaluation the reliability of the measurements. One can see that RMSEs with reliability test are much smaller than those without test, which means that the reliability evaluation method is effective and necessary in estimation.

For comparison, Table 5.2 shows the time-averaged RMSEs without reliability test. Compare Table 5.1 with Table 5.2, one can draw the same conclusions as those from Fig. 5.2. This indicates that the algorithm presented in this chapter is effective.

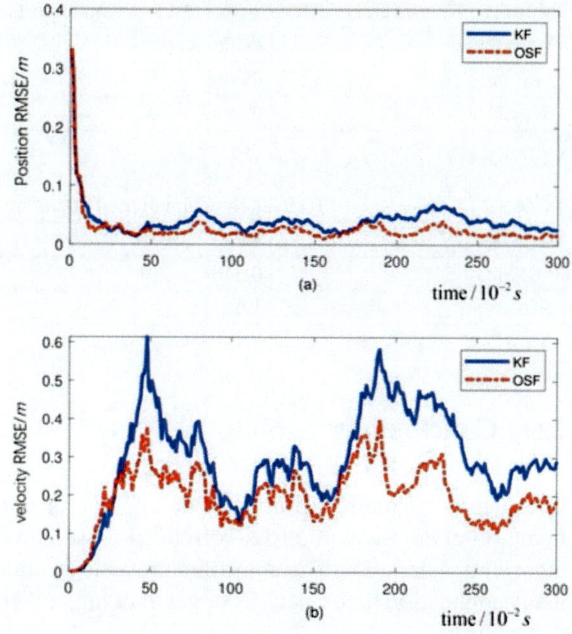

Fig. 5.1 RMSEs for position and velocity by different algorithms

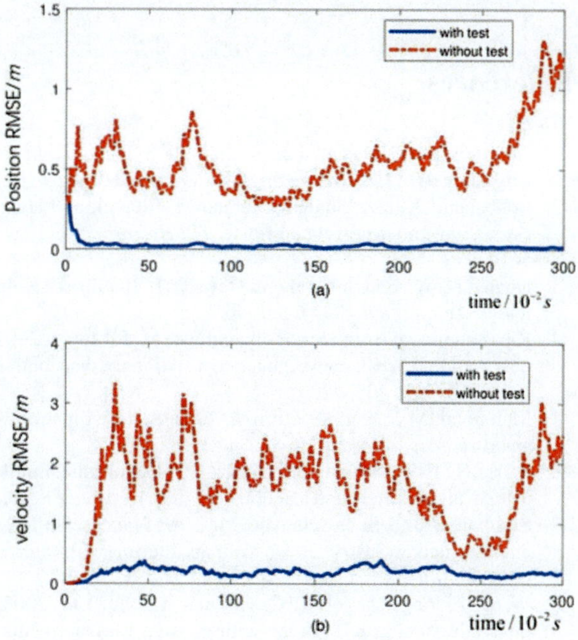

Fig. 5.2 RMSEs with reliability test and without test for estimation of position and velocity

Table 5.1 Time-averaged RMSEs with test and without test

Algorithm	Without test	With test
Position	0.0409	0.0259
Velocity	0.2989	0.2007

Table 5.2 Time-averaged RMSEs for estimation of position and velocity

Algorithm	KF	OSF
Position	0.5466	0.5457
Velocity	1.6477	1.6299

5.5 Conclusions

The state estimation algorithm by data fusion of multi-rate sensors based on unreliable observations with cross-correlated noise is presented. The sequential fusion algorithm is deduced. The algorithm presented in this chapter has potential value in many application fields, such as target tracking, integrated navigation, fault detection and control.

References

1. Atherton, Derek P., and J.A. Bather. 2005. Data fusion for several Kalman filters tracking a single target. In *IEE Proceedings – Radar, Sonar and Navigation*, volume 152, pages 372–C376.
2. Behbahani, Alireza Shahan, Ahmed M. Eltawil, and Hamid Jafarkhani. 2014. Decentralized estimation under correlated noise. *IEEE Transactions on Signal Processing* 62 (21): 5603–5614.
3. Huang, Minyi, and Subhrakanti Dey. 2007. Stability of Kalman filtering with markovian packet losses. *Automatica* 43 (4): 598–607.
4. Kar, Soummya, Bruno Sinopoli, and José M. F. Moura. 2009. Kalman filtering with intermittent observations: Weak convergence to a stationary distribution. *IEEE Transactions on Automatic Control* 57 (2): 405–420.
5. Mahmoud, M.S., and M.F. Emzir. 2012. State estimation with asynchronous multi-rate multi-smart sensors. *Information Sciences* 196: 15–27.
6. Safari, S., F. Shabani, and D. Simon. 2014. Multirate multisensor data fusion for linear systems using Kalman filters and a neural network. *Aerospace Science and Technology* 39 (1): 465–471.
7. Sinopoli, Bruno, Luca Schenato, Massimo Franceschetti, Kameshwar Poolla, Michael I. Jordan, and S. Shankar Sastry. 2004. Kalman filtering with intermittent observations. *IEEE Transactions on Automatic Control* 49 (9): 1453–1464.
8. Wang, Zidong, Daniel W. C. Ho, and Xiaohui Liu. 2003. Variance-constrained filtering for uncertain stochastic systems with missing measurements. *IEEE Transactions on Automatic Control*, 48 (7): 1254–1258.
9. Yan, Liping, X. Rong Li, and Yuanqing Xia. 2015. Modeling and estimation of asynchronous multirate multisensor system with unreliable measurements. *IEEE Transactions on Aerospace and Electronic Systems* 51 (3): 2012–2026.

10. Yan, Liping, X. Rong Li, Yuanqing Xia, and Mengyin Fu. 2013. Optimal sequential and distributed fusion for state estimation in cross-correlated noise. *Automatica* 49 (12): 3607–3612.
11. Yan, Liping, Yuanqing Xia, Baosheng Liu, and Mengyin Fu. 2015. *Multisensor Optimal Estimation Theory and its Application*. Beijing: The Science Publishing House.

Chapter 6
Distributed Fusion Estimation for Systems with Network Delays and Uncertainties

In this chapter, the problem of the distributed weighted Kalman filter fusion (DWKFF) for a class of multi-sensor unreliable networked systems (MUNSs) with correlated noises is studied. We assume that the process noise and the measurement noises are one-step, two-step cross-correlated and one-step autocorrelated, and the measurement noises of each sensor are one-step cross-correlated. The correlated multiplicative noises are used to describe the stochastic uncertainties in the state and measurements. It can cause the MUNSs suffer measurement delay or loss due to their system. To deal with measurement delay or loss, we propose buffers of finite length and derive an optimal local Kalman filter estimator with a buffer of finite length for each subsystem. We develop the DWKFF algorithm with finite length buffers with a stronger fault-tolerance ability according to the new optimal local Kalman filter estimator. A numerical example is given to show the effectiveness of the proposed algorithm.

6.1 Introduction

Multi-sensor data fusion technology is a method to solve the problem of dynamic target tracking, and it is also one of the hot spots in the current control field. There are still some challenges due to the inherent characteristics of multi-sensor unreliable networked systems (MUNSs), fading and loss, such as data transmission delay, fading and loss, the mathematical complexity of uncorrelated, autocorrelated and cross-correlated noises, etc. The development of distributed data fusion is gaining more attention from researchers due to the direct limitations of these issues. In the distributed data fusion systems, there is an important and practical issue: finding an optimal local state estimator to fuse so as to evaluate the system's performance.

One of the most widely used methodology in literatures is multi-sensor distributed fusion to meet the requirements of higher precision of estimation in navigation and dynamic target tracking, such as location, velocity and acceleration. Up to now, a

L. Yan et al., *Multisensor Fusion Estimation Theory and Application*,
https://doi.org/10.1007/978-981-15-9426-7_6

variety of distributed fusion studied in the literatures have been published, such as distributed weighted fusion [13, 20, 25, 26], distributed robust fusion [6, 14], federated filter [4, 5], distributed Kalman filtering fusion [8, 24], distributed covariance intersection fusion [27], optimal sequential distributed fusion [29], etc. Considering the advantage of Kalman filtering technology, some distributed fusion methods based on it were studied, such as [2, 10, 19]. In [25, 26], the researchers proposed a two-layer fusion structure and a three-layer fusion structure with fault tolerant property and reliability. The study in [13] discussed a problem of distributed weighted robust Kalman filter fusion for a class of uncertain systems with autocorrelated and cross-correlated noises. In [20], by using the principle of minimax robust estimation and the optimal estimation rule of unbiased linear minimum variance, a distributed robust weighted fusion Kalman filters for multi-sensor time-varying systems with uncorrelated noise was designed. Reference [24] researched the problem of the multi-sensor linear dynamic systems with cross-correlated observation noises, considered that the distributed Kalman filtering fusion problem without feedback from the fusion center (FC) to local sensors (LEs), and proved that under a mild condition the fused state estimate was equivalent to the centralized Kalman filtering using all sensor measurements. In [8], the researchers investigated the distributed finite-horizon fusion Kalman filtering problem for a class of networked multi-sensor fusion systems in a bandwidth and energy constrained wireless sensor network. The study in [27] discussed the information fusion estimation problems for multi-sensor stochastic uncertain systems with correlated noises. However, they did not consider the multiplicative noise in the measurement equation and measurement delay or loss. Reference [29] focuses on the problem of the optimal sequential and distributed fusion estimation problems for multi-sensor linear systems under the conditions that the noises are cross-correlated and coupled with the system noise in the previous step.[13] did not involve the unreliability of systems, and [20] considered uncorrelated the noises of the system and did not consider the unreliability of systems. In [8, 24, 26, 29], the cross-correlated noises were considered.

For uncertain systems with autocorrelated and cross-correlated noises, [12] designed a novel optimal robust non-fragile Kalman-type recursive filter for a class of uncertain systems with finite-step autocorrelated noises. The study in [32] discussed the robust recursive filtering problem for a class of uncertain systems with finite-step correlated noises, stochastic nonlinearities and autocorrelated missing measurements. The optimal filtering problem was addressed in [16] for a class of uncertain dynamical systems with finite-step correlated observation noises and multiple packet dropouts. In [11], the researchers designed a recursive Kalman-type filter for systems with one-step auto-correlated observation noises and multiple packet dropouts. Reference [15] concerned a recursive filtering problem with random parameter matrices, stochastic nonlinearity, multiple fading measurements, autocorrelated noises and cross-correlated noises. Reference [3] addressed the optimal least-squares state estimation problem for a class of discrete-time multisensor linear stochastic systems with state transition and measurement random parameter matrices and correlated noises. The optimal least-squares linear estimation problem was discussed for a class of discrete-time stochastic systems with random parameter matrices and correlated additive noises in [17]. In [3, 15, 17], the measurement noises are one-step autocorrelated and one-step correlated among different sensors, while the process noise is one-step autocorrelated and two-step correlated with the measurement noises for different sensors.

To deal with the delays and available data packets, various forms of processing methods have been used in literatures. Reference [30] proposed the key ideas of "nearest-neighbour compensation" to completely compensate for the unknown time delays. The research in [7, 9] proposed a novel stochastic model to describe the transmission delays and packet dropouts. Reference [23] performed the problem of Kalman filtering with intermittent observations to model the arrival of the observation as a random process. In [22, 33, 34], the researchers presented finite length buffers to deal with measurements delay or loss. Reference [22] considered the state estimate problem on the single channel with a finite buffer, and analyzed the relationship between the stability and the packet arrival process. References [33, 34] extended the state estimation algorithm in single channel in [22] to multi-channel, and presented the centralized and the distributed fusion estimation algorithms for MUNSs. Based on the two papers, [28] proposed the centralized and the distributed fusion methods using the federated filter criteria for MUNSs with uncorrelated noises. In [21], the researchers modeled delay and loss of measurement in the transmission from sensor to filter by a Bernoulli distributed random sequence, and researched the problem of robust filtering for a class of MUNSs. A discrete-time Markov chain to capture the signal losses, to design and to show the convergence of the distributed estimation algorithms under various uncertainties was introduced in [31]. However, the stochastic uncertainties and the autocorrelated and cross-correlated noises of systems were not taken into account in [28, 33, 34].

Motivated by the above problems, resolving the distributed weighted Kalman filter fusion (DWKFF) problem for a class of MUNSs with autocorrelated and cross-correlated noises is the objective of this chapter. In this chapter, we assume that the process noise and the measurement noises are one-step, two-step cross-correlated and one-step auto-correlated, and the measurement noises of each sensor are one-step cross-correlated. The stochastic uncertainties in the state and measurements are formulated by correlated multiplicative noises. A distributed data fusion algorithm for the MUNS is presented, in which all measurements of LEs are time stamped, and are sent to local estimators through communication channels, and then local optimal estimates are transmitted to the fusion center directly. The system suffers measurement delay or loss because of unreliability of MUNSs. For the derived model, we address an optimal local weighted Kalman filter with a finite length buffer for each subsystem. Based on the new optimal local weighted Kalman filter, we propose a multi-sensor optimal DWKFF with finite length buffers.

The rest of this chapter is organized as follows. The multi-sensor linear discrete-time stochastic control system model and some problem statements are described in Sect. 6.2. Section 6.3 describes an optimal local Kalman filter estimator with a finite length buffer of MUNSs. Section 6.4 is dedicated to the DWKFF with finite length buffers of MUNSs. Simulation results are provided in Sect. 6.5. Section 6.6 is the summary.

6.2 Model and Problem Statements

Consider a discrete-time linear stochastic system with N sensors, the model is as follows

$$\begin{cases} x_{k+1} = (A_k + F_k \varepsilon_k)x_k + B_k w_k \\ z_{i,k} = (C_{i,k} + \Gamma_{i,k}\epsilon_{i,k})x_k + D_{i,k}v_{i,k}, i = 1, 2, \ldots, N \end{cases} \tag{6.1}$$

where $x_k \in R^n$ is the state vector, $z_{i,k} \in R^{m_i}$ $(i = 1, 2, \ldots, N)$ are measurement vectors, $w_k \in R^r$ is the process noise, $v_{i,k} \in R^{m_i}$, $i = 1, 2, \ldots, N$ are the measurement noise of the ith sensor, A_k, F_k, B_k, $C_{i,k}$, $H_{i,k}$ and $D_{i,k}$ are known time-varying matrices with appropriate dimensions.

Assumption 6.1 $\varepsilon_k \in R$ and $\epsilon_{i,k} \in R$ are zero-mean correlated multiplicative noises and uncorrelated with other signals, which satisfy:

$$E\{\varepsilon_k\} = 0,\ E\{\epsilon_{i,k}\} = 0,\ E\{\varepsilon_k \varepsilon_l^T\} = Q_{\varepsilon,k}\delta_{kl} \tag{6.2}$$

$$E\{\epsilon_{i,k}\epsilon_{j,l}^T\} = R_{\epsilon,k}^{ij}\delta_{kl},\ E\{\varepsilon_k \epsilon_{i,l}^T\} = S_{\varepsilon\epsilon,k}^{i}\delta_{kl} \tag{6.3}$$

where $Q_{\varepsilon,k} = 1$ and $R_{\epsilon,k}^{ii} > 0$ are noise covariances, and δ_{kj} is the Kronecker delta function.

Assumption 6.2 w_k and $v_{i,k}$ $(i = 1, 2, \ldots, N)$ are autocorrelated and cross-correlated white noises with zero mean, which meet:

$$\begin{cases} E\{w_k\} = 0,\ E\{v_{i,k}\} = 0 \\ E\{w_k w_l^T\} = Q_{w,k}\delta_{kl} + Q_{w,(k,l)}\delta_{(k-1)l} + Q_{w,(k,l)}\delta_{(k+1)l} \\ E\{v_{i,k}v_{j,l}^T\} = R_{v,k}^{ij}\delta_{kl} + R_{v,(k,l)}^{ij}\delta_{(k-1)l} + R_{v,(k,l)}^{ij}\delta_{(k+1)l} \\ E\{w_k v_{i,l}^T\} = S_{wv,k}^{i}\delta_{kl} + S_{wv,(k,l)}^{i}\delta_{(k+1)l} + S_{wv,(k,l)}^{i}\delta_{(k+2)l} \end{cases} \tag{6.4}$$

where $Q_{w,k} \geq 0$ and $R_{v,k}^{ii} > 0$ are noise covariances, respectively.

Remark 6.1 From Assumption 6.2 , it is easy to see: (1) the relationship between the process noise and measurement noises is one-step autocorrelated. For example, the process noise and measurement noises at time k are correlated with themselves at time $k-1$ and $k+1$ with the covariance matrices $Q_{w,(k,k-1)}$, $Q_{w,(k,k+1)}$ and $R_{v,(k,k-1)}^{ii}$, $R_{v,(k,k+1)}^{ii}$, respectively. (2) the relationship between the measurement noises of different sensors is one-step forward and backward cross-correlated. For instance, the measurement noise at time k is correlated with another measurement noise at time $k-1$ and $k+1$ with the covariance matrices $R_{v,(k,k-1)}^{ij}$ and $R_{v,(k,k+1)}^{ij}$, respectively. (3) the relationship between the process noise and measurement noises is one-step and two-step forward cross-correlated. For example, the measurement noises at time k are correlated with the process noise at time $k-1$ and $k-2$ with the covariances $S_{wv,(k,k-1)}^{i}$ and $S_{wv,(k,k-2)}^{i}$, respectively.

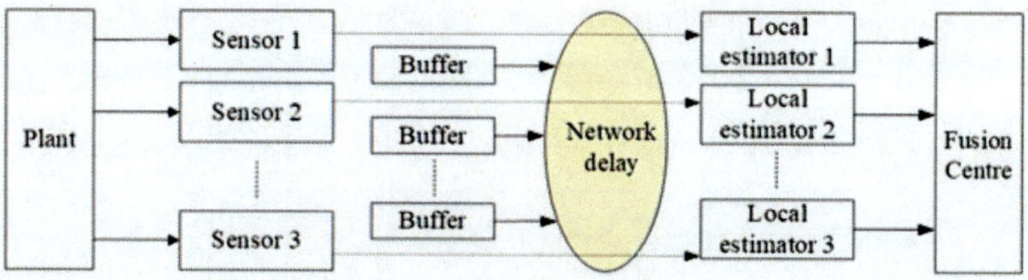

Fig. 6.1 Distributed fusion structure of the MUNS with buffers

Assumption 6.3 The initial state x_0 is independent of w_k, ε_k, $\epsilon_{i,k}$ and $v_{i,k}$ $(i = 1, 2, \ldots, N)$, and

$$\begin{cases} E\{x_0\} = \bar{x}_0 \\ E\{[x_0 - \bar{x}_0][x_0 - \bar{x}_0]^T\} = P_0 \end{cases} \tag{6.5}$$

Assumption 6.4 $(A_k + F_k\varepsilon_k, C_{i,k} + \Gamma_{i,k}\epsilon_{i,k})$ is observable, and $(A_k + F_k\varepsilon_k, Q_{w,k}^{1/2})$ is controllable.

It is assumed that all measurements are sent to local estimators through communication channels, and the time stamp of the measurements are known. Suppose that measurements are sent to local estimators from sensors with negligible quantization effects, we propose buffers of finite length to deal with measurement delay or a term of measurement noises [23] processes measurement loss. Then, the fusion center accepts the information of all local estimators (see Fig. 6.1). Thus, a distributed method of computing the optimal estimate and the error covariance based on this distributed fusion architecture is presented.

For the model in this chapter, each $z_{i,k}$ is delayed by $d_{i,k}$ times, where $d_{i,k}$ is a random variable described by a probability mass function f_i:

$$f_i(j) = \text{Prob}[d_{i,k} = j],\ j = 0, 1, 2, \ldots,\ i = 1, 2, \ldots, N \tag{6.6}$$

For simplicity, it is assumed that d_{i_1,k_1} and d_{i_2,k_2} are independent if $i_1 \neq i_2$ or $k_1 \neq k_2$. It is further assumed that $d_{i,k}$ is independent of ε_k, w_k, $v_{i,k}$ and the initial state x_0.

It is assumed that maintaining the length of the buffer of the local estimator at each time k as L $(L \geq 2)$, and all available measurement data packets up to time $k - L + 1$ can be retrieved. The measurements are not delayed if $L = 1$ while delayed if $L \geq 2$ and $k \geq L$. Let $\gamma_{t,k}^i$ be an indicator function for $z_{i,t}$ at time k $(t = k - L + 1, \ldots, k)$ as follows

$$\gamma_{t,k}^i = \begin{cases} 1, & \text{if } z_{i,t} \text{ arrives before or at time } k \\ 0, & \text{otherwise.} \end{cases} \tag{6.7}$$

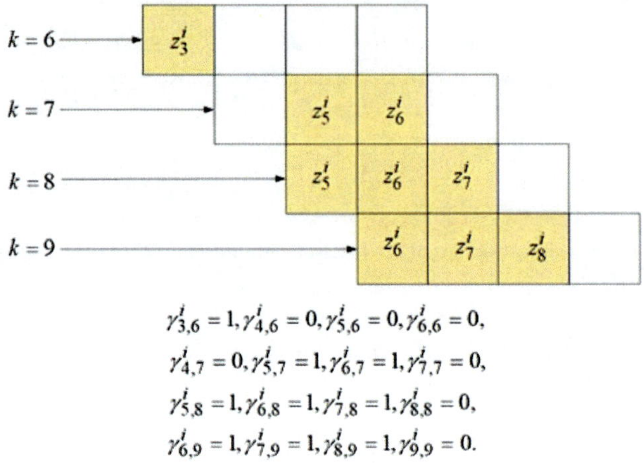

Fig. 6.2 The value of $\gamma_{t,k}^i$ in the ith channel when $L = 4$

More formally, $z_{t,k}^i$ $(i = 1, 2, \ldots, N)$ are stored in the $(t + L - k)$th slot of the buffer at time k, and $z_{t,k}^i$ can be written as

$$z_{t,k}^i = \gamma_{t,k}^i z_{i,t} = \gamma_{t,k}^i (C_{i,t} + \Gamma_{i,t}\epsilon_{i,t})x_t + \gamma_{t,k}^i D_{i,t} v_{i,t} \tag{6.8}$$

Let $\gamma_{t,k}^i v_{i,t} = v_{t,k}^i$, the measurement noise $v_{t,k}^i$ is defined as follows [23]:

$$p(v_{t,k}^i | \gamma_{t,k}^i) = \begin{cases} N(0, R_{v,t}^{ii}), \ \gamma_{t,k}^i = 1 \\ N(0, \rho^2 I), \ \rho \to \infty, \gamma_{t,k}^i = 0 \end{cases} \tag{6.9}$$

Remark 6.2 In this chapter, $z_{i,t}$ will be considered to be lost, if it is not received by local estimator i before or at k. If $\gamma_{t,k}^i = 1$, then $\gamma_{t,k+l}^i = 1 \ \forall l \in N$, which indicates that $z_{i,t}$ will be present for all later time, if it is received by local estimator i before or at time k. An example of $\gamma_{t,k}^i$ is graphically illustrated in Fig. 6.2.

Remark 6.3 In Fig. 6.2, a yellow cell means that the corresponding measurement is stored in the buffer, i.e., $z_{i,3}$ is received before or at time 6, $z_{i,4}$ is lost, $z_{i,5}$ and $z_{i,6}$ is received at time 7, $z_{i,7}$ is received at time 8 and $z_{i,8}$ is received at time 9.

There are two goals of this research described as follows:

1. Firstly, an optimal local Kalman filter estimator with a finite length buffer of MUNSs will be deduced;

2. Secondly, a DWKFF algorithm with finite length buffers of MUNSs with stronger fault-tolerance ability will be presented.

Lemma 6.1 ([27]) *For system (6.1) under Assumptions 6.1–6.4, the state second-order moment matrix X_{t-1} is given by:*

$$X_{t-1} = A_{t-2}X_{t-2}A_{t-2}^T + Q_{\varepsilon,k}F_{t-2}X_{t-2}F_{t-2}^T + B_{t-2}Q_{w,t-2}B_{t-3}^T + A_{t-2}B_{t-3}Q_{w,(t-3,t-2)}B_{t-2}^T + B_{t-2}Q_{w,(t-2,t-3)}B_{t-3}^T A_{t-2}^T \tag{6.10}$$

with the initial value $X_0 = \hat{x}_0\hat{x}_0^T + P_0$.

6.3 Optimal Local Kalman Filter Estimator with a Buffer of Finite Length

An optimal local Kalman filter estimator i with a buffer of finite length is designed by the optimal Kalman filter criterion [1], and its form is as follows:

$$\hat{x}_{i,k|k-1} \triangleq E\{x_{i,k}\,|Z_{k-1}, \gamma_{i,k-1}\} = A_{k-1}\hat{x}_{i,k-1|k-1} \tag{6.11}$$

$$\begin{aligned}\hat{x}_{i,k|k} &\triangleq E\{x_{i,k}\,|Z_{i,k}, \gamma_{i,k}\}\\ &= \hat{x}_{i,k|k-1} + \gamma_{i,k}K_{i,k}(z_{i,k} - C_{i,k}\hat{x}_{i,k|k-1})\end{aligned} \tag{6.12}$$

where $Z_{i,k} = [z_{1,k}^i, z_{2,k}^i, \ldots, z_{k,k}^i]$ and $\gamma_{i,k} = [\gamma_{1,k}^i, \gamma_{2,k}^i, \ldots, \gamma_{k,k}^i]$.

Without loss of generality, we define the error, the error covariance, one-step prediction error, one-step prediction error covariance, and some variables as follows:

$$\tilde{x}_{i,k|k} \triangleq x_k - \hat{x}_{i,k|k} \tag{6.13}$$

$$P_{i,k|k} \triangleq E\{\tilde{x}_{i,k|k}(\tilde{x}_{i,k|k})^T\,|Z_{i,k}, \gamma_{i,k}\} \tag{6.14}$$

$$\tilde{x}_{i,k|k-1} \triangleq x_k - \hat{x}_{i,k|k-1} \tag{6.15}$$

$$P_{i,k|k-1} \triangleq E\{\tilde{x}_{i,k|k-1}(\tilde{x}_{i,k|k-1})^T\,|Z_{i,k-1}, \gamma_{i,k-1}\} \tag{6.16}$$

$$\hat{x}_{t|l,k}^i \triangleq E\{x_t\,|z_{1,k}^i, z_{2,k}^i, \ldots, z_{l,k}^i, \gamma_{1,k}^i, \gamma_{2,k}^i, \ldots, \gamma_{l,k}^i\} \tag{6.17}$$

$$\tilde{x}_{t|l,k}^i \triangleq x_t - \hat{x}_{t|l,k}^i \tag{6.18}$$

$$P_{t|l,k}^i \triangleq E\{\tilde{x}_{t|l,k}^i(\tilde{x}_{t|l,k}^i)^T\,|z_{1,k}^i, z_{2,k}^i, \ldots, z_{l,k}^i, \gamma_{1,k}^i, \gamma_{2,k}^i, \ldots, \gamma_{l,k}^i\} \tag{6.19}$$

Then, we depict the set of lengths of buffers in all channels as

$$\{L_i\} = \{L_i\,|\text{if } \exists L_i \text{ s.t. } \text{Prob}[P_{i,k|k}(L_i, f_i) \le M] \ge 1 - \epsilon\}$$

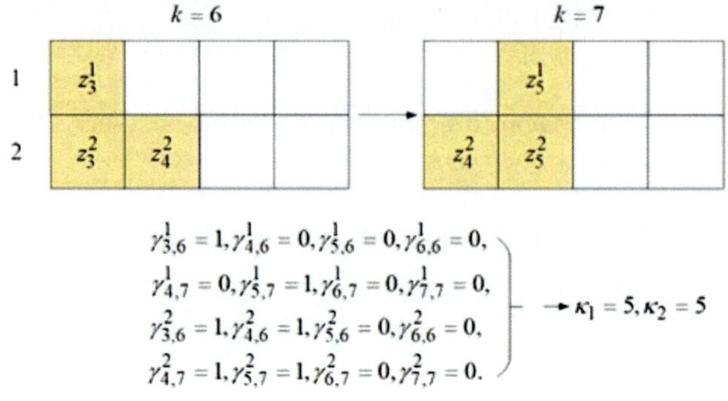

Fig. 6.3 An example of κ_i with $N = 2$, $L = 4$

where L_i is the length of the given buffer in the ith channel and its computation can be found in [22]. f_i is defined in Eq. (6.6) and $\text{Prob}[P_{i,k|k}(L_i, f_i) \leq M]$ is the probability of $P_{i,k|k} \leq M$ under the condition of L_i, f_i.

An earliest measurement updated time for the ith channel is defined as follows

$$\kappa_i = \begin{cases} \min\{t \mid \text{if } \exists t \text{ s.t. } \gamma_{t,k}^i = 1,\ \gamma_{t,k-1}^i = 0, \\ \qquad\qquad 0 < k - L_i + 1 \leq t < k\} \\ k, \quad \text{otherwise} \end{cases} \tag{6.20}$$

From the definition of κ_i, one can see that $z_{\kappa_i,k}^i$ is the earliest measurement vector, in which $z_{\kappa_i,k}^i$ is received by the ith local estimator at time k. Figure 6.3 declares an example of κ_i.

Remark 6.4 In Fig. 6.3, a yellow cell means the corresponding measurement stored in the buffer at present, i.e., z_3^1 is received before or at time 6, z_3^2 and z_4^2 are received before or at time 6, z_5^1 and z_5^2 are received at time 7.

In the following section, we give two useful theorems to obtain the optimal local Kalman filter estimator i with a buffer of finite length. The ith optimal local Kalman filter estimate $\hat{x}_{k|k,k}^i$, the error covariance of the ith subsystem $P_{k|k,k}^i$ and the error cross-covariance of the ith and jth subsystems $P_{k|k,k}^{ij}$ are provided by Theorems 6.1 and 6.2, respectively. Obviously, $\hat{x}_{i,k|k} = \hat{x}_{k|k,k}^i$, $P_{i,k|k} = P_{k|k,k}^i$ and $P_{k|k}^{ij} = P_{k|k,k}^{ij}$.

Theorem 6.1 *Consider the ith subsystem ($i = 1, 2, \ldots, N$) of system (6.1), based on the standard Kalman filter and Eqs. (6.11)–(6.19), the ith optimal local Kalman filter estimate $\hat{x}_{k|k,k}^i$ and the error covariance $P_{k|k,k}^i$ are given by*

$$\hat{x}^i_{\kappa_i-1|\kappa_i-1,k} = \hat{x}^i_{\kappa_i-1|\kappa_i-1,k-1} \tag{6.21}$$

$$P^i_{\kappa_i-1|\kappa_i-1,k} = P^i_{\kappa_i-1|\kappa_i-1,k-1} \tag{6.22}$$

$$\hat{x}^i_{t|t-1,k} = A_{t-1}\hat{x}^i_{t-1|t-1,k} \tag{6.23}$$

$$\hat{x}^i_{t|t,k} = \hat{x}^i_{t|t-1,k} + \gamma^i_{t,k}K^i_{t,k}(z_{i,t} - C_{i,t}\hat{x}^i_{t|t-1,k}) \tag{6.24}$$

$$\begin{aligned} K^i_{t,k} = {} & (P^i_{t|t-1,k}C^T_{i,t} + H^i_t D^T_{i,k})(C_{i,t}P^i_{t|t-1,k}C^T_{i,t} + R^{ii}_{\epsilon,t}\Gamma_{i,t}P^i_{t|t-1,k}\Gamma^T_{i,t} \\ & + D_{i,t}R^{ii}_{v,t}D^T_{i,t} + C_{i,t}H_{i,t}D^T_{i,t} + D_{i,t}(H_{i,t})^T C^T_{i,t})^{-1} \end{aligned} \tag{6.25}$$

$$\begin{aligned} P^i_{t|t-1,k} = {} & A_{t-1}P^i_{t-1|t-1,k}A^T_{t-1} + Q_{\varepsilon,k}F_{t-1}X_{t-1}F^T_{t-1} \\ & + B_{t-1}Q_{w,t-1}B^T_{t-1} + A_{t-1}M_{i,t-1}B^T_{t-1} \\ & + B_{t-1}(M_{i,t-1})^T A^T_{t-1} \end{aligned} \tag{6.26}$$

$$P^i_{t|t,k} = P^i_{t|t-1,k} - \gamma^i_{t,k}K^i_{t,k}(C_{i,t}P^i_{t|t-1,k} + D_{i,t}(H_{i,t})^T) \tag{6.27}$$

where

$$\begin{aligned} M_{i,t-1} = {} & B_{t-2}Q_{w,(t-2,t-1)} - \gamma^i_{t-1,k}K^i_{t-1,k}C_{i,t-1}B_{t-2} \\ & \cdot Q_{w,(t-2,t-1)} - \gamma^i_{t-1,k}K^i_{t-1,k}D_{i,t-1}(S^i_{wv,t-1})^T \end{aligned} \tag{6.28}$$

$$\begin{aligned} H_{i,t} = {} & A_{t-1}B_{t-2}S^i_{wv,(t-2,t)} + B_{t-1}S^i_{wv,(t-1,t)} - \gamma^i_{t-1,k}A_{t-1}K^i_{t-1,k} \\ & \cdot C_{i,t-1}B_{t-2}S^i_{wv,(t-2,t)} - \gamma^i_{t-1,k}A_{t-1}K^i_{t-1,k}R^{ii}_{v,(t-1,t)} \end{aligned} \tag{6.29}$$

where, $\hat{x}^i_{0|0,k} = x_0$, $P^i_{0|0,k} = P_0$, *and* $t = \kappa_i, \ldots, k$. *The* i*th optimal local Kalman filter estimate* $\hat{x}_{i,k|k}$ *and the error covariance* $P_{i,k|k}$ *are obtained by iterating* $k - \kappa_i + 1$ *times from* $\hat{x}^i_{\kappa_i-1|\kappa_i-1,k}$ *and* $P^i_{\kappa_i-1|\kappa_i-1,k}$, *respectively.*

Proof According to the definition of κ_i in Eq. (6.20), one can obviously see that κ_i is the earliest time when the measurement of the ith communication channel is updated at least, which means that no new measurement $z_{i,t}$ $(t = k - L + 1, \ldots, \kappa_i - 1,\ i = 1, 2, \ldots, N)$ can be received at time t. Therefore, we can easily get Eqs. (6.21) and (6.22). Then, Eqs. (6.23) and (6.24) follow directly from Eqs. (6.11), (6.12) and (6.17).

From Eq. (6.18), it is easy to get that $\tilde{x}^i_{t|t-1,k} = x_t - \hat{x}^i_{t|t-1,k}$. Then, we can calculate the ith one-step prediction error covariance $P^i_{t|t-1,k}$ based on Assumptions 6.1–6.2 and Lemma 6.1:

$$\begin{aligned}
P^i_{t|t-1,k} &= E\{\tilde{x}^i_{t|t-1,k}(\tilde{x}^i_{t|t-1,k})^T\,|z^i_{1,k}, z^i_{2,k}, \ldots, z^i_{t-1,k}, \gamma^i_{1,k}, \gamma^i_{2,k}, \ldots, \gamma^i_{t-1,k}\} \\
&= E\{(x_t - \hat{x}^i_{t|t-1,k})(x_k - \hat{x}^i_{t|t-1,k})^T\} \\
&= E\{(A_{t-1}\tilde{x}^i_{t-1|t-1,k} + F_{t-1}\varepsilon_{t-1}x_{t-1} + B_{t-1}\omega_{t-1}) \\
&\quad \cdot (A_{t-1}\tilde{x}^i_{t-1|t-1,k} + F_{t-1}\varepsilon_{t-1}x_{t-1} + B_{t-1}\omega_{t-1})^T\} \\
&= A_{t-1}P^i_{t-1|t-1,k}A^T_{t-1} + Q_{\varepsilon,k}F_{t-1}X_{t-1}F^T_{t-1} + B_{t-1} \\
&\quad \cdot Q_{w,t-1}B^T_{t-1} + A_{t-1}M_{t-1}B^T_{t-1} + B_{t-1}M^T_{t-1}A^T_{t-1}
\end{aligned} \tag{6.30}$$

where $M_{t-1} = E\{\tilde{x}^i_{t-1|t-1,k}\,w^T_{t-1}\}$. Note that Eq. (6.30) yields Eq. (6.26).

Similarly, M_{t-1} is computed by:

$$M_{t-1} = B_{t-2}Q_{w,(t-2,t-1)} - E\{\hat{x}^i_{t-1|t-1,k}\,w^T_{t-1}\} \tag{6.31}$$

Using Eqs. (6.2)–(6.7), (6.17) and noting Assumptions 6.1–6.3, $E\{\hat{x}^i_{t-1|t-1,k}\,w^T_{t-1}\}$ can be obtained by

$$\begin{aligned}
E\{\hat{x}^i_{t-1|t-1,k}\,w^T_{t-1}\} &= \gamma^i_{t-1,k}K^i_{t-1,k}C_{i,t-1}B_{t-2}Q_{w,(t-2,t-1)} \\
&\quad + \gamma^i_{t-1,k}K^i_{t-1,k}D_{i,t-1}(S^i_{wv,t-1})^T
\end{aligned} \tag{6.32}$$

Substituting Eq. (6.32) into Eq. (6.31), we have Eq. (6.28).

Note that $\tilde{x}^i_{t|t-1,k} = x_t - \hat{x}^i_{t|t-1,k}$ from Eq. (6.18) and according to Eq. (6.12), $\tilde{x}^i_{t|t,k}$ is derived as follows:

$$\begin{aligned}
\tilde{x}^i_{t|t,k} &= x_t - \hat{x}^i_{t|t-1,k} - \gamma^i_{t,k}K^i_{t,k}(z_{i,t} - C_{i,t}\hat{x}^i_{t|t-1,k}) \\
&= x_t - \hat{x}^i_{t|t-1,k} - \gamma^i_{t,k}K^i_{t,k}((C_{i,t} + \Gamma_{i,t}\epsilon_{i,t})x_t + D_{i,t}v_{i,t} - C_{i,t}\hat{x}^i_{t|t-1,k}) \\
&= (I - \gamma^i_{t,k}K^i_{t,k}(C_{i,t} + \Gamma_{i,t}\epsilon_{i,t}))\tilde{x}^i_{t|t-1,k} - \gamma^i_{t,k}K^i_{t,k}D_{i,t}v_{i,t}
\end{aligned} \tag{6.33}$$

Also, we can calculate the expectation $\tilde{x}^i_{t|t,k}$ by

$$\begin{aligned}
E\{\tilde{x}^i_{t|t,k}\} &= (I - \gamma^i_{t,k}K^i_{t,k}C_{i,k})E\{\tilde{x}^i_{t|t-1,k}\} \\
&= (I - \gamma^i_{t,k}K^i_{t,k}C_{i,t})E\{x_t - \hat{x}^i_{t|t-1,k}\} \\
&= (I - \gamma^i_{t,k}K^i_{t,k}C_{i,t})E\{(A_{t-1} + F_{t-1}\varepsilon_{t-1})x_{t-1} + B_{t-1}w_{t-1} \\
&\quad - A_{t-1}\hat{x}^i_{t-1|t-1,k}\} \\
&= (I - \gamma^i_{t,k}K^i_{t,k}C_{i,t})A_{t-1}E\{\tilde{x}^i_{t-1|t-1,k}\}
\end{aligned} \tag{6.34}$$

Denoting

$$
\begin{aligned}
G_{i,t} &= I - \gamma_{t,k}^{i} K_{t,k}^{i} C_{i,t} \\
\Phi_{i,t} &= \gamma_{t,k}^{i} K_{t,k}^{i} \Gamma_{i,t} \\
L_{i,t} &= \gamma_{t,k}^{i} K_{t,k}^{i} D_{i,t} \\
H_{i,t} &= E\{\tilde{x}_{t|t-1,k}^{i} v_{i,t}^{T}\}
\end{aligned}
$$

and according to Assumptions 6.1–6.3, we can determine $H_{i,t}$ by

$$
\begin{aligned}
H_{i,t} &= E\{x_t v_{i,t}^T\} - \{\hat{x}_{t|t-1,k}^{i} v_{i,t}^{T}\} \\
&= A_{t-1} B_{t-2} S_{wv,(t-2,t)}^{i} + B_{t-1} S_{wv,(t-1,t)}^{i} - \gamma_{t-1,k}^{i} A_{t-1} K_{t-1,k}^{i} C_{i,t-1} \\
&\quad \cdot B_{t-2} S_{wv,(t-2,t)}^{i} - \gamma_{t-1,k}^{i} A_{t-1} K_{t-1,k}^{i} R_{v,(t-1,t)}^{ii}
\end{aligned} \tag{6.35}
$$

Note that Eq. (6.35) is consistent with Eq. (6.29). Thus, we can compute the local Kalman error covariance $P_{t|t,k}^{i}$ by

$$
\begin{aligned}
&P_{t|t,k}^{i} \\
&= E\{\tilde{x}_{t|t,k}^{i} (\tilde{x}_{t|t,k}^{i})^T \,|z_{1,k}^{i}, z_{2,k}^{i}, \ldots, z_{t,k}^{i}, \gamma_{1,k}^{i}, \gamma_{2,k}^{i}, \ldots, \gamma_{t,k}^{i}\} \\
&= E[(G_{i,t} \tilde{x}_{t|t-1,k}^{i} - \Phi_{i,t} \epsilon_{i,t} \tilde{x}_{t|t-1,k}^{i} - L_{i,t} v_{i,t}) \\
&\quad \cdot (G_{i,t} \tilde{x}_{t|t-1,k}^{i} - \Phi_{i,t} \epsilon_{i,t} \tilde{x}_{t|t-1,k}^{i} - L_{i,t} v_{i,t})^T] \\
&= G_{i,t} P_{t|t-1,k}^{i} (G_{i,t})^T + R_{\epsilon,t}^{ii} \Phi_{i,t} P_{t|t-1,k}^{i} (\Phi_{i,t})^T \\
&\quad + L_{i,t} R_{v,t}^{ii} (L_{i,t})^T - G_{i,t} H_{i,t} (L_{i,t})^T - L_{i,t} (H_{i,t})^T (G_{i,t})^T
\end{aligned} \tag{6.36}
$$

According to [13], we can solve Eq. (6.36) with the first variation and obtain the following form:

$$
\begin{aligned}
&- 2\gamma_{t,k}^{i} P_{t|t-1,k}^{i} C_{i,t}^{T} + 2(\gamma_{t,k}^{i})^2 K_{t,k}^{i} C_{i,t} P_{t|t-1,k}^{i} C_{i,t}^{T} \\
&+ 2R_{\epsilon,t}^{ii} (\gamma_{t,k}^{i})^2 K_{t,k}^{i} \Gamma_{i,t} P_{t|t-1,k}^{i} \Gamma_{i,t}^{T} + 2(\gamma_{t,k}^{i})^2 K_{t,k}^{i} D_{i,t} R_{v,t}^{ii} D_{i,t}^{T} \\
&- 2\gamma_{t,k}^{i} H_{i,t} D_{i,t}^{T} + 2(\gamma_{t,k}^{i})^2 K_{t,k}^{i} C_{i,t} H_{i,t} D_{i,t}^{T} \\
&+ 2(\gamma_{t,k}^{i})^2 K_{t,k}^{i} D_{i,t} (H_{i,t})^T C_{i,t}^{T} = 0
\end{aligned} \tag{6.37}
$$

where $\gamma_{t,k}^{i} = 0$ or 1. $K_{t,k}^{i}$ refers to any value if $\gamma_{t,k}^{i} = 0$ and $K_{t,k}^{i}$ can be calculated by Eq. (6.37) if $\gamma_{t,k}^{i} = 1$. Therefore, we can determine the optimal local Kalman filter gain $K_{t,k}^{i}$ by

$$
\begin{aligned}
K_{t,k}^{i} &= (P_{t|t-1,k}^{i} C_{i,t}^{T} + H_{i,t} D_{i,t}^{T})(C_{i,t} P_{t|t-1,k}^{i} C_{i,t}^{T} + R_{\epsilon,t}^{ii} \Gamma_{i,t} P_{t|t-1,k}^{i} \Gamma_{i,t}^{T} \\
&\quad + D_{i,t} R_{v,t}^{ii} D_{i,t}^{T} + C_{i,t} H_{i,t} D_{i,t}^{T} + D_{i,t} H_{i,t}^{T} C_{i,t}^{T})^{-1}
\end{aligned} \tag{6.38}
$$

Then, substituting Eq. (6.38) into Eq. (6.36), we can rewrite the optimal local Kalman filter error covariance $P_{k|k}^{i}$ as follows:

$$P^i_{t|t,k} = P^i_{t|t-1,k} - \gamma^i_{t,k} K^i_{t,k} (C_{i,t} P^i_{t|t-1,k} + D_{i,t} (H_{i,t})^T) \tag{6.39}$$

Note that Eqs. (6.38) and (6.39) are the equation in Eqs. (6.25) and (6.27), respectively. Repeating Eqs. (6.21)–(6.27), we can obtain the ith optimal local Kalman filter estimate $\hat{x}_{i,k|k}$ and the error covariance $P_{i,k|k}$ by iterating $k - \kappa_i + 1$ times from $\hat{x}^i_{\kappa_i-1|\kappa_i-1,k}$ and $P^i_{\kappa_i-1|\kappa_i-1,k}$, respectively. Thus the proof is completed.

Theorem 6.2 *Consider the ith and jth subsystems ($i, j = 1, 2, \ldots, N$) of system (6.1), based on Eqs. (6.21)–(6.25) in Theorem 6.1, we have the optimal local Kalman filter one-step prediction error cross-covariance $P^{ij}_{t|t-1,k}$ and error cross-covariance $P^{ij}_{t|t,k}$ between any two subsystems (i.e. the ith and jth subsystems) at time instant k by:*

$$P^{ij}_{\kappa_i-1|\kappa_i-1,k} = P^{ij}_{\kappa_i-1|\kappa_i-1,k-1} \tag{6.40}$$

$$\begin{aligned} P^{ij}_{t|t-1,k} &= A_{t-1} P^{ij}_{t-1|t-1,k} A^T_{t-1} + Q_{\varepsilon,k} F_{t-1} X_{t-1} F^T_{t-1} \\ &\quad + B_{t-1} Q_{w,t-1} B^T_{t-1} + A_{t-1} M_{i,t-1} B^T_{t-1} \\ &\quad + B_{t-1} M^T_{j,t-1} A^T_{t-1} \end{aligned} \tag{6.41}$$

$$\begin{aligned} P^{ij}_{t|t,k} &= G_{i,t} P^i_{t|t-1,k} G^T_{j,t} + R^{ij}_{\epsilon,t} \Phi_{i,t} P^{ij}_{t|t-1,k} \Phi^{j,T}_t \\ &\quad + L_{i,t} R^{ij}_{v,t} L^T_{j,t} - G_{i,t} H_{ij,t} L^T_{j,t} \\ &\quad - L_{i,t} H^T_{ij,t} G^T_{j,t} \end{aligned} \tag{6.42}$$

where

$$\begin{aligned} H_{ij,t} &= A_{t-1} B_{t-2} S^j_{wv,(t-2,t)} + B_{t-1} S^j_{wv,(t-1,t)} \\ &\quad - \gamma^i_{t-1,k} A_{t-1} K^i_{t-1,k} C_{i,t-1} B_{t-2} S^j_{wv,(t-2,t)} \\ &\quad - \gamma^i_{t-1,k} A_{t-1} K^i_{t-1,k} R^{ij}_{v,(t-1,t)} \end{aligned} \tag{6.43}$$

where we can calculate $G_{i,t} = I - \gamma^i_{t,k} K^i_{t,k} C_{i,t}$, $\Phi_{i,t} = \gamma^i_{t,k} K^i_{t,k} \Gamma_{i,t}$, $L_{i,t} = \gamma^i_{t,k} K^i_{t,k} D_{i,t}$, $P^i_{0|0,k} = P_0$, $t = \kappa_i, \ldots, k$. $M_{i,t-1}$ can be computed by Eq. (6.28) in Theorem 6.1. Then, we can obtain the optimal local Kalman filter error cross-covariance $P_{ij,k|k}$ between any two subsystems (i.e. the ith and jth subsystems) at time instant k by iterating $k - \kappa_i + 1$ times from $P^{ij}_{\kappa_i-1|\kappa_i-1,k}$.

Proof According to the definition of κ_i in Eqs. (6.20) and (6.22) in Theorem 6.1, we can easily obtain Eq. (6.40).

For the ith subsystem, we can write the optimal local Kalman filter error $\tilde{x}^i_{t|t,k}$ and one-step prediction error $\tilde{x}^i_{t|t-1,k}$ as the following forms from Eqs. (6.30) and (6.36) in Theorem 6.1, respectively.

$$\tilde{x}^i_{t|t,k} = G_{i,t} \tilde{x}^i_{t|t-1,k} - \Phi_{i,t} \epsilon_{i,t} \tilde{x}^i_{t|t-1,k} - L_{i,t} v_{i,t} \tag{6.44}$$

$$\tilde{x}^i_{t|t-1,k} = A_{t-1} \tilde{x}^i_{t-1|t-1,k} + F_{t-1} \varepsilon_{t-1} x_{t-1} + B_{t-1} w_{t-1} \tag{6.45}$$

Similarly, for the jth subsystem, we have

$$\tilde{x}^{j}_{t|t,k} = G_{j,t}\tilde{x}^{j}_{t|t-1,k} - \Phi^{j}_{t}\epsilon_{j,t}\tilde{x}^{j}_{t|t-1,k} - L_{j,t}v_{j,t} \tag{6.46}$$

$$\tilde{x}^{j}_{t|t-1,k} = A_{t-1}\tilde{x}^{j}_{t-1|t-1,k} + F_{t-1}\varepsilon_{t-1}x_{t-1} + B_{t-1}w_{t-1} \tag{6.47}$$

According to Eqs. (6.45), (6.47), (6.30) and Assumptions 6.1–6.3, we can calculate the one-step prediction error cross-covariance $P^{ij}_{t|t-1,k}$ between any two subsystems (i.e. the ith and jth subsystems) by

$$\begin{aligned}
P^{ij}_{t|t-1,k} &= E\{(A_{t-1}\tilde{x}^{i}_{t-1|t-1,k} + F_{t-1}\varepsilon_{t-1}x_{t-1} + B_{t-1}w_{t-1}) \\
&\quad \cdot (A_{t-1}\tilde{x}^{j}_{t-1|t-1,k} + F_{t-1}\varepsilon_{t-1}x_{t-1} + B_{t-1}w_{t-1})^{T}\} \\
&= A_{t-1}P^{ij}_{t-1|t-1,k}A^{T}_{t-1} + Q_{\varepsilon,k}F_{t-1}X_{t-1}F^{T}_{t-1} \\
&\quad + B_{t-1}Q_{w,t-1}B^{T}_{t-1} + A_{t-1}M_{i,t-1}B^{T}_{t-1} \\
&\quad + B_{t-1}M^{T}_{j,t-1}A^{T}_{t-1}
\end{aligned} \tag{6.48}$$

Note that Eq. (6.48) yields Eq. (6.41).

Denoting $H_{ij,t} = E\{\tilde{x}^{i}_{t|t,k}v^{T}_{j,t}\}$ and noting Assumptions 6.1–6.3, we can compute $H_{ij,t}$ by

$$\begin{aligned}
H_{ij,t} &= E\{x_{t}v^{T}_{j,t}\} - \{\hat{x}^{i}_{t|t-1,k}v^{T}_{j,t}\} \\
&= A_{t-1}B_{t-2}S^{j}_{wv,(t-2,t)} + B_{t-1}S^{j}_{wv,(t-1,t)} \\
&\quad - \gamma^{i}_{t-1,k}A_{t-1}K^{i}_{t-1,k}C_{i,t-1}B_{t-2}S^{j}_{wv,(t-2,t)} \\
&\quad - \gamma^{i}_{t-1,k}A_{t-1}K^{i}_{t-1,k}R^{ij}_{v,(t-1,t)}
\end{aligned} \tag{6.49}$$

Note that Eq. (6.49) is just the equation in Eq. (6.40). Therefore, we can compute the local Kalman error cross-covariance $P^{ij}_{t|t,k}$ between any two subsystems (i.e. the ith and jth subsystems) by

$$\begin{aligned}
P^{ij}_{t|t,k} &= E\{(G_{i,t}\tilde{x}^{i}_{t|t-1,k} - \Phi_{i,t}\epsilon_{i,t}\tilde{x}^{i}_{t|t-1,k} - L_{i,t}v_{i,t}) \\
&\quad \cdot (G_{j,t}\tilde{x}^{j}_{t|t-1,k} - \Phi^{j}_{t}\epsilon_{j,t}\tilde{x}^{j}_{t|t-1,k} - L_{j,t}v_{j,t})^{T}\} \\
&= G_{i,t}P^{ij}_{t|t-1,k}G^{T}_{j,t} + R^{ij}_{\epsilon,t}\Phi_{i,t}P^{ij}_{t|t-1,k}\Phi^{T}_{j,t} \\
&\quad + L_{i,t}R_{ij,t}L^{T}_{j,t} - G_{i,t}H_{ij,t}L^{T}_{j,t} - L_{i,t}H^{T}_{ij,t}(G_{j,t})^{T}
\end{aligned} \tag{6.50}$$

Note that Eq. (6.50) is Eq. (6.42). Repeating Eqs. (6.21)–(6.25) and Eqs. (6.40)–(6.42), one can obtain the optimal local Kalman filter error cross-covariance $P_{ij,k|k}$ between any two subsystems (i.e. the ith and jth subsystems) by iterating $k - \kappa_i + 1$ times from $P^{ij}_{\kappa_i-1|\kappa_i-1,k}$. The proof is complete.

Remark 6.5 From the above iteration steps of Theorems 6.1 and 6.2, one can notice that the correlated multiplicative noises, autocorrelated noises and cross-correlated noises are considered, and buffers of finite length are used to process measurement delay. When ε_k, $\epsilon_{i,k}$, $Q_{w,(k,l)}$, $Q_{w,(k,l)}$, $R^{ij}_{v,(k,l)}$, $R^{ij}_{v,(k,l)}$, $S^i_{wv,k}$, $S^i_{wv,(k,l)}$, $S^i_{wv,(k,l)} = 0$ and $L = 1$, it degrades to the Kalman filtering with intermittent observations in [23].

6.4 Distributed Weighted Kalman Filter Fusion with Buffers of Finite Length

In this section, based on the optimal local Kalman filter estimate $\hat{x}^i_{k|k}$ from Eq. (6.24) and error cross-covariance $P_{ij,k|k}$ from Eq. (6.42), the DWKFF with buffers for a class of MUNSs with measurement delay or loss and autocorrelated and cross-correlated noises is investigated. Based on Theorems 6.1 and 6.2, we can easily obtain the following theorem from the optimal weighted fusion algorithm in the linear minimum-variance sense [26].

Theorem 6.3 *For system (6.1), based on Assumptions 6.1–6.3 and Theorems 6.1–6.2, it can be calculated by the following form, which is the DWKFF estimate* $\hat{x}^{fusion}_{k|k}$

$$\hat{x}^{fusion}_{k|k} = \Xi_1 \hat{x}_{1,k|k} + \Xi_2 \hat{x}_{2,k|k} + \cdots + \Xi_N \hat{x}_{N,k|k} \tag{6.51}$$

where, we can calculate the weighted matrices Ξ_i, $i = 1, 2, \ldots, N$ *by*

$$\Xi = \Omega^{-1} e (e^T \Omega^{-1} e)^{-1} \tag{6.52}$$

where $\Omega = E\{(\hat{x}_{1,k|k}, \ldots, \hat{x}_{N,k|k})^T (\hat{x}_{1,k|k}, \ldots, \hat{x}_{N,k|k})\} = (P_{ij,k|k})$, $(i, j = 1, 2, \ldots, N)$ *is an* $nN \times nN$ *symmetric positive definite matrix,* $\Xi = [\Xi_1, \ldots, \Xi_N]^T$ *and* $e = [I_n, \ldots, I_n]^T$ *are both* $nN \times n$ *matrices. It can be computed by the following form, which is the corresponding variance of the DWKFF estimate error*

$$P^{fusion}_{k|k} = (e^T \Omega^{-1} e)^{-1} \tag{6.53}$$

where $P^{fusion}_{k|k} \le P_{i,k|k} = P_{ii,k|k}$, $i = 1, 2, \ldots, N$.

Based on Theorem 1 in [26], we can directly obtain the proof of Theorem 6.3.

According to Theorems 6.1–6.3, we can easily obtain the DWKFF algorithm of MUNSs with buffers, as follows:

DWKFF algorithm of MUNSs with buffers
Step 1: Initialization
$\hat{x}^i_{0\|0,k} = x_0$, $P^i_{0\|0,k} = P_0$ and $P^{ij}_{0\|0,k} = P_0$
Step 2: Buffer process
(a) Compute $\hat{x}^i_{k\|k,k}$ and $P^i_{k\|k,k}$ by iterating $k - \kappa_i + 1$ times following (6.24) and (6.27) from $\hat{x}^i_{\kappa_i-1\|\kappa_i-1,k}$ and $P^i_{\kappa_i-1\|\kappa_i-1,k}$, respectively
(b) Compute $P^{ij}_{k\|k,k}$ by iterating $k - \kappa_i + 1$ times following (6.42) from $P^{ij}_{\kappa_i-1\|\kappa_i-1,k}$
Step 3: Weighted fusion
(a) $\hat{x}_{i,k\|k} = \hat{x}^i_{k\|k,k}$, $P_{i,k\|k} = P^i_{k\|k,k}$ and $P_{ij,k\|k} = P^{ij}_{k\|k,k}$
(b) Compute $\hat{x}^{fusion}_{k\|k}$ and $P^{fusion}_{k\|k}$ by (6.51) and (6.53), respectively
Step 4: Set $k = k + 1$ and go to Step 2

Remark 6.6 In this chapter, the multiplicative noises both in the state equation and the measurement equation, the autocorrelated noises and the cross-correlated noises are considered. The probabilistic metric in this chapter is similar to that of [7, 9]. Although the expressions are not the same, they are essentially similar. In this chapter, the packet delay is modeled as a Poisson distribution with mean d_i, i.e., the probability density function satisfies:

$$\text{Prob}\{d_{i,k} = j\} = f_i(j) = \frac{(d_i)e^{-d_i}}{j!}, \quad j = 0, 1, \ldots$$

where $d_i = E\{d_{i,k}\}$ denotes the mean value of the packet delay in the ith channel, and $\sum_{j=0}^{\infty} f_i(j) = 1$. In [7, 9], the occurrence probabilities of delays are known as a priori through statistical test, that is

$$\text{Prob}\{d_i(t) = \ell\} = \pi_{i,\ell}, \ \ell = 0, 1, \ldots, d_i$$

where $\pi_{i,\ell}$ is a positive scalar, and $\sum_{\ell=0}^{d_i} \pi_{i,\ell} = 1$.

6.5 Simulation Results

In this section, the following discrete-time linear stochastic system with 2 local sensors in [13, 27] is considered:

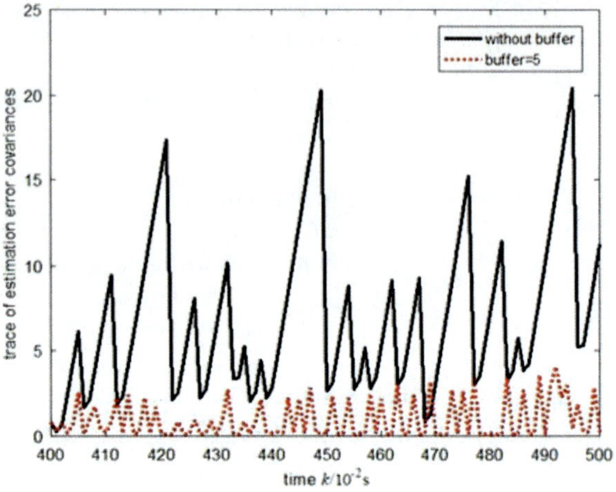

Fig. 6.4 The traces of error covariances of the fusion center (FC) with buffers $L_1 = L_2 = 5$ and without buffer

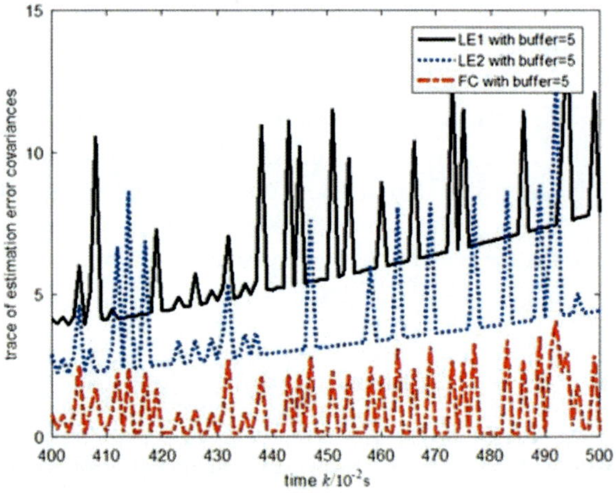

Fig. 6.5 The traces of error covariances of the local estimators (LEs) and FC with buffers $L_1 = L_2 = 5$

$$x_{k+1} = \left(\begin{bmatrix} 0.95 & T \\ 0 & 0.95 \end{bmatrix} + \begin{bmatrix} 0.01 & 0 \\ 0 & 0.01 \end{bmatrix} \varepsilon_k \right) x_k + \begin{bmatrix} 0.8 \\ 0.6 \end{bmatrix} w_k \tag{6.54}$$

$$z_{i,k} = (C_i + \Gamma_i \epsilon_{i,k}) x_k + D_i v_{i,k}, \ i = 1, 2 \tag{6.55}$$

$$\epsilon_{i,k} = \tau_i \varepsilon_k + \varphi_k, \ i = 1, 2 \tag{6.56}$$

$$w_k = \phi_k + \phi_{k-1} \tag{6.57}$$

$$v_{i,k} = c_i \phi_{k-1} + c_i \phi_{k-2}, \ i = 1, 2 \tag{6.58}$$

where $T = 0.01s$ is the sample period, two components of x_k is the position and velocity of the target. The variables $\varepsilon_k \in R$, $\varphi_k \in R$ and $\phi_k \in R$ are zero-mean Gaussian white noises with the covariances 1, 0.5 and 0.5, respectively. Without loss of generality, suppose the measurement multiplicative noises $\epsilon_{i,k} (i = 1, 2)$ sat-

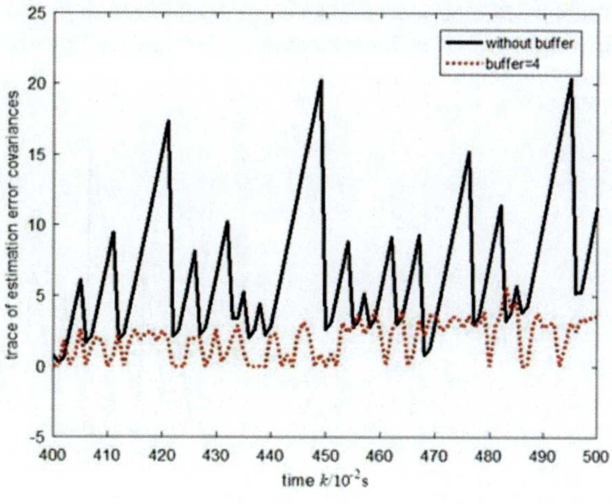

Fig. 6.6 The traces of error covariances of FC with buffers $L_1 = L_2 = 4$ and without buffer

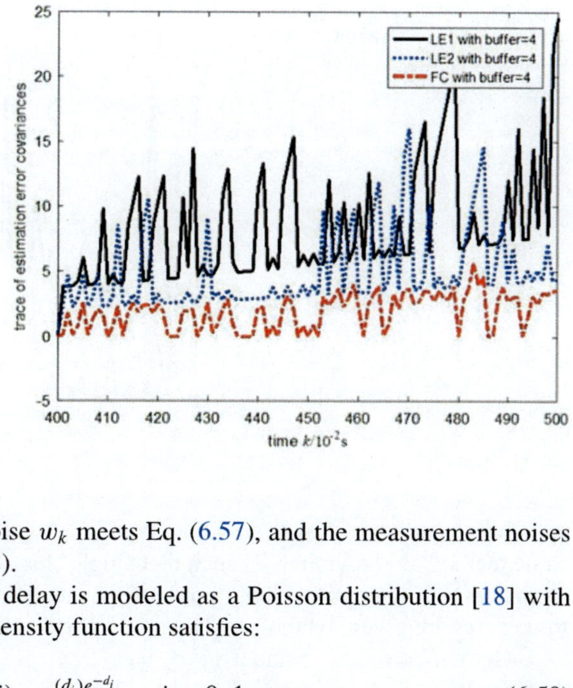

Fig. 6.7 The traces of error covariances of LEs and FC with buffers $L_1 = L_2 = 4$

isfy Eq. (6.56), the process noise w_k meets Eq. (6.57), and the measurement noises $v_{i,k}(i = 1, 2)$ satisfy Eq. (6.58).

Similar to [22], the packet delay is modeled as a Poisson distribution [18] with mean d_i, i.e., the probability density function satisfies:

$$f_i(j) = \frac{(d_i)e^{-d_i}}{j!}, \quad j = 0, 1, \ldots \tag{6.59}$$

where $d_i = E\{d_{i,k}\}$ denotes the mean value of the packet delay in the ith channel, and it is supposed that $d_1 = d_2 = 3$ or $d_1 = d_2 = 2$.

In the simulation, we select the initial conditions as $x_0 = [200, 10]^T$ and $P_0 = \text{diag}(0.5, 0.1)$. $c_i (i = 1, 2)$ are chosen as 1 and 0.5. $\tau_1 = 0.6$, $\tau_2 = 0.5$, and the measurement matrices are set as $C_1 = [1, 1]$, $C_2 = [1, 1.5]$, $\Gamma_1 = [0.1, 0.1]$, $\Gamma_2 = [0.2, 0.1]$ and $D_1 = D_2 = 1$. Suppose $d_1 = d_2 = 3$ or $d_1 = d_2 = 2$, and the corre-

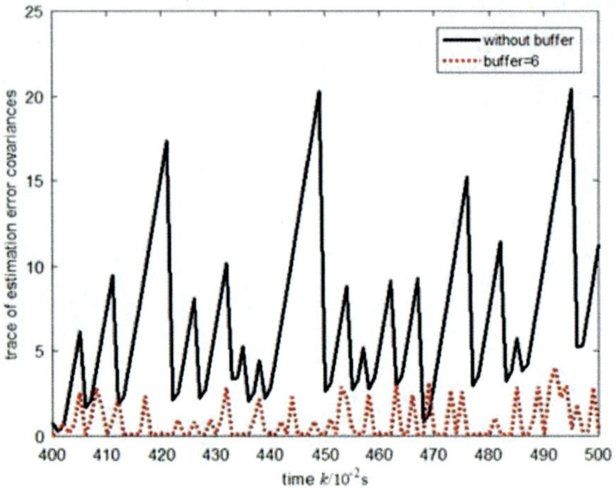

Fig. 6.8 The traces of error covariances of FC with buffers $L_1 = L_2 = 6$ and without buffer

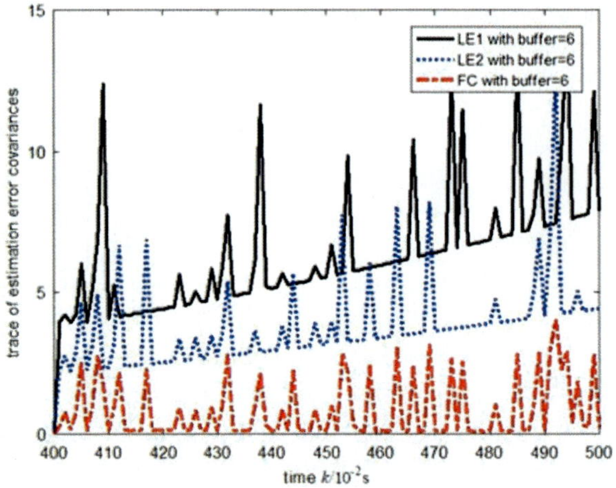

Fig. 6.9 The traces of error covariances of LEs and FC with buffers $L_1 = L_2 = 6$

sponding $L_1 = L_2 = 5$ or $L_1 = L_2 = 4$ or $L_1 = L_2 = 6$ are obtained by utilizing the same method as shown in [22] such that single channel keeps stable. To make a fair comparison of the optimal length of buffers, the simulation results of the following three cases are given, which are shown as follows.

Case A: $L_1 = L_2 = 5$ and $d_1 = d_2 = 3$.

Case B: $L_1 = L_2 = 4$ and $d_1 = d_2 = 2$.

Case C: $L_1 = L_2 = 6$ and $d_1 = d_2 = 3$.

As shown in Figs. 6.4, 6.6 and 6.8, the trace of the error covariance of FC is smaller than that without buffer. From Figs. 6.5, 6.7 and 6.9, one can see that the trace of the error covariance of FC is smaller than that of each LE. There is no relationship with the length of buffers within a certain length. From Fig. 6.10, it can be seen that the performance of FC with buffers $L_1 = L_2 = 5$ and buffers $L_1 = L_2 = 4$ are better than that without buffer. As shown in Fig. 6.11, the trace of the error covariance of

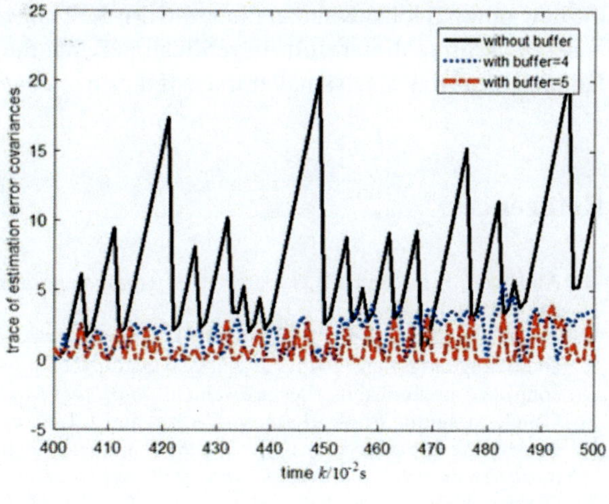

Fig. 6.10 The traces of error covariances of FC with buffers $L_1 = L_2 = 4$, buffers $L_1 = L_2 = 5$ and without buffer

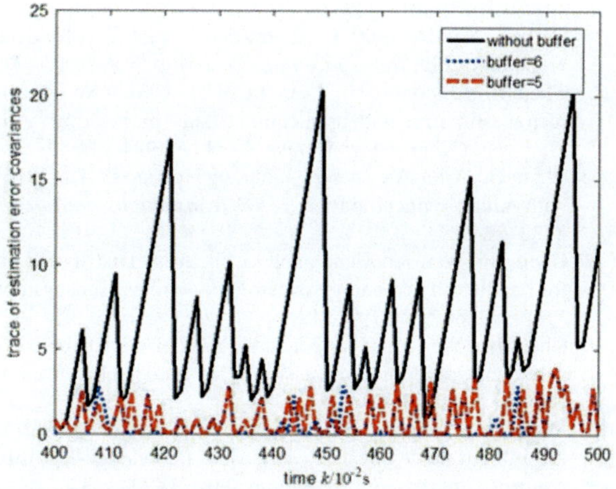

Fig. 6.11 The traces of error covariances of FC with buffers $L_1 = L_2 = 6$, buffers $L_1 = L_2 = 5$ and without buffer

FC with buffers $L_1 = L_2 = 5$ and buffers $L_1 = L_2 = 6$ are much smaller than that without buffer. Meanwhile, from Fig. 6.11, it is also clear that buffers $L_1 = L_2 = 6$ lead to little difference on FC, but increase burden of calculations.

6.6 Conclusion

In this chapter, we have investigated the problem of DWKFF for a class of MUNSs with network delays, stochastic uncertainties and correlated noises. A distributed multi-sensor data fusion system suffers measurement delay or loss due to the unreliability of the network. By using finite length buffers to deal with measurement delay or loss, an optimal local Kalman filter for each subsystem has been derived. Based on the new optimal local Kalman filter, the DWKFF algorithm with finite length buffers is derived. Simulation results have illustrated that the DWKFF algorithm with finite length buffers has much smaller trace of the error covariance than that without buffer.

References

1. Anderson, B.D., and J.B. Moore. 1979. *Optimal Filtering*. Englewood Cliffs, New Jersey: Prentice-Hall.
2. Annamalai, Andy S. K., Amit Motwani, Sanjay Sharma, Robert Sutton, Phil F. Culverhouse, and Chenguang Yang. 2015. A robust navigation technique for integration in the guidance and control of an uninhabited surface vehicle. *Journal of Navigation* 68 (4): 750–768.
3. Caballero-águila, R., A. Hermoso-Carazo, and J. Linares-Pérez. 2015. Optimal state estimation for networked systems with random parameter matrices, correlated noises and delayed measurements. *International Journal of General Systems* 44 (1–2): 142–154.
4. Carlson, N.A., and M.P. Berarducci. 1994. Federated Kalman filter simulation results. *Navigation* 41 (3): 297–322.
5. Carlson, Neal A. 1990. Federated square root filter for decentralized parallel processors. *IEEE Transactions on Aerospace and Electronic Systems* 26 (3): 517–525.
6. Chen, Bo, Guoqiang Hu, Daniel W. C. Ho, Wen-An Zhang, and Li Yu. 2017. Distributed robust fusion estimation with application to state monitoring systems. *IEEE Transactions on Systems, Man, and Cybernetics: Systems* 47 (11): 2994–3005.
7. Chen, Bo, Wen-An Zhang, Guoqiang Hu, and Li Yu. 2015. Networked fusion kalman filtering with multiple uncertainties. *IEEE Transactions on Aerospace and Electronic Systems* 51 (3): 2232–2249.
8. Chen, Bo, Wen-An Zhang, and Li Yu. 2014. Distributed finite-horizon fusion Kalman filtering for bandwidth and energy constrained wireless sensor networks. *IEEE Transactions on Signal Processing* 62 (4): 797–812.
9. Chen, Bo, Wen-An Zhang, and Li Yu. 2014. Distributed fusion estimation with missing measurements, random transmission delays and packet dropouts. *IEEE Transactions on Automatic Control* 59 (7): 1961–1967.
10. Feng, Bo, Hongbin Ma, Mengyin Fu, and Chenguang Yang. 2015. Real-time state estimator without noise covariance matrices knowledge-fast minimum norm filtering algorithm. *Iet Control Theory and Applications* 9 (9): 1422–1432.
11. Feng, Jianxin, Tingfeng Wang, and Jin Guo. 2014. Recursive estimation for descriptor systems with multiple packet dropouts and correlated noises. *Aerospace Science and Technology* 32 (1): 200–211.
12. Feng, Jianxin, Zidong Wang, and Ming Zeng. 2011. Optimal robust non-fragile Kalman-type recursive filtering with finite-step autocorrelated noises and multiple packet dropouts. *Aerospace Science and Technology* 15 (6): 486–494.
13. Feng, Jianxin, Zidong Wang, and Ming Zeng. 2013. Distributed weighted robust Kalman filter fusion for uncertain systems with autocorrelated and cross-correlated noises. *Information Fusion* 14 (1): 78–86.
14. Gao, Yongxin, X. Rong Li, and Enbin Song. 2016. Robust linear estimation fusion with allowable unknown cross-covariance. *IEEE Transactions on Systems Man and Cybernetics Systems* 46 (9): 1314–1325.
15. Hu, Jun, Zidong Wang, and Huijun Gao. 2013. Recursive filtering with random parameter matrices, multiple fading measurements and correlated noises. *Automatica* 49 (11): 3440–3448.
16. Li, Fan, Jie Zhou, and Duzhi Wu. 2013. Optimal filtering for systems with finite-step autocorrelated noises and multiple packet dropouts. *Dialogues in Cardiovascular Medicine Dcm* 24 (1): 255–263.

17. Linares-Perez, J., R. Caballero-Aguila, and I. Garcia-Garrido. 2014. Optimal linear filter design for systems with correlation in the measurement matrices and noises: recursive algorithm and applications. *International Journal of Systems Science* 45 (7): 1548–1562.
18. Liu, Kun, Emilia Fridman, Karl Henrik Johansson, and Yuanqing Xia. 2016. Generalized jensen inequalities with application to stability analysis of systems with distributed delays over infinite time-horizons. *Automatica* 69: 222–231.
19. Ma, Hongbin, Yini Lv, Chenguang Yang, and Mengyin Fu. 2014. Decentralized adaptive filtering for multi-agent systems with uncertain couplings. *IEEE/CAA Journal of Automatica Sinica* 1 (1): 101–112.
20. Qi, Wenjuan, Peng Zhang, and Zili Deng. 2014. Robust weighted fusion Kalman filters for multisensor time-varying systems with uncertain noise variances. *Signal Process* 99: 185–200.
21. Rezaei, Hossein, Reza Mahboobi Esfanjani, and Milad Farsi. 2015. Robust filtering for uncertain networked systems with randomly delayed and lost measurements. *IET Signal Processing* 9 (4): 320–327.
22. Shi, Ling, Lihua Xie, and R.M. Murray. 2009. Kalman filtering over a packet-delaying network: A probabilistic approach. *Automatica* 45 (9): 2134–2140.
23. Sinopoli, Bruno, Luca Schenato, Massimo Franceschetti, Kameshwar Poolla, Michael I. Jordan, and S. Shankar Sastry. 2004. Kalman filtering with intermittent observations. *IEEE Transactions on Automatic Control* 49 (9): 1453–1464.
24. Song, Enbin, Yunmin Zhu, Jie Zhou, and Zhisheng You. 2007. Optimal Kalman filtering fusion with cross-correlated sensor noises. *Automatica* 43 (8): 1450–1456.
25. Sun, Shuli. 2004. Multisensor optimal information fusion input white noise deconvolution estimators. *IEEE transactions on systems, man, and cybernetics. Part B, Cybernetics : a publication of the IEEE Systems, Man, and Cybernetics Society* 34 (4): 1886–1893.
26. Sun, Shuli, and Zili Deng. 2004. Multi-sensor optimal information fusion Kalman filter. *Automatica* 40 (6): 1017–1023.
27. Tian, Tian, Shuli Sun, and Na Li. 2016. Multi-sensor information fusion estimators for stochastic uncertain systems with correlated noises. *Information Fusion* 27: 126–137.
28. Xia, Yuanqing, Mengyin Fu, and Guoping Liu. 2010. *Multi-Channel Networked Data Fusion with Intermittent Observations*. Springer Science and Business Media.
29. Yan, Liping, X. Rong Li, Yuanqing Xia, and Mengyin Fu. 2013. Optimal sequential and distributed fusion for state estimation in cross-correlated noise. *Automatica* 49 (12): 3607–3612.
30. Yang, Chenguang, Hongbin Ma, and Mengyin Fu. 2013. Adaptive predictive control of periodic non-linear auto-regressive moving average systems using nearest-neighbour compensation. *Control Theory and Applications Iet* 7 (7): 936–951.
31. Zhang, Qiang, and Jifeng Zhang. 2012. Distributed parameter estimation over unreliable networks with markovian switching topologies. *IEEE Transactions on Automatic Control* 57 (10): 2545–2560.
32. Zhang, Shuo, Yan Zhao, Falin Wu, and Jianhui Zhao. 2014. Robust recursive filtering for uncertain systems with finite-step correlated noises, stochastic nonlinearities and autocorrelated missing measurements. *Aerospace Science and Technology* 39: 272–280.
33. Zhu, Cui, Yuanqing Xia, Liping Yan, and Mengyin Fu. 2010. Multi-channel networked data fusion with intermittent observations. In *Proceedings of the 29th Chinese Control Conference*, pages 4317–4322.
34. Zhu, Cui, Yuanqing Xia, Liping Yan, and Mengyin Fu. 2012. Centralised fusion over unreliable networks. *International Journal of Control* 85 (4): 409–418.

Chapter 7
State Estimation of Asynchronous Multirate Multisensor Systems

In the above chapters, the sampling of different sensors is assumed to be synchronous. In this chapter, the problem of optimal state estimation is studied for fusion of asynchronous multirate multiscale sensors with unreliable measurements and correlated noises. The noise of different sensors is cross-correlated and coupled with the system noise of the previous step and the same time step. The system is described at the highest sampling rate with different sensors observing a single target independently with multiple sampling rates. An optimal state estimation algorithm based on iterative estimation of white noise estimator is presented, which makes full use of the observation information effectively, overcomes the packet loss, data fault, unreliable factors, and improves the precision and the robustness of the system state estimation. A numerical example is used to illustrate the effectiveness of the presented algorithm.

7.1 Introduction

Data fusion technique combines data from multiple sensors and related information to achieve more specific inferences than could be achieved by using a single, independent sensor. Multisensor data fusion seeks to combine information from multiple sources (including sensors, human reports, and data from the internet) to achieve inferences that cannot be obtained from a single sensor or source, or whose quality exceeds that of an inference drawn from any single source [12]. Estimation fusion deals with the problem of how to best utilize useful information contained in multiple sets of data for the purpose of estimating a quantity [2].

When multiple sensors observe a common thing at the same sampling rate synchronously, the corresponding data fusion problem is easier and has been studied extensively [2, 4–6]. However, sensors may be asynchronous and sample nonuniformly with multiple rates and at multiscales, such as the case of target tracking systems [13, 32], fault detection systems [11], and predictive control systems [24]. Publications on multirate data fusion seldom consider the asynchronous or

L. Yan et al., *Multisensor Fusion Estimation Theory and Application*,
https://doi.org/10.1007/978-981-15-9426-7_7

multiscale observations, such as [18, 22, 23]. Measurements from different sensors maybe at different scales or different resolutions. For example, in image fusion, the source images taken by different sensors may have different resolutions. Literatures on asynchronous multiscale data fusion for discrete time systems are usually based on multiscale system theory, where the asynchronous multiscale observations are linked by the bridge of multiscale models [3, 13, 32, 33, 35]. The difficulty lies in the establishment of the proper multiscale models which could handle multisensor data effectively without making the original problem complicated. In [13, 35], the multiscale models are established by use of wavelet, which are used to fuse the multiscale data with the sampling rate ratio being the power of two. In [16, 32–34], moving average and its variations are used for multiscale modeling, and centralized fusion, recursive fusion, sequential fusion, and distributed fusion algorithms are presented, respectively, for fusing asynchronous data with multirate sampling at multiscales.

The research on asynchronous sampling systems seldom consider the correlation of noise. However, many multisensor systems have correlated noise in practical applications, for example, when the dynamic process is observed in a common noisy environment [25]. For synchronous sampling systems with the same sampling rate observations, there are quite a few state estimation algorithms. When the sensor noise is cross-correlated, Wen et al. propose a systematic way to handle the recursive fusion problem based on a unified data model in which the measurement noise across sensors at the same time is correlated [28]. By using the Cholesky factorization, the coupled noises can be decoupled and the optimal state estimation is derived by Duan et al. [9]. Reference [27] considers the correlation between the measurement noises and the system noise at the same time step. In the process of discretizing the continuous systems, the measurement noise is shown to be coupled with the system noise of the previous step, and several optimal state estimation algorithms are presented in [31]. For multirate multisensor data fusion with cross-correlated noise, recursive fusion and distributed fusion algorithms are presented in [19, 20], respectively. The white noise estimation theory is introduced systematically in [8].

The literatures listed above do not consider the date missing or unreliability of measurements, which may occur, for example, in networked environment due to the limitation of communication and sensor fault, etc. When the missing of measurements is Bernulli distributed, based on a discrete-time linear system, Kalman filtering with intermittent observations is studied in [17]. By using peak covariance as the estimation of filtering deterioration caused by packet losses, the stability of Kalman filtering with Markovian packet losses is studied in [15] for a linear time-invariant system. Xia et al. study networked data fusion with packet losses and variable delays, and an optimal state estimator is obtained [29]. When packets are not time-stamped, by giving a Riccati difference equation and a Lyapunov difference equation, Sun et al. study an optimal linear filter dependent on the probabilities of possible random delays and losses of measurements by the innovation analysis approach [26]. Recursive filtering with multiple fading measurements and correlated noises are studied in [10, 14]. For multirate sensor fusion, an effective multirate distributed estimation fusion algorithm for sensor networks with packet losses is presented in [36]. For asynchronous fusion with packet losses, a distributed estimation algorithm is presented in [7]. A state estimation algorithm for fusing asynchronous multirate multi–smart sensors is presented in [21]. For the above literatures, the multiscale sensor fusion, and the unreliability evaluation of the measurements are not concerned.

Some data may be lost or unreliable due to communication problems or sensor faults, and if the discrete-time linear system was obtained from discretization of the continuous-time system, then the measurement noise was correlated with the system noise of the previous time step and the same time step. How to fuse such unreliable, asynchronous, multirate, multiscale, multisensor data for optimal state estimation is a challenge and is the topic of this chapter. From the above, clearly fusion of asynchronous multirate multiscale sensor data is difficult, especially in the networked environment. The classical Kalman filter is optimal when the linear-Gaussian system of dynamics and measurements is known. This chapter assumes that the dynamics is known, but the measurement system is known only partially: Except for Sensor 1 that at the finest scale, for sensor $i = 2, 3, \ldots, N$, the sensors' measurement equations in this scenario are not exactly known. Besides, a measurement is unreliable—it may be a real observation of the system state or something else due to communication problems or sensor faults. In such a scenario, the classical Kalman filter cannot be used without a major modification. This chapter essentially presents such a major modification. For sensors having different sampling rates at different scales, multiscale system theory is a useful tool. Our work stems from multiscale system theory, but provides a new multiscale modeling architecture that is not limited to wavelet [32–34]. In [32], the state at a coarser scale is modeled by the moving average of states at the finest scale; in [33, 34], the backward model and the forward model are presented, respectively, through certain modifications of moving average model. In [30], the multiscale multirate asynchronous data fusion problem is reformulated and data is fused more effectively. In this chapter, the multiscale model is the same as in [30]. The difference in between lies the cross–correlation of the noise is considered in this chapter. Based on iterative estimation of white noise estimator, an optimal state estimation algorithm is generated in this chapter for a class of asynchronous multirate multisensor dynamic systems with unreliable measurements, where the noise of different sensors is cross–correlated and coupled with the system noise of the previous time step and the same time step. The data is examined before putting to use, so the state estimation algorithm is more robust and can cope with (but not limited to) networked data fusion subject to random data dropouts. Simulation on a target tracking example is given to demonstrate the effectiveness of the presented algorithm.

This chapter is organized as follows. In Sect. 7.2, the problem is formulated. Section 7.3 presents the optimal state estimation algorithm. Section 7.4 is the simulation and Sect. 7.5 draws the conclusions.

7.2 Problem Formulation

A class of linear dynamic systems with multiple sensors are considered in this chapter. There are N sensors observing a single target during the same period of time with different sampling rates and could encounter randomly data missing and unreliable observations. The observations are obtained asynchronously at multiple scales, and the dynamic system at the finest scale is known. We use $i = 1, 2, \ldots, N$ to denote the sensors as well as the scales. The sensors that observe the target with higher sampling rates are at finer scales, and the sensors that observe with lower sampling rates are at coarser scales. Namely, Sensor 1 has the highest sampling rate, then is the Sensor $2, 3, \ldots, N$, and Sensor N has the lowest sampling rate. The sampling period of Sensor i is n_i times of Sensor 1, namely, the kth measurement of Sensor 1 is obtained at time k, the kth measurement of sensor i is sampled during time $(n_i(k-1), n_i k]$, $i = 1, 2, \ldots, N$. So, the measurements of Sensor 1 are sampled uniformly and the measurements of Sensor $2, \ldots, N$ are sampled non-uniformly [30].

Three sensors with different sampling rates are illustrated in Fig. 7.1, where Sensor 1 has the highest sampling rate and samples uniformly. As for Sensors 2 and 3, their sampling rates are one third and one fourth of Sensor 1, respectively. $y_{2,k}$ samples at any time between $(3(k-1), 3k]$, and $y_{3,k}$ samples between $(4(k-1), 4k]$. As shown in this example, $y_{2,1}$, $y_{2,2}$, $y_{2,3}$ and $y_{2,4}$ are obtained at time 1.5, 5, 7.7, 11, respectively, and $y_{3,1}$, $y_{3,2}$ and $y_{3,3}$ are sampled at time 3.2, 6.5, 11.

The discrete-time linear dynamic system with N sensors can be described by

$$\begin{aligned} x_{k+1} &= A_k x_k + w_k \\ z_{i,k_i} &= \gamma_{i,k_i}(C_{i,k_i} x_{i,k_i} + v_{i,k_i}), i = 1, 2, \ldots, N \end{aligned} \tag{7.1}$$

where $x_k \in \mathbb{R}^n$ is the system state at time kT_s at the finest scale, scale 1, and T_s is the sampling period of Sensor 1. $A_k \in \mathbb{R}^{n \times n}$ is the system transition matrix. w_k is the system noise and is assumed to be Gaussian distributed.

$$\begin{cases} E\{w_k\} = 0 \\ E\{w_k w_j^T\} = Q_k \delta_{kj} \end{cases} \tag{7.2}$$

where $Q_k \geq 0$, δ_{kj} is the Kronecker δ function.

$z_{i,k_i} \in \mathbb{R}^{m_i}$ is the k_ith measurement of Sensor i sampled at time t_i. The sampling period of Sensor i is n_i times of Sensor 1, where n_i is a given positive integer. $C_{i,k_i} \in \mathbb{R}^{m_i \times n}$ is the measurement matrix. x_{i,k_i} is the coarser projection of x_k, $k \in (n_i(k_i - 1), n_i k_i]$ from scale 1 to scale i. When $i = 1$, we have $k_1 = k$ and $x_{1,k_1} = x_k$. v_{i,k_i} is the measurement noise and assumed to be white Gaussian

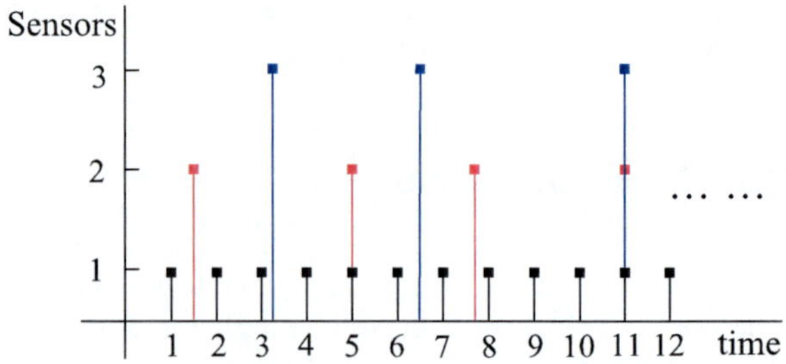

Fig. 7.1 Illustration of time (sampling instant) versus sensor (Scale)

$$\begin{cases} E\{v_{i,k_i}\} = 0 \\ E\{v_{i,k_i} v_{i,k_i}^T\} = R_{i,k_i} \end{cases} \tag{7.3}$$

$$E\{v_{i,k_i} v_{l,k_l}^T\} = \begin{cases} R_{il,k}, & \text{if } t_i, t_l \in (k-1, k] \\ 0, & \text{otherwise} \end{cases} \tag{7.4}$$

$$E\{w_k v_{i,k_i}^T\} = \begin{cases} S_{i,k_i}, & \text{if } t_i \in (k-1, k] \\ 0, & \text{otherwise} \end{cases} \tag{7.5}$$

$$E\{w_{k-1} v_{i,k_i}^T\} = \begin{cases} S_{i,k_i}^*, & \text{if } t_i \in (k-1, k] \\ 0, & \text{otherwise} \end{cases} \tag{7.6}$$

where $i, l = 1, 2, \ldots, N$, $k = n_i(k_i - 1) + j$, $1 \le j \le n_i$. From Eq. (7.4), it can be seen that the noises of different sensors are cross-correlated in the same sampling period at the finest scale. From Eqs. (7.5) and (7.6), it can be seen that the measurement noise is correlated with the system noise.

The initial state x_0 is independent of w_k and v_{i,k_i}, for $i = 1, 2, \ldots, N$ and is assumed to be Gaussian distributed

$$\begin{cases} E\{x_0\} = \bar{x}_0 \\ E\{(x_0 - \bar{x}_0)(x_0 - \bar{x}_0)^T\} = \bar{P}_0 \end{cases} \tag{7.7}$$

$\gamma_{i,k_i} \in R$ is a stochastic sequence that takes values on *unit* and *non-unit* with Bernoulli distribution, which is used to describe the missing or the unreliability of the measurements, and is supposed to be independent of w_k, v_{i,k_i} and x_0, $i = 1, 2, \ldots, N$.

The objective of this chapter is to generate the optimal estimation of state x_k by an optimal batched fusion algorithm based on the above description.

7.3 The Optimal State Fusion Estimation Algorithm

7.3.1 Modeling of Asynchronous, Multirate, Multisensor Systems

Without any prior information, the state variable vector x_{i,k_i} at scale i may be established by the state vector $x_{n_i k_i}$ at the finest scale 1 through the following linear transformation [34]

$$x_{i,k_i} = \frac{1}{n_i}\left(\sum_{m=0}^{n_i-1} A^{-m}\right) x_{n_i k_i} \tag{7.8}$$

where n_i is the sampling ratio of the sensor that with the highest sampling rate to Sensor i, $i = 1, 2, \ldots, N$.

In the sequel, we first establish the state space model at scale $i\,(1 \leq i \leq N)$. Then, we generate the optimal estimation of state x_k by fusing the asynchronous multirate multisensor data.

Similar to Theorem 1 in [30], we have the following theorem.

Theorem 7.1 *From multiscale system theory and the problem formulation, the dynamic system (7.1) and (7.2) could be rewritten as*

$$x_{k+1} = A_k x_k + w_k \tag{7.9}$$

$$z_{i,k_i} = \gamma_{i,k_i}(\bar{C}_{i,k} x_k + v_{i,k_i}), \tag{7.10}$$

where $t_i \in (k-1, k]$, $i = 1, 2, \ldots, N_k$, $1 \leq N_k \leq N$ *denotes the sampling time of* z_{i,k_i}, N_k *is the number of sensors that have observations between* $(k-1, k]$, *and*

$$\bar{C}_{i,k} = C_{i,k_i} A_{i,k_i} \tag{7.11}$$

$$A_{i,k_i} = \frac{1}{n_i}[I + \sum_{m=1}^{j-1} \prod_{l=m}^{j-1} A^{-1}_{n_i(k_i-1)+l} + \sum_{m=0}^{n_i-j-1} \prod_{l=j+m}^{j} A_{n_i(k_i-1)+l}] \tag{7.12}$$

where $j = k - n_i(k_i - 1)$.

Proof As Sensor i samples non-uniformly, z_{i,k_i} could be obtained at any time during $(n_i(k_i-1), n_i k_i]$. Suppose z_{i,k_i} is sampled at $t_i \in (n_i(k_i-1)+j-1, n_i(k_i-1)+j]$, $1 \leq j \leq n_i$. Then, from the multiscale system theory, x_{i,k_i}, i.e., the coarser projection of x_k, $k \in (n_i(k_i-1), n_i k_i)$ from scale 1 to scale i, could be computed by

$$x_{i,k_i} = A_{i,k_i} x_{n_i(k_i-1)+j} \tag{7.13}$$

Substituting (7.13) into (7.2) yields

$$z_{i,k_i} = \gamma_{i,k_i}(C_{i,k_i} A_{i,k_i} x_{n_i(k_i-1)+j} + v_{i,k_i}) \tag{7.14}$$

Let $n_i(k_i-1)+j = k$, then

$$z_{i,k_i} = \gamma_{i,k_i}(\bar{C}_{i,k} x_k + v_{i,k_i}) \tag{7.15}$$

where $j = k - n_i(k_i-1)$, $(n_i(k_i-1)+j-1, n_i(k_i-1)+j] = (k-1, k]$. There are N data sampled in the period of $(n_i(k_i-1), n_i k_i]$ from the system description and N_k observe data are expected to be sampled during $(k-1, k]$, where $1 \leq N_k \leq N$. Specially, $z_{1,k}$ is sampled at time k.

In the sequel, we generate the state estimation of x_k by fusing the measurements from multiple sensors in Sect. 7.3.2 for normal measurements and in Sect. 7.3.3 for unreliable measurements.

Remark 7.1 Inverse of A_k are used in the model. If A_k is not full rank, A_k^{-1} should be replaced by the Moore-Penrose inverse A_k^{-}.

7.3.2 Data Fusion with Normal Measurements

When it comes to data fusion, the most intuitive method is to collect information from all sensors for centralized estimation fusion. The batch fusion algorithm in cross-correlated noise with normal measurements i.e. $\gamma_{i,k_i} = 1$ for all $i = 1, 2, \ldots, N$, and $k_i = 1, 2, \ldots$, is given by the following theorem.

Theorem 7.2 (The optimal centralized batch fusion (BF)) *For system (7.1)–(7.2), the estimation of x_k by the centralized batch fusion is given by*

$$\begin{cases} \hat{x}_{b,k|k} &= \hat{x}_{b,k|k-1} + K_{b,k}(z_k - \bar{C}_k\hat{x}_{b,k|k-1}) \\ P_{b,k|k} &= P_{b,k|k-1} - K_{b,k}P_{\tilde{z},k|k-1}K_{b,k}^T \\ \hat{x}_{b,k|k-1} &= A_{k-1}\hat{x}_{b,k-1|k-1} + \hat{w}_{b,k-1|k-1} \\ P_{b,k|k-1} &= A_{k-1}P_{b,k-1|k-1}A_{k-1}^T + A_{k-1}P^b_{\tilde{x}\tilde{w},k-1|k-1} \\ &\quad + P^b_{\tilde{w}\tilde{x},k-1|k-1}A_{k-1}^T + P^b_{\tilde{w},k-1|k-1} \\ K_{b,k} &= (P_{b,k|k-1}\bar{C}_k^T + S_k^*)(\bar{C}_kP_{b,k|k-1}\bar{C}_k^T + R_k + \bar{C}_kS_k^* + S_k^{*,T}\bar{C}_k^T)^{-1} \\ P_{\tilde{z},k|k-1} &= \bar{C}_kP_{b,k|k-1}\bar{C}_k^T + R_k + \bar{C}_kS_k^* + S_k^{*,T}\bar{C}_k^T \end{cases} \tag{7.16}$$

where the white noise estimator is computed by

$$\begin{cases} \hat{w}_{b,k|k} = K^b_{w,k}(z_k - \bar{C}_k\hat{x}_{b,k|k-1}) \\ K^b_{w,k} = S_k(\bar{C}_kP_{b,k|k-1}\bar{C}_k^T + R_k + \bar{C}_kS_k^* + S_k^{*,T}\bar{C}_k^T)^{-1} \\ P^b_{\tilde{w},k|k} = Q_k - K^b_{w,k}S_k^T \end{cases} \tag{7.17}$$

The filtering error cross-covariance matrix between the state and the system noise is computed by

$$\begin{cases} P^b_{\tilde{x}\tilde{w},k|k} = -K_{b,k}S_k^T \\ P^b_{\tilde{w}\tilde{x},k|k} = -S_kK_{b,k}^T \end{cases} \tag{7.18}$$

The subscript b and superscript b in (7.16)–(7.18) stands for the batch fusion and

$$z_k = \left[\, z_{1,k_1}^T \; \cdots \; z_{i,k_i}^T \; \cdots \; z_{N_k,k_{N_k}}^T \,\right]^T \tag{7.19}$$

$$\bar{C}_k = [\, \bar{C}_{1,k}^T \; \cdots \; \bar{C}_{i,k}^T \; \cdots \; \bar{C}_{N_k,k}^T \,]^T \tag{7.20}$$

$$v_k = \begin{bmatrix} v_{1,k_1}^T & \cdots & v_{i,k_i}^T & \cdots & v_{N_k,k_{N_k}}^T \end{bmatrix}^T \tag{7.21}$$

$$R_k = \begin{bmatrix} R_{1,k_1} & \cdots & R_{1i,k} & \cdots & R_{1N_k,k} \\ \vdots & & \vdots & & \vdots \\ R_{i1,k} & \cdots & R_{i,k_i} & \cdots & R_{iN_k,k} \\ \vdots & & \vdots & & \vdots \\ R_{N_k 1,k} & \cdots & R_{N_k i,k} & \cdots & R_{N_k,k_{N_k}} \end{bmatrix} \tag{7.22}$$

$$S_k^* = \begin{bmatrix} S_{1,k_1}^* & \cdots & S_{i,k_i}^* & \cdots & S_{N_k,k_{N_k}}^* \end{bmatrix} \tag{7.23}$$

$$S_k = \begin{bmatrix} S_{1,k_1} & \cdots & S_{i,k_i} & \cdots & S_{N_k,k_{N_k}} \end{bmatrix} \tag{7.24}$$

where z_{i,k_i}, $i = 1, 2, \ldots, N_k$, $k_i = 1, 2, \ldots$ is the k_ith measurement of Sensor i in time $(k-1, k]$. N_k measurements of sensors arrived during time $(k-1, k]$, $R_{ij,k} = E\{v_{i,k_i} v_{j,k_j}^T\}$, if $t_i, t_j \in (k-1, k]$.

Proof We use projection theorem and inductive method to prove this theorem.

From Theorem 7.1, where $\gamma_{i,k_i} = 1$, we have

$$x_{k+1} = A_k x_k + w_k \tag{7.25}$$

$$z_{i,k_i} = \bar{C}_{i,k} x_k + v_{i,k_i}, \quad i = 1, 2, \ldots, N_k \tag{7.26}$$

By expanding z_{i,k_i}, v_{i,k_i} and $\bar{C}_{i,k}$, we have

$$z_k = \begin{bmatrix} z_{1,k_1}^T & \cdots & z_{i,k_i}^T & \cdots & z_{N_k,k_{N_k}}^T \end{bmatrix}^T \tag{7.27}$$

$$\bar{C}_k = \begin{bmatrix} \bar{C}_{1,k}^T & \cdots & \bar{C}_{i,k}^T & \cdots & \bar{C}_{N_k,k}^T \end{bmatrix}^T \tag{7.28}$$

$$v_k = \begin{bmatrix} v_{1,k_1}^T & \cdots & v_{i,k_i}^T & \cdots & v_{N_k,k_{N_k}}^T \end{bmatrix}^T \tag{7.29}$$

If all the data are reliable, by using the projection theory, the innovation sequence $\tilde{z}_{k|k-1}$ is defined as follows

$$\tilde{z}_{k|k-1} = z_k - \hat{z}_{k|k-1} \tag{7.30}$$

where

$$\hat{z}_{k|k-1} = \bar{C}_k \hat{x}_{b,k|k-1} \tag{7.31}$$

Then, substituting (7.31) to (7.30) yields

$$\tilde{z}_{k|k-1} = z_k - \bar{C}_k \hat{x}_{b,k|k-1} \tag{7.32}$$

where

$$\hat{x}_{b,k|k} = \hat{x}_{b,k|k-1} + K_{b,k} \tilde{z}_{k|k-1} \tag{7.33}$$

Next, based on the projection theory, we have the following filtering and one step prediction error equations for the state

$$\begin{aligned} \tilde{x}_{b,k|k} &= x_k - \hat{x}_{b,k|k} \\ &= \tilde{x}_{b,k|k-1} - K_{b,k} \tilde{z}_{k|k-1} \end{aligned} \tag{7.34}$$

$$\begin{aligned} \tilde{x}_{b,k|k-1} &= x_k - \hat{x}_{b,k|k-1} \\ &= A_{k-1} \tilde{x}_{b,k-1|k-1} + \tilde{w}_{b,k-1|k-1} \end{aligned} \tag{7.35}$$

Then, we can get another form of the innovation sequence as follows

$$\tilde{z}_{k|k-1} = \bar{C}_k \tilde{x}_{b,k|k-1} + v_k \tag{7.36}$$

From the projection theory [1], we have the white noise filter

$$\begin{aligned} \hat{w}_{b,k|k} &= \hat{w}_{b,k|k-1} + K_{w,k}^{b} \tilde{z}_{k|k-1} \\ &= K_{w,k}^{b} \tilde{z}_{k|k-1} \end{aligned} \tag{7.37}$$

$$\begin{aligned} \tilde{w}_{b,k|k} &= w_k - \hat{w}_{b,k|k} \\ &= w_k - K_{w,k}^{b} \tilde{z}_{k|k-1} \end{aligned} \tag{7.38}$$

$$\begin{aligned} E\{\tilde{x}_{b,k|k-1} v_k^T\} &= E\{(A_k \tilde{x}_{b,k-1|k-1} + \tilde{w}_{b,k-1|k-1}) v_k^T\} \\ &= E\{\tilde{w}_{b,k-1|k-1} v_k^T\} \\ &= E\{(w_{k-1} - K_{w,k-1}^{b} \tilde{z}_{k-1|k-2}) v_k^T\} \\ &= E\{w_{k-1} v_k^T\} \\ &= S_k^* \end{aligned} \tag{7.39}$$

We have the covariance matrix of $\tilde{z}_{k|k-1}$

$$\begin{aligned}
P_{\tilde{z},k|k-1} &= E\{\tilde{z}_{k|k-1}\tilde{z}^T_{k|k-1}\} \\
&= E\{(\bar{C}_k\tilde{x}_{b,k|k-1} + v_k)(\bar{C}_k\tilde{x}_{b,k|k-1} + v_k)^T\} \\
&= \bar{C}_k P_{k|k-1}\bar{C}^T_k + R_k + \bar{C}_k E\{\tilde{x}_{b,k|k-1}v^T_k\} + E\{v_k\tilde{x}^T_{b,k|k-1}\}\bar{C}^T_k \\
&= \bar{C}_k P_{k|k-1}\bar{C}^T_k + \bar{C}_k S^*_k + S^{*,T}_k\bar{C}^T_k + R_k
\end{aligned} \tag{7.40}$$

The filtering gain for the white noise is computed by [8]

$$\begin{aligned}
K^b_{w,k} &= E\{w_k\tilde{z}^T_{k|k-1}\}P^{-1}_{\tilde{z},k|k-1} \\
&= E\{w_k(\bar{C}_k\tilde{x}_{b,k|k-1} + v_k)^T\}P^{-1}_{\tilde{z},k|k-1} \\
&= E\{w_k v^T_k\}P^{-1}_{\tilde{z},k|k-1} \\
&= S_k P^{-1}_{\tilde{z},k|k-1}
\end{aligned} \tag{7.41}$$

Similarly, we define the filtering gain for the state as follows

$$K_{b,k} = E\{x_k\tilde{z}^T_{k|k-1}\}P^{-1}_{\tilde{z},k|k-1} \tag{7.42}$$

where

$$\begin{aligned}
E\{x_k\tilde{z}^T_{k|k-1}\} &= E\{x_k(\bar{C}_k\tilde{x}_{b,k|k-1} + v_k)^T\} \\
&= E\{x_k\tilde{x}^T_{b,k|k-1}\bar{C}^T_k\} + E\{x_k v^T_k\} \\
&= P_{b,k|k-1}\bar{C}^T_k + E\{(A_{k-1}x_{k-1} + w_{k-1})v^T_k\} \\
&= P_{b,k|k-1}\bar{C}^T_k + E\{w_{k-1}v^T_k\} \\
&= P_{b,k|k-1}\bar{C}^T_k + S^*_k
\end{aligned} \tag{7.43}$$

then, we have

$$K_{b,k} = (P_{b,k|k-1}\bar{C}^T_k + S^*_k)P^{-1}_{\tilde{z},k|k-1} \tag{7.44}$$

Substituting Eq. (7.34) into $P_{b,k|k} = E\{\tilde{x}_{b,k|k}\tilde{x}^T_{b,k|k}\}$, we have the filtering error covariance matrix for the state

$$\begin{aligned}
P_{b,k|k} &= P_{b,k|k-1} - E\{\tilde{x}_{b,k|k-1}\tilde{z}^T_{k|k-1}\}K^T_{b,k} - K_{b,k}E\{\tilde{z}_{k|k-1}\tilde{x}^T_{b,k|k-1}\} \\
&\quad + K_{b,k}P_{\tilde{z},k|k-1}K^T_{b,k} \\
&= P_{b,k|k-1} - K_{b,k}E\{\tilde{z}_{k|k-1}\tilde{x}^T_{b,k|k-1}\} \\
&= P_{b,k|k-1} - K_{b,k}(S^{*,T}_k + \bar{C}_k P_{b,k|k-1}) \\
&= P_{b,k|k-1} - K_{b,k}P_{\tilde{z},k|k-1}K^T_{b,k}
\end{aligned} \tag{7.45}$$

Then, substituting $P_{b,k|k-1} = E\{\tilde{x}_{b,k|k-1}\tilde{x}_{b,k|k-1}^T\}$, we have the prediction error covariance matrix

$$
\begin{aligned}
P_{b,k|k-1} &= A_{k-1}P_{b,k-1|k-1}A_{k-1}^T + A_{k-1}E\{\tilde{x}_{b,k-1|k-1}\tilde{w}_{b,k-1|k-1}^T\} \\
&\quad + E\{\tilde{w}_{b,k-1|k-1}\tilde{x}_{b,k-1|k-1}^T\}A_{k-1}^T + E\{\tilde{w}_{b,k-1|k-1}\tilde{w}_{b,k-1|k-1}^T\} \\
&= A_{k-1}P_{b,k-1|k-1}A_{k-1}^T + A_{k-1}P_{\tilde{x}\tilde{w},k-1|k-1}^b + P_{\tilde{w}\tilde{x},k-1|k-1}^b A_{k-1}^T \\
&\quad + P_{\tilde{w},k-1|k-1}^b
\end{aligned} \tag{7.46}
$$

where

$$
\begin{aligned}
P_{\tilde{w},k|k}^b &= E\{\tilde{w}_{b,k|k}\tilde{w}_{b,k|k}^T\} \\
&= E\{(w_k - K_{w,k}^b\tilde{z}_{k|k-1})(w_k - K_{w,k}^b\tilde{z}_{k|k-1})^T\} \\
&= Q_k - K_{w,k}^b P_{\tilde{z},k|k-1}K_{w,k}^{b,T} - K_{w,k}^b P_{\tilde{z},k|k-1}K_{w,k}^{b,T} + K_{w,k}^b P_{\tilde{z},k|k-1}K_{w,k}^{b,T} \\
&= Q_k - K_{w,k}^b P_{\tilde{z},k|k-1}K_{w,k}^{b,T} \\
&= Q_k - K_{w,k}^b S_k^T
\end{aligned} \tag{7.47}
$$

and

$$
\begin{aligned}
P_{\tilde{x}\tilde{w},k|k}^b &= E\{\tilde{x}_{b,k|k}\tilde{w}_{b,k|k}^T\} \\
&= E\{(\tilde{x}_{b,k|k-1} - K_{b,k}\tilde{z}_{k|k-1})(w_k - K_{w,k}^b\tilde{z}_{k|k-1})^T\} \\
&= -E\{\tilde{x}_{b,k|k-1}\tilde{z}_{k|k-1}^T\}K_{w,k}^{b,T} - K_{b,k}E\{\tilde{z}_{k|k-1}w_k^T\} \\
&\quad + K_{b,k}E\{\tilde{z}_{k|k-1}\tilde{z}_{k|k-1}^T\}K_{w,k}^{b,T} \\
&= -K_{b,k}P_{\tilde{z},k|k-1}K_{w,k}^{b,T} \\
&= -K_{b,k}S_k^T
\end{aligned} \tag{7.48}
$$

7.3.3 Data Fusion with Unreliable Measurements

From the problem formulation and the Kalman filter, it can be proven that when the measurement is reliable, $\tilde{z}_{i,k_i}$ is Gaussian distributed, with zero mean and variance being $P_{N_k,k_{N_k}}$, i.e., $\tilde{z}_{i,k_i}^T \sim \mathcal{N}(0, P_{N_k,k_{N_k}})$, where

$$
\begin{aligned}
P_{N_k,k_{N_k}} &= \mathrm{var}\{\tilde{z}_{i,k_i}\} \\
&= \bar{C}_{i,k}P_{\tilde{z}_i,k_i|k_i-1}\bar{C}_{i,k}^T + R_{i,k_i} + \bar{C}_{i,k}S_{i,k_i}^* + S_{i,k_i}^{*,T}\bar{C}_{i,k}^T
\end{aligned} \tag{7.49}
$$

where $S_{i,k_i}^{*,T} = [S_{i,k_i}^*]^T$. Denote

$$
\rho_{i,k_i} = \tilde{z}_{i,k_i}^T P_{N_k,k_{N_k}}^{-1}\tilde{z}_{i,k_i} \tag{7.50}
$$

then, $\rho_{i,k_i} \sim \chi^2(m_i)$ is standard chi-square distribution with m_i degrees of freedom, whose mean and variance is m_i and $2m_i$, respectively, where m_i is the dimension of z_{i,k_i}. Hence, we can use ρ_{i,k_i} as a measure to evaluate whether the measurement z_{i,k_i} is normal or faulty.

From the above analysis, when unreliable measurements are available, the optimal batch fusion algorithm in cross-correlated noise with random packet dropout is given by the following theorem.

Theorem 7.3 (The optimal batch fusion with unreliable measurements (OBF)) *For system (7.1)–(7.2), the optimal estimation of x_k by the centralized batch fusion is given by*

$$\begin{cases} \hat{x}_{k|k} & = \hat{x}_{k|k-1} + K_k(z_k - \bar{C}_k\hat{x}_{k|k-1}) \\ P_{k|k} & = P_{k|k-1} - K_k P_{\tilde{z},k|k-1} K_k^T \\ \hat{x}_{k|k-1} & = A_{k-1}\hat{x}_{k-1|k-1}\hat{w}_{k-1|k-1} \\ P_{k|k-1} & = A_{k-1}P_{k-1|k-1}A_{k-1}^T + A_{k-1}P_{\tilde{x}\tilde{w},k-1|k-1} \\ & \quad +P_{\tilde{w}\tilde{x},k-1|k-1}A_{k-1}^T + P_{\tilde{w},k-1|k-1} \\ K_k & = (P_{k|k-1}\bar{C}_k^T + S_k^*)(\bar{C}_k P_{k|k-1}\bar{C}_k^T + R_k + \bar{C}_k S_k^* + S_k^{*,T}\bar{C}_k^T)^{-1} \cdot \delta_k \\ P_{\tilde{z},k|k-1} & = (\bar{C}_k P_{k|k-1}\bar{C}_k^T + R_k + \bar{C}_k S_k^* + S_k^{*,T}\bar{C}_k^T)^{-1} \cdot \delta_k \\ \delta_k & = \mathrm{diag}\{\delta_{1,k_1} I_{m_1}, \ldots, \delta_{i,k_i} I_{m_i}, \ldots, \delta_{N_k,k_{N_k}} I_{m_{N_k}}\} \end{cases} \tag{7.51}$$

where the white noise estimator is computed by

$$\begin{cases} \hat{w}_{k|k} & = K_{w,k}(z_k - \bar{C}_k\hat{x}_{k|k-1}) \\ K_{w,k} & = S_k(\bar{C}_k P_{k|k-1}\bar{C}_k^T + R_k + \bar{C}_k S_k^* + S_k^{*,T}\bar{C}_k^T)^{-1} \cdot \delta_k \\ P_{\tilde{w},k|k} & = Q_k - K_{w,k}S_k^T \end{cases} \tag{7.52}$$

The filtering error cross-covariance matrix between the state and the system noise is computed by

$$\begin{cases} P_{\tilde{x}\tilde{w},k|k} = -K_k S_k^T \\ P_{\tilde{w}\tilde{x},k|k} = -S_k K_k^T \end{cases} \tag{7.53}$$

where z_k, $\bar{C}_k$, v_k, R_k, S_k^* and S_k could be computed by Eqs. (7.19)–(7.24), z_{i,k_i}, $i = 1, 2, \ldots, N_k$, $k_i = 1, 2, \ldots$ is the k_ith measurement of Sensor i in time $(k-1, k]$. N_k measurements of sensors arrived during time $(k-1, k]$, $R_{ij,k} = E[v_{i,k_i}v_{j,k_j}^T]$, if $t_i, t_j \in (k-1, k]$. δ_{i,k_i} is a sequence that takes values on 0 and 1. $\delta_{i,k_i} = 0$ if z_{i,k_i} is deemed unreliable, otherwise $\delta_{i,k_i} = 1$, and m_i is the dimension of z_{i,k_i}.

Remark 7.2 In Theorem 7.3, when the measurement is randomly unreliable, we use the residual as a statistic to test its reliability. Namely, in this testing problem, the original hypothesis is $H_0 : \gamma_{i,k_i} = 1$, and the antithetic hypothesis is $H_1 : \gamma_{i,k_i} \neq 1$. The rejection interval is $(\chi_\alpha^2(m_i), +\infty)$, where $\chi_\alpha^2(m_i)$ is the one-side χ^2 value with confidence α, $1 \le i \le N_k$. Basically, if $\rho_{i,k_i} > \chi_\alpha^2(m_i)$, z_{i,k_i} will be deemed faulty, and will not be used in the generation of the state estimation, therefore we set $\delta_{i,k_i} = 0$. Accordingly, if $\rho_{i,k_i} \leq \chi^2_\alpha(m_i)$, z_{i,k_i} will be deemed as reliable measurement and will be used to update the estimations, where one has $\delta_{i,k_i} = 1$.

7.4 Numerical Example

We use an example to illustrate the effectiveness of the presented algorithm in this section. A tracking system with two sensors can be described by

$$x_{k+1} = \begin{bmatrix} 1 & T_s \\ 0 & 1 \end{bmatrix} x_k + \Gamma_k \xi_k \tag{7.54}$$

$$z_{1,k} = \gamma_{1,k} C_1 x_k + v_{1,k}, k = 1, 2, \ldots, L \tag{7.55}$$

$$z_{2,k_2} = \gamma_{2,k_2} C_2 x_{2,k_2} + v_{2,k_2} \tag{7.56}$$

$$v_{1,k} = \eta_{1,k} + \beta_{11}\xi_{k-1} + \beta_{12}\xi_k \tag{7.57}$$

$$v_{2,k} = \eta_{2,k} + \beta_{21}\xi_{k-1} + \beta_{22}\xi_k \tag{7.58}$$

where $n_2 = 3$, $k_2 = 1, 2, \ldots, \frac{L}{n_2}$, and $L = 300$ is the length of the signal x to be estimated. $T_s = 0.01$ is the sampling period. The state $x_k = [s_k \ \dot{s}_k]^T$, where s_k and $\dot{s}_k$ are the position and velocity of the target at time kT_s, respectively. $\xi_k \in R$ is the system noise, assumed to be white and Gaussian distributed with zero mean and variance σ^2_ξ. $\Gamma_k = [T_s \ \ 1]$ is the noise transition matrix. z_{i,k_i} $(i = 1, 2)$ are the observation vector of the two sensors at time t_i, which observe the position and velocity, respectively, i.e., $C_1 = [1 \ 0]$, $C_2 = [0 \ 1]$. $v_{i,k}$ $(i = 1, 2)$ are the measurement noises of Sensor i, and are cross- correlated and coupled with the system noise ξ_{k-1} and ξ_k. The strength of the correlation is determined by β_{i1} and β_{i2}. v_{2,k_2} and x_{2,k_2} are obtained by the non-uniform sampling of $v_{2,k}$ and x_k, respectively. η_{i,t_i} $(i = 1, 2)$ are zero mean white Gaussian noises with variances $\sigma^2_{\eta_i}$ and are independent of ξ_k, $k = 1, 2, \ldots$. The initial values are $\bar{x}_0 = [10 \ 1]^T$, $\bar{P}_0 = I_2$.

From Eq. (7.54), we have $Q_k = \Gamma_k \Gamma_k^T \sigma^2_\xi$, which is the system noise covariance corresponding to $w_k = \Gamma_k \xi_k$. From Eqs. (7.57) and (7.58), the measurement noise covariance is computed by

$$R_k = \begin{bmatrix} (\beta_{11}^2 + \beta_{12}^2)\sigma^2_\xi + \sigma^2_{\eta_1} & (\beta_{11}\beta_{21} + \beta_{12}\beta_{22})\sigma^2_\xi \\ (\beta_{21}\beta_{11} + \beta_{22}\beta_{12})\sigma^2_\xi & (\beta_{21}^2 + \beta_{22}^2)\sigma^2_\xi + \sigma^2_{\eta_2} \end{bmatrix} \tag{7.59}$$

The covariance between w_{k-1} and v_{i,k_i} is

$$S_k^* = [\beta_{11}\sigma^2_\xi \Gamma_{k-1} \ \ \beta_{21}\sigma^2_\xi \Gamma_{k-1}] \tag{7.60}$$

and the covariance between w_k and v_{i,k_i} is

$$S_k = [\beta_{12}\sigma_\xi^2 \Gamma_k \;\; \beta_{22}\sigma_\xi^2 \Gamma_k] \tag{7.61}$$

In the sequel, we will give the state estimation of x_k by fusing information from the two sensors, and compare the differences between the estimated results obtained by different estimation algorithms in case of noise correlation and uncorrelation. In the case of noise correlation, we will analyze the influence on fusion results when noise correlation is ignored.

Denote $\sigma_\xi^2 = 0.01$, $\sigma_{\eta_1}^2 = 0.25$ and $\sigma_{\eta_2}^2 = 0.16$. The unreliable ratio of each sensor is 0.1. When testing the reliability of the observations, we set confidence level $\alpha = 0.01$. For β_{i1} and $\beta_{i2}, i = 1, 2$, we set two group of values. In the first group, we make β_{i1} and $\beta_{i2}, i = 1, 2$ equal to zero, which means there is no correlation between the sensor noise, the system noise and the sensor noise; in the second group, we set $\beta_{11} = 3, \beta_{12} = 2, \beta_{21} = 2, \beta_{22} = 3$, consequently the measurement noises are cross-correlated and coupled with the system noise. We select 300 sampling time for 100 Monte Carlo simulations, and observe the effectiveness of the presented algorithm. The simulation results of case 1 are shown in Fig. 7.2 and Table 7.1; the simulation results of case 2 are shown in Figs. 7.3, 7.4 and Table 7.2.

Figure 7.2 shows the statistical simulation curves of Root Mean Square Errors (RMSEs) for position and velocity of the state estimations of different algorithms

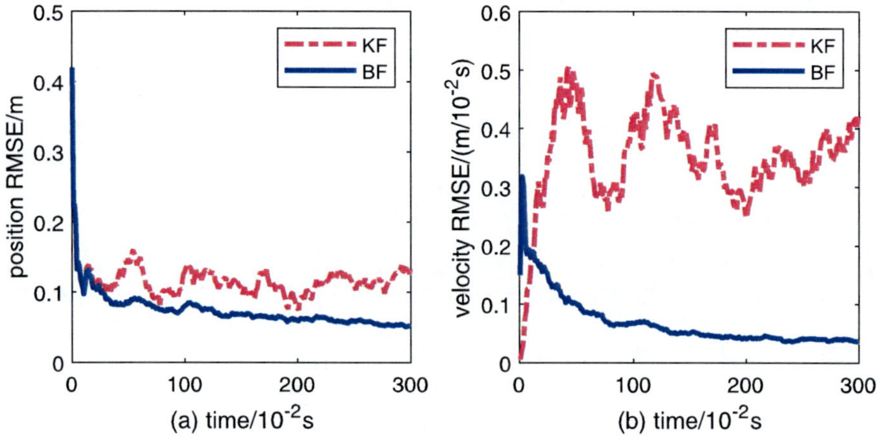

Fig. 7.2 RMSEs for position and velocity of case 1

Table 7.1 Time-averaged RMSEs for position and velocity of case 1

	KF	BF
Position	0.1153	0.0737
Velocity	0.3486	0.0684

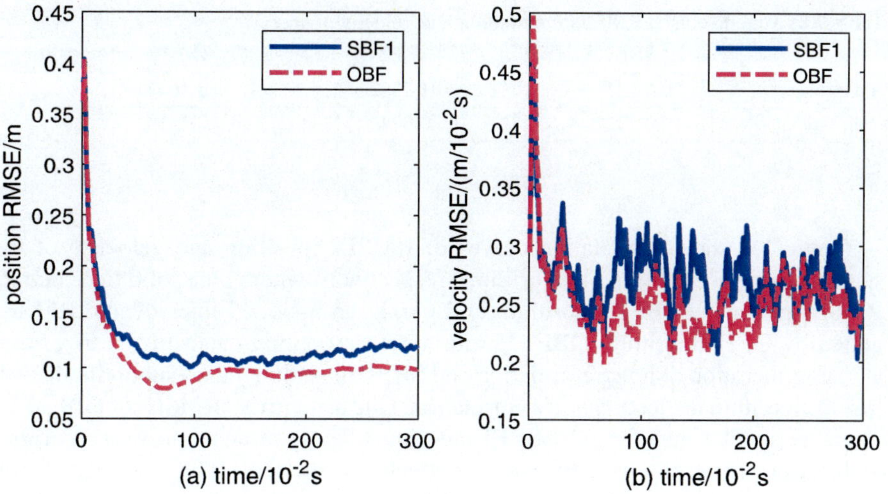

Fig. 7.3 RMSEs for position and velocity of case 2

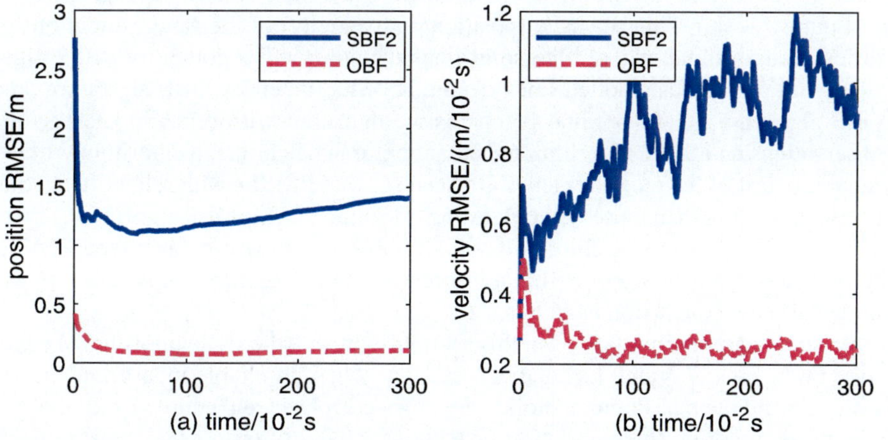

Fig. 7.4 RMSEs with reliability test and without test of the observations in case 2

in case 1, respectively, where the purple dash–dotted lines denote Kalman filtering algorithm (KF) and blue solid lines denote batch fusion algorithm (BF). One can see that solid lines are closer to zero, followed by the dash-dotted lines. This indicates that batch fusion algorithm is more effective than Kalman filtering algorithm. Table 7.1 shows time-averaged RMSEs for position and velocity of case 1. We can draw the same conclusion as in Fig. 7.2. This indicates that the batch fusion algorithm is effective.

Figures 7.3, 7.4 and Table 7.2 show simulation results of case 2.

Table 7.2 Time-averaged RMSEs for position and velocity of case 2

	SBF1	SBF2	OBF
Position	0.121	1.2382	0.1047
Velocity	0.2722	0.9743	0.2514

Figure 7.3 shows simulation curves of RMSEs (position and velocity) of the state estimations of different algorithms, respectively, where blue solid lines denote RMSEs gotten by SBF1 algorithm, and purple dash–dotted lines denote RMSEs gotten by OBF algorithm. OBF algorithm means optimal batch fusion algorithm with consideration of noise correlation and SBF1 algorithm means suboptimal batch fusion algorithm neglecting noise correlation. One can see that RMSEs of OBF algorithm are smaller than those by SBF1 algorithm. These show that fusion algorithm with consideration of noise correlation is effective while neglecting noise correlation can reduce estimation precision even with batch fusion. The first and third columns of Table 7.2 show time-averaged RMSEs for position and velocity of SBF1 algorithm and OBF algorithm in case 2. We can draw the same conclusion as in Fig. 7.3.

Figure 7.4 shows the RMSEs (position and velocity) of the state estimations of different algorithms, where blue solid lines denote RMSEs gotten by SBF2 algorithm and purple dash–dotted lines denote RMSEs gotten by OBF algorithm, and OBF algorithm means optimal batch fusion algorithm with reliability test to the observations and SBF2 algorithm means suboptimal batch fusion algorithm without reliability test of the measurements. One can see that RMSEs with reliability testing are much smaller than those without testing, meaning that the OBF algorithm is more effective. The second and third columns of Table 7.2 show time-averaged RMSEs for position and velocity of SBF2 algorithm and OBF algorithm in case 2. We can draw the same conclusion as in Fig. 7.4.

In summary, the simulations in this section show that the OBF algorithm is effective. From Figs. 7.3 and 7.4, one can see that when unreliable measurements are available and the measurement noises are cross-correlated and coupled with the system noise, ignoring the correlation or without reliability test of the measurements will reduce the estimation accuracy. And a divergence in Kalman filter performance is noted for SBF2, which is not observed in SBF1. It seems to provide an interesting inference that skipping reliability test of the measurements can be more damaging than ignoring noise cross-correlation.

7.5 Summary

This chapter studied the optimal state estimation for an asynchronous multirate multisensor dynamic systems over unreliable measurements with correlated noise. The main contributions of this chapter are: (1) For an asynchronous multirate multisensor dynamic system whose sensor noise is cross-correlated and coupled with the system noise of the previous time step and the same time step, an optimal white noise estimator is generated. (2) When the measurements are randomly unreliable, a rule for evaluating whether the observation is reliable or not is given, which could be used on line. (3) When unreliable measurements are available for the asynchronous multirate multisensor dynamic system with correlated noise, an optimal state estimation algorithm is presented. The performance of the algorithm is compared through an example in terms of root mean square error. From theoretical analysis and simulation results, it can be concluded that the proposed algorithm is effective and has potential value especially in the application fields including target tracking, integrated navigation, network transportation, fault tolerance control and so on.

References

1. Anderson, B.D., and J.B. Moore. 1979. *Optimal Filtering*. Englewood Cliffs, New Jersey: Prentice-Hall.
2. Bar-Shalom, Yarkov, X. Rong Li, and T. Kirubarajan. 2001. *Estimation with Applications to Tracking and Navigation: Theory, Algorithms and Software*. New York: Wiley.
3. Basseville, M., A. Benveniste, K.C. Chou, S.A. Golden, R. Nikoukhah, and A.S. Willsky. 1992. Modeling and estimation of multiresolution stochastic processes. *IEEE Transactions on Information Theory* 38 (2): 766–784.
4. Carlson, N.A. 1990. Federated square root filter for decentralized parallel processors. *IEEE Transactions on Aerospace and Electronic Systems* 26 (3): 517–525.
5. Chang, Kuochu C., Rajat K. Saha, and Yaakov Bar-Shalom. 1997. On optimal track-to-track fusion. *IEEE Transactions on Aerospace and Electronic Systems* 33 (4): 1271–1276.
6. Chong, Chee-Yee, and Mori Shozo. 2000. Architectures and algorithms for track association and fusion. *IEEE Aerospace and Electronic Systems Magazine* 15 (1): 5–13.
7. Chu, Tianpeng, Guoqing Qi, and Yinya Li. 2014. Distributed asynchronous fusion algorithm for sensor networks with packet losses. *Discrete Dynamics in Nature and Society* 5 (1): 1–10.
8. Deng, Zili, Xin Wang, and Yuan Gao. 2007. *Modeling and Estimation*. Beijing: Science Press.
9. Duan, Zhansheng, Chongzhao Han, and Tangfei Tao. 2004. Optimal multi-sensor fusion target tracking with correlated measurement noises. *IEEE International Conference on Systems*, vol. 2, 1272–1278. Man and Cybernetics Netherlands: Hague.
10. Feng, Jianxin, Tingfeng Wang, and Jin Guo. 2014. Recursive estimation for descriptor systems with multiple packet dropouts and correlated noises. *Aerospace Science and Technology* 32 (1): 200–211.
11. Geng, Hang, Yan Liang, Feng Yang, Linfeng Xu, and Quan Pan. 2016. The joint optimal filtering and fault detection for multi-rate sensor fusion under unknown inputs. *Information Fusion* 29 (1): 57–67.
12. Hall, David L., and James Llinas. 2001. *Multisensor Data Fusion*. Boca Raton: CRC Press.
13. Hong, Long. 1993. Multiresolutional filtering using wavelet transform. *IEEE Transactions on Aerospace and Electronic Systems* 29 (4): 1244–1251.
14. Hu, Jun, Zidong Wang, and Huijun Gao. 2013. Recursive filtering with random parameter matrices, multiple fading measurements and correlated noises. *Automatica* 49 (11): 3440–3448.
15. Huang, Minyi, and S Dey. 2007. Stability of Kalman filtering with marikovian packet losses. *Automatica* 43 (4): 598–607.
16. Jiang, Lu, Liping Yan, Bo Xiao, Yuanqing Xia, and Mengyin Fu. 2014. Sequential fusion and state estimation for asynchronous multirate multisensor dynamic systems. In *Proceedings of the 33th Chinese Control Conference*, 291–296. Nanjing, China.

17. Kar, Soummya, Bruno Sinopoli, and José M. F. Moura. 2009. Kalman filtering with intermittent observations: Weak convergence to a stationary distribution. *IEEE Transactions on Automatic Control* 57 (2): 405–420.
18. Liang, Yan, Tongwen Chen, and Quan Pan. 2009. Multi-rate optimal state estimation. *International Journal of Control* 82 (11): 2059–2076.
19. Liu, Yulei, Liping Yan, Yuanqing Xia, Mengyin Fu, and Bo Xiao. 2013. Multirate multisensor distributed data fusion algorithm for state estimation with cross-correlated noises. In *Proceedings of the 32th Chinese Control Conference*, 4682–4687. Xi'an, China.
20. Liu, Yulei, Liping Yan, Bo Xiao, Yuanqing Xia, and Mengyin Fu. 2013. Multirate multisensor data fusion algorithm for state estimation with cross-correlated noises. *Advances in Intelligent Systems and Computing* 214 (1): 19–29.
21. Mahmoud, M.S., and M. F. Emzir. 2012. State estimation with asynchronous multi-rate multi-smart sensors. *Information Sciences* 196: 15–27.
22. Peng, Fangfang, and Shuli Sun. 2014. Distributed fusion estimation for multisensor multirate systems with stochastic observation multiplicative noises. *Mathematical Problems in Engineering*, 1–9.
23. Safari, S., F. Shabani, and D. Simon. 2014. Multirate multisensor data fusion for linear systems using Kalman filters and a neural network. *Aerospace Science and Technology* 39 (1): 465–471.
24. Sheng, Jie, Tongwen Chen, and Sirish L. Shah. 2002. Generalized predictive control for non-uniformly sampled systems. *Journal of Process Control* 12 (8): 875–885.
25. Song, Eenbin, Yunmin Zhu, Jie Zhou, and Zhisheng You. 2007. Optimal Kalman filtering fusion with cross-correlated sensor noises. *Automatica* 43 (8): 1450–1456.
26. Sun, Shuli. 2013. Optimal linear filters for discrete-time systems with randomly delayed and lost measurements with/without time stamps. *IEEE Transactions on Automatic Control* 58 (6): 1551–1556.
27. Wang, Xiaoxu, Yan Liang, Quan Pan, and Feng Yang. 2012. A gaussian approximation recursive filter for nonlinear systems with correlated noises. *Automatica* 48 (9): 2290–2297.
28. Wen, Chuanbo, Yunze Cai, Chenglin Wen, and Xiaoming Xu. 2013. Optimal sequential Kalman filtering with cross-correlated measurement noises. *Aerospace Science and Technology* 26 (1): 153–159.
29. Xia, Yuanqing, Jizong Shang, Jie Chen, and Guoping Liu. 2009. Networked data fusion with packet losses and variable delays. *IEEE Transactions on Systems, Man and Cybernetics, Part B (Cybernetics)* 39 (5): 1107–1120.
30. Yan, Liping, X. Rong Li, and Yuanqing Xia. 2015. Modeling and estimation of asynchronous multirate multisensor system with unreliable measurements. *IEEE Transactions on Aerospace and Electronic Systems* 51 (3): 2012–2026.
31. Yan, Liping, X. Rong Li, and Yuanqing Xia. 2013. Optimal sequential and distributed fusion for state estimation in cross-correlated noise. *Automatica* 49 (12): 3607–3612.
32. Yan, Liping, Baosheng Liu, and Donghua Zhou. 2006. The modeling and estimation of asynchronous multirate multisensor dynamic systems. *Aerospace Science and Technology* 10 (1): 63–71.
33. Yan, Liping, Baosheng Liu, and Donghua Zhou. 2007. An asynchronous multirate multisensor information fusion algorithm. *IEEE Transactions on Aerospace and Electronic Systems* 43 (3): 1135–1146.
34. Yan, Liping, Donghua Zhou, Mengyin Fu, and Yuanqing Xia. 2010. State estimation for asynchronous multirate multisensor dynamic systems with missing measurements. *IET Signal Processing* 4 (6): 728–739.
35. Zhang, Lei, Xiaolin Wu, Quan Pan, and Hongcai Zhang. 2004. Multiresolution modeling and estimation of multisensor data. *IEEE Transactions on Signal Processing* 52 (11): 3170–3182.
36. Zhang, Wen-An, Gang Feng, and Li Yu. 2012. Multi-rate distributed fusion estimation for sensor networks with packet losses. *Automatica* 58 (9): 2016–2028.

Part III
Fusion Estimation Under Event-Triggered Mechanisms

Chapter 8
Event-Triggered Centralized Fusion for Correlated Noise Systems

For decades, with the rapid development of sensing and communication technology, the scale of information collection and transmission speed have reached an unprecedented level, which puts a lot of pressure on the transmission channel bandwidth, transmission costs and the energy consumption in the transmission process. Hence, minimizing the sensor-to-estimator communication rate is practically important, and reducing the sensor-to-estimator communication rate while guarantee a certain level of the estimation performance has great significance. Based on the above reasons, event-based control and data transmission strategies have been widely used, ranging from control systems, signal processing, to various network physical systems, and gradually become the focus of research at domestic and abroad.

In this part, we will introduce a novel transformation strategy, i.e., event-triggered strategy. Based on the design of event-triggered strategies, the state fusion estimation algorithms will be introduced in three chapters, namely, the event-triggered centralized fusion, the event-triggered distributed fusion and the event-triggered sequential fusion algorithms, respectively.

8.1 Introduction

In the early 1980s, Ho and Cassandras first proposed the concept of event-triggered sampling strategy in the study of discrete systems [5]. The event-triggered concept was first applied to the signal acquisition and processing of the dynamic system by Åström [3] and Årzén [2] et al. in 1999. Rabi et al. considered the efficient sampling for state estimation of continuous-time linear systems in [10]. An event-based sensor data scheduler for linear systems and derived the corresponding minimum squared error estimator, and a desired balance between the sensor-to-estimator communication rate and the estimation quality was achieved was proposed by [19]. An event based state-estimator was presented in [14], which was applicable to various types of event sampling strategies, and an event-based estimator with a hybrid update was

L. Yan et al., *Multisensor Fusion Estimation Theory and Application*,
https://doi.org/10.1007/978-981-15-9426-7_8

proposed by using a sum of Gaussian approach to reduce the computational complexity. In [12], the maximum likelihood estimation problem for an event-triggering scheme quantifying the magnitude of the innovation of the estimator at each time instant was studied, and the computation of upper and lower bounds for the communication rate was discussed. A stochastic representation was employed for modeling noises and the state estimation was obtained by minimizing the maximum possible mean squared error in [15]. Based on the Send-on-Delta (SoD) strategy, the distributed filtering problem about a class of discrete time-varying systems based on event-triggered communication mechanism was studied by Liu et al.[8]. Shi et al. used the "set-valued" filtering method to achieve event-based state estimation [13]. Wang et al. studied the problem of event-triggered state estimation under mixed delay and nonlinear interference by using the convex optimization method, and the upper bound of the estimation error was given [17]. Reference [9] studied the problem of event-triggered filter design for a class of linear discrete-time systems under the condition of loss of measured value in communication process.

For multisensor systems, multiple event schedulers were configured to trigger measurement transmissions from multiple sensors assembled together to monitor the same system. This naturally leads to the multi-sensor fusion problem under event-triggered mechanism. Two event-triggered fusion estimation algorithms under sequential and parallel fusion structures were proposed in [6], respectively, where each sensor sent its observations to the fusion center only when its event-triggering condition was satisfied. Reference [11] studied the optimal fusion estimation problem of event-based mixed measurement information (i.e., set-value measurements and point-value measurements). In [20], the distributed state estimation problem over sensor networks with event-triggered communication was addressed for nonlinear discrete time-delay systems. Trimpe et al. proposed an event-triggered distributed data fusion algorithm, in which the event triggering mechanism was based on the estimated error covariance, and each sensor was individually triggered [16]. The event-triggered state estimator was deduced in detail by mining the information behind the event based on the residual triggering conditions [18]. Liu et al. given an distributed adaptive filtering algorithm for determining the triggering threshold with stochastically fading phenomenon, which could achieve the desired average transmission rate in the finite wireless channel resources [7]. Under the condition that the noise-dependent state and the sensor network were stochastically uncertain, a class of event-triggered robust distributed state estimation problems were proposed in [4].

The correlation of noise is seldom considered in the research on event-triggering sampling systems. However, multisensor systems usually have the problem of correlated noise in practical applications. As mentioned in [21], the measurement noise was correlated with the system noise of the previous time step and the same time step, if the discrete-time linear system was obtained from discretization of the continuous-time system. The difference in between lies the cross–correlation of the noise is considered in this chapter. In this chapter, an optimal event-triggered state estimation algorithm based on iterative estimation of white noise estimator is generated for a class of multisensor dynamic systems is proposed, where the noise of different sensors is cross–correlated and coupled with the system noise of the previous time step and the same time step.

This chapter is organized as follows. Section 8.2 formulates the problem. In Sect. 8.3, the event-triggered multisensor state estimation algorithm is presented. In Sect. 8.4, the simulation results is given and Sect. 8.5 draws the conclusions.

8.2 Problem Formulation

8.2.1 System Model Characterization

Consider the following linear dynamic system

$$x_{k+1} = A_k x_k + w_k, \quad k = 0, 1, \ldots \tag{8.1}$$

$$z_{i,k} = C_{i,k} x_k + v_{i,k}, \quad i = 1, 2, \ldots, N \tag{8.2}$$

where $x_k \in R^n$ is the system state, $A_k \in R^{n \times n}$ is the state transition matrix, w_k is the system noise assumed to be white Gaussian distributed and satisfies

$$E\{w_k\} = 0 \tag{8.3}$$

$$E\{w_k w_l^T\} = Q_k \delta_{kl} \tag{8.4}$$

where δ_{kl} is the Kronecker delta function. $z_{i,k} \in R^{m_i}$ is the measurement of sensor i at time k, and $C_{i,k} \in R^{m_i \times n}$ is the measurement matrix. The measurement noise $v_{i,k}$ is assumed to be white Gaussian and satisfies

$$E\{v_{i,k}\} = 0 \tag{8.5}$$

$$E\{v_{i,k} v_{j,l}^T\} = R_{i,k} \delta_{kl} \delta_{ij} \tag{8.6}$$

$$E\{w_{k-1} v_{i,k}^T\} = S_{i,k}^* \tag{8.7}$$

$$E\{w_k v_{i,k}^T\} = S_{i,k} \tag{8.8}$$

Note that the measurement noises are correlated with the system noise—$v_{i,k}$ are dependent on w_{k-1} and w_k for $k = 1, 2, \ldots$ and $i = 1, 2, \ldots, N$; the measurement noises of different sensors $v_{i,k}$ and $v_{j,k}$ are cross-correlated at time k with $E[v_{i,k} v_{j,k}^T] = R_{ij,k} \neq 0$ for $i, j = 1, 2, \ldots, N$. For simplicity, we denote $R_{i,k} := R_{ii,k} > 0$ in the sequel for $i = 1, 2, \ldots, N$.

The initial state x_0 is independent of w_k and $v_{i,k}$ for $k = 1, 2, \ldots$ and $i = 1, 2, \ldots, N$, and is assumed to be Gaussian distributed with mean $\bar{x}_0$ and covariance $\bar{P}_0$.

The purpose of this chapter is to obtain the optimal estimation of state x_k by effectively fusing the observations under limited communication resources.

8.2.2 *Event-Triggered Mechanism of Sensors*

An event-triggered mechanism similar to that in [19] is equipped with each sensor to reduce the sensor-to-estimator communication and prolong the lifetime of sensors. The event-triggering condition of the ith sensor scheduler is associated with the Kalman filtering innovation

$$\tilde{z}_{i,k|k-1} = z_{i,k} - C_{i,k}\hat{x}_{i,k|k-1} \tag{8.9}$$

where $\hat{x}_{i,k|k-1}$ is the prediction of x_k associated with sensor i. The covariance of $\tilde{z}_{i,k|k-1}$ is $P_{\tilde{z}_i,k|k-1} = C_{i,k}P_{i,k|k-1}C_{i,k}^T + R_{i,k} + C_{i,k}S_{i,k}^* + S_{i,k}^{*,T}C_{i,k}^T$, where $P_{i,k|k-1}$ is the state prediction covariance of $\hat{x}_{i,k|k-1}$ [21].

Owing to $P_{\tilde{z}_i,k|k-1}$ is a positive semidefinite matrix, we can obtain a unitary matrix $U_{i,k} \in R^{m_i \times m_i}$ to get the diagonalization matrix:

$$U_{i,k}^T P_{\tilde{z}_i,k|k-1} U_{i,k} = \Lambda_{i,k} \tag{8.10}$$

where the matrix

$$\Lambda_{i,k} = \text{diag}(\lambda_{i1,k}, \ldots, \lambda_{im_i,k}) \in R^{m_i \times m_i} \tag{8.11}$$

and these diagonal scalar elements $\lambda_{i1,k}, \ldots, \lambda_{im_i,k} \in R^{m_i \times m_i}$ are the eigenvalues of $P_{\tilde{z}_i,k|k-1}$. Define $H_{i,k} \in R^{m_i \times m_i}$ by

$$H_{i,k} = U_{i,k}\Lambda_{i,k}^{-1/2} \tag{8.12}$$

Evidently, $H_{i,k}H_{i,k}^T = P_{\tilde{z}_i,k|k-1}^{-1}$. Then, $\tilde{z}_{i,k|k-1}$ is normalized and decorrelated by:

$$\bar{z}_{i,k|k-1} = H_{i,k}^T \tilde{z}_{i,k|k-1} \tag{8.13}$$

Self-evidently, elements of the normalized innovation $\bar{z}_{i,k|k-1}$ are standard Gaussian distributed and independent with each other. Similar to [19], we define the event-triggering condition of the ith sensor scheduler as

$$\gamma_{i,k} = \begin{cases} 0, & \text{if } ||\bar{z}_{i,k|k-1}||_\infty \le \theta_i \\ 1, & \text{otherwise} \end{cases} \tag{8.14}$$

where $\theta_i \ge 0$ is a predetermined threshold, which is related to the estimation accuracy and the communication rate. $||\cdot||_\infty$ represents the infinity-norm of a vector, $|\bar{z}_{im_1,k|k-1}|, \ldots, |\bar{z}_{im_i,k|k-1}|$ are the absolute values of the first to the m_ith dimensional elements of $\bar{z}_{i,k|k-1}$, respectively. Consequently, we know that when $\gamma_{i,k} = 0$, the fusion center can not get the exact measurement, and we only have the knowledge that $\max\{|\bar{z}_{im_1,k|k-1}|, \ldots, |\bar{z}_{im_i,k|k-1}|\} \le \theta_i$; Otherwise, when $\gamma_{i,k} = 1$, the raw sensor measurement $z_{i,k}$ is transmitted to the fusion center.

8.3 The State Fusion Estimation Algorithm with Event-Triggered Mechanism

After $z_{i,k}$ is taken, the sensor decides whether it shall be transmitted to a remote estimator for further processing by event-triggered mechanism. Let $\gamma_{i,k} = 1$ or 0 be the decision variable whether $z_{i,k}$ shall be sent or not.

Define $I_{i,k} \equiv \{\gamma_{i,0} z_{i,0}, \ldots, \gamma_{i,k} z_{i,k}\} \cup \{\gamma_{i,0}, \ldots, \gamma_{i,k}\}$ with $I_{i,-1} = \emptyset$, $\hat{I}_{i,k} = I_{i,k-1} \cup \{\gamma_{i,k} = 0\}$. Define the set $\Omega \subset R^{m_i}$ as

$$\Omega = \{\bar{z}_{i,k|k-1} \in R^{m_i} : ||\bar{z}_{i,k|k-1}||_\infty \leq \theta_i\} \tag{8.15}$$

where $\tilde{z}_{i,k|k-1}$ is zero-mean Gaussian distribution conditioned on $I_{i,k-1}$.

From

$$\tilde{z}_{i,k|k-1} = C_{i,k} \tilde{x}_{i,k|k-1} + v_{i,k} \tag{8.16}$$

We obtain

$$\begin{aligned} P_{\tilde{z}_i,k|k-1} &= E\{\tilde{z}_{i,k|k-1} \tilde{z}_{i,k|k-1}^T | I_{i,k-1}\} \\ &= E\{(C_{i,k} \tilde{x}_{i,k|k-1} + v_{i,k})(C_{i,k} \tilde{x}_{i,k|k-1} + v_{i,k})^T | I_{i,k-1}\} \\ &= C_{i,k} P_{i,k|k-1} C_{i,k}^T + R_{i,k} + C_{i,k} E\{\tilde{x}_{i,k|k-1} v_{i,k}^T | I_{i,k-1}\} \\ &\quad + E\{v_{i,k} \tilde{x}_{i,k|k-1}^T | I_{i,k-1}\} C_{i,k}^T \\ &= C_{i,k} P_{i,k|k-1} C_{i,k}^T + C_{i,k} S_{i,k}^* + S_{i,k}^{*,T} C_{i,k}^T + R_{i,k} \end{aligned} \tag{8.17}$$

From (8.13) and (8.17)

$$\begin{aligned} E\{\bar{z}_{i,k|k-1} \bar{z}_{i,k|k-1}^T | I_{i,k-1}\} &= H_{i,k}^T E\{\tilde{z}_{i,k|k-1} \tilde{z}_{i,k|k-1}^T | I_{i,k-1}\} H_{i,k} \\ &= I_{m_i} \end{aligned} \tag{8.18}$$

Therefore, $\bar{z}_{i,k|k-1}$ is a zero-mean Gaussian multivariate random variable with unit variance. Denote $\bar{z}_{i,k|k-1}^m$ as the mth element of $\bar{z}_{i,k|k-1}$. Thus $\bar{z}_{i,k|k-1}^n$ and $\bar{z}_{i,k|k-1}^m$ are mutually independent if $n \neq m$. Notice that $\gamma_i = 0$ suggests that the event $||\bar{z}_{i,k|k-1}||_\infty \leq \theta_i$ happens. We then have:

Lemma 8.1 ([19]) *Let $x \in R$ be a Gaussian r.v. with zero mean and variance $E[x^2] = \sigma^2$. Denoting $\Delta = \delta\sigma$, then $E[x^2 | |x| \leq \Delta] = \sigma^2(1 - \beta_\delta)$, where*

$$\beta_\delta = \frac{2}{\sqrt{2\pi}} \delta e^{-\frac{\delta^2}{2}} [1 - 2Q_\delta]^{-1} \tag{8.19}$$

$$Q_\delta = \int_\delta^\infty \frac{1}{\sqrt{2\pi}} e^{-\frac{x^2}{2}} dx \tag{8.20}$$

Lemma 8.2 ([19]) *We have*

$$H_{i,k}^T E\{\tilde{z}_{i,k|k-1}\tilde{z}_{i,k|k-1}^T|\hat{I}_{i,k}\}H_{i,k} = E\{\bar{z}_{i,k|k-1}\bar{z}_{i,k|k-1}^T|\hat{I}_{i,k}\} \\ = [1-\beta_{\theta_i}]I_{m_i} \tag{8.21}$$

Lemma 8.3 ([19]) *Let* $\theta_i \geq 0$. *Then*

$$\text{Prob}(||\bar{z}_{i,k|k-1}||_\infty \leq \theta_i|I_{i,k-1}) = [1-2Q_{\theta_i}]^{m_i} \tag{8.22}$$

where Prob$(\cdot)$ *denotes the probability of a random event.*

The average sensor communication rate is defined as

$$\gamma_i = \limsup_{T\longrightarrow\infty} \frac{1}{T+1}\sum_{k=0}^{T} E[\gamma_{i,k}] \tag{8.23}$$

then

$$\gamma_i = 1-(1-2Q_{\theta_i})^{m_i} \tag{8.24}$$

8.3.1 Event-Triggered Kalman Filter with Correlated Noise

Theorem 8.1 (The Kalman filter (KF)) *For system (8.1)–(8.2), the estimation of* x_k *by the Kalman filter is given by*

$$\begin{cases} \hat{x}_{i,k|k} &= \hat{x}_{i,k|k-1} + K_{i,k}\gamma_{i,k}(z_{i,k} - C_{i,k}\hat{x}_{i,k|k-1}) \\ P_{i,k|k} &= P_{i,k|k-1} - g_{i,k}K_{i,k}P_{\bar{z}_i,k|k-1}K_{i,k}^T \\ \hat{x}_{i,k|k-1} &= A_{k-1}\hat{x}_{i,k-1|k-1} + \hat{w}_{i,k-1|k-1} \\ P_{i,k|k-1} &= A_{k-1}P_{i,k-1|k-1}A_{k-1}^T + A_{k-1}P_{\tilde{x}\tilde{w},k-1|k-1}^i \\ &\quad + P_{\tilde{w}\tilde{x},k-1|k-1}^i A_{k-1}^T + P_{\tilde{w},k-1|k-1}^i \\ K_{i,k} &= (P_{i,k|k-1}C_{i,k}^T + S_{i,k}^*)(C_{i,k}P_{i,k|k-1}C_{i,k}^T + R_{i,k} \\ &\quad + C_{i,k}S_{i,k}^* + S_{i,k}^{*,T}C_{i,k}^T)^{-1} \end{cases} \tag{8.25}$$

where, the white noise estimator is computed by

$$\begin{cases} \hat{w}_{i,k|k} &= K_{w,k}^i\gamma_{i,k}(z_{i,k} - C_{i,k}\hat{x}_{i,k|k-1}) \\ K_{w,k}^i &= S_{i,k}(C_{i,k}P_{i,k|k-1}C_{i,k}^T + R_{i,k} + C_{i,k}S_{i,k}^* + S_{i,k}^{*,T}C_{i,k}^T)^{-1} \\ P_{\tilde{w},k|k}^i &= Q_{i,k} - K_{w,k}^i\gamma_{i,k}S_{i,k}^T \end{cases} \tag{8.26}$$

The filtering error cross-covariance matrix between the state and the system noise is computed by

$$\begin{cases} P^i_{\tilde{x}\tilde{w},k|k} = -K_{i,k}\gamma_{i,k}S^T_{i,k} \\ P^i_{\tilde{w}\tilde{x},k|k} = -S_{i,k}K^T_{i,k}\gamma_{i,k} \end{cases} \tag{8.27}$$

The subscript i and superscript i in (8.25)–(8.27) stands for the sensor i and

$$g_{i,k} = \begin{cases} \beta_{\theta_i}, \text{ if } \gamma_{i,k} = 0 \\ 1, \quad \text{if } \gamma_{i,k} = 1 \end{cases} \tag{8.28}$$

$$\beta_{\theta_i} = \frac{2}{\sqrt{2\pi}}\theta_i e^{-\frac{\theta_i^2}{2}}[1 - 2Q_{\theta_i}]^{-1} \tag{8.29}$$

$$Q_{\theta_i} = \int_{\theta_i}^{\infty} \frac{1}{\sqrt{2\pi}} e^{-\frac{x^2}{2}} dx \tag{8.30}$$

where, $z_{i,k}, i = 1, 2, \ldots, N, k = 1, 2, \ldots$ *is the kth measurement of sensor i in time* $(k-1, k]$.

Proof Projection theorem and inductive method is used to prove this theorem.

For whether $z_{i,k}$ shall be sent or not, we have the following two cases:

(1) $\gamma_{i,k} = 1$: For $\gamma_{i,k} = 1$, the estimator can get the exact measurement $z_{i,k}$. In the circumstance, the filter estimation is the same with the time-driven schedule, and we can get

$$\begin{cases} \hat{x}_{i,k|k} &= \hat{x}_{i,k|k-1} + K_{i,k}(z_{i,k} - C_{i,k}\hat{x}_{i,k|k-1}) \\ P_{i,k|k} &= P_{i,k|k-1} - K_{i,k}P_{\tilde{z}_i,k|k-1}K^T_{i,k} \\ \hat{x}_{i,k|k-1} &= A_{k-1}\hat{x}_{i,k-1|k-1} + \hat{w}_{i,k-1|k-1} \\ P_{i,k|k-1} &= A_{k-1}P_{i,k-1|k-1}A^T_{k-1} + A_{k-1}P^i_{\tilde{x}\tilde{w},k-1|k-1} \\ &\quad + P^i_{\tilde{w}\tilde{x},k-1|k-1}A^T_{k-1} + P^i_{\tilde{w},k-1|k-1} \\ K_{i,k} &= (P_{i,k|k-1}C^T_{i,k} + S^*_{i,k})(C_{i,k}P_{i,k|k-1}C^T_{i,k} + R_{i,k} \\ &\quad + C_{i,k}S^*_{i,k} + S^{*,T}_{i,k}C^T_{i,k})^{-1} \end{cases} \tag{8.31}$$

where, the white noise estimator is computed by

$$\begin{cases} \hat{w}_{i,k|k} = K^i_{w,k}(z_{i,k} - C_{i,k}\hat{x}_{i,k|k-1}) \\ K^i_{w,k} = S_{i,k}(C_{i,k}P_{i,k|k-1}C^T_{i,k} + R_{i,k} + C_{i,k}S^*_{i,k} + S^{*,T}_{i,k}C^T_{i,k})^{-1} \\ P^i_{\tilde{w},k|k} = Q_{i,k} - K^i_{w,k}S^T_{i,k} \end{cases} \tag{8.32}$$

The filtering error cross-covariance matrix between the state and the system noise is computed by

$$\begin{cases} P^i_{\tilde{x}\tilde{w},k|k} = -K_{i,k}S^T_{i,k} \\ P^i_{\tilde{w}\tilde{x},k|k} = -S_{i,k}K^T_{i,k} \end{cases} \tag{8.33}$$

(2) $\gamma_{i,k} = 0$: The remote estimator can not get the exact measurement $z_{i,k}$ and we only know that $||\bar{z}_{i,k|k-1}||_\infty \le \theta_i$.

Given $I_{i,k-1}$, $\tilde{z}_{i,k|k-1}$ is Gaussian with zero mean and unit covariance, so we can define $p_{\theta_i} = \text{Prob}(||\bar{z}_{i,k|k-1}||_\infty \le \theta_i | I_{i,k-1})$. Using the conditional probability density function (pdf):

$$f_{\bar{z}_{i,k|k-1}}(\bar{z}_i|\hat{I}_{i,k}) = \begin{cases} \frac{f_{\bar{z}_{i,k|k-1}}(\bar{z}_i|I_{i,k-1})}{p_{\theta_i}}, & \text{if } ||\bar{z}_{i,k|k-1}||_\infty \le \theta_i \\ 0, & \text{otherwise} \end{cases} \tag{8.34}$$

We have

$$\begin{aligned} \hat{x}_{i,k|k} &= E[x_k|\hat{I}_{i,k}] \\ &= \frac{1}{p_{\theta_i}} \int_\Omega E\{x_k|I_{i,k-1}, ||\bar{z}_{i,k|k-1}||_\infty \le \theta_i, \tilde{z}_{i,k|k-1} = H_{i,k}^{T,-1}\bar{z}_i\} \\ &\quad \cdot f_{\bar{z}_{i,k|k-1}}(\bar{z}_i|I_{i,k-1})d\bar{z}_i \\ &= \frac{1}{p_{\theta_i}} \int_\Omega (\hat{x}_{i,k|k-1} + K_{i,k}H_{i,k}^{T,-1}\bar{z}_i) f_{\bar{z}_{i,k|k-1}}(\bar{z}_i|I_{i,k-1})d\bar{z}_i \\ &= \hat{x}_{i,k|k-1} + \frac{K_{i,k}H_{i,k}^{T,-1}}{p_{\theta_i}} \int_\Omega \bar{z}_i f_{\bar{z}_{i,k|k-1}}(\bar{z}_i|I_{i,k-1})d\bar{z}_i \\ &= \hat{x}_{i,k|k-1} \end{aligned} \tag{8.35}$$

where, the last equality is due to

$$\int_\Omega \bar{z}_i f_{\bar{z}_{i,k|k-1}}(\bar{z}_i|I_{i,k-1})d\bar{z}_i = 0 \tag{8.36}$$

then,

$$\begin{aligned} & E\{\tilde{x}_{i,k|k-1}\tilde{z}_{i,k|k-1}^T|\hat{I}_{i,k}\} \\ &= \frac{1}{p_{\theta_i}} \int_\Omega E\{\tilde{x}_{i,k|k-1}|I_{i,k-1}, ||\bar{z}_{i,k|k-1}||_\infty \le \theta_i, \\ &\quad \tilde{z}_{i,k|k-1} = H_{i,k}^{T,-1}\bar{z}_i\}\bar{z}_i^T H_{i,k}^{-1} f_{\bar{z}_{i,k|k-1}}(\bar{z}_i|I_{i,k-1})d\bar{z}_i \\ &= \frac{1}{p_{\theta_i}} \int_\Omega E\{x_k - \hat{x}_{i,k|k-1}|I_{i,k-1}, ||\bar{z}_{i,k|k-1}||_\infty \le \theta_i, \\ &\quad \tilde{z}_{i,k|k-1} = H_{i,k}^{T,-1}\bar{z}_i\}\bar{z}_i^T H_{i,k}^{-1} f_{\bar{z}_{i,k|k-1}}(\bar{z}_i|I_{i,k-1})d\bar{z}_i \\ &= \frac{1}{p_{\theta_i}} \int_\Omega \{E\{x_k|I_{i,k-1}, ||\bar{z}_{i,k|k-1}||_\infty \le \theta_i, \tilde{z}_{i,k|k-1} = H_{i,k}^{T,-1}\bar{z}_i\} \\ &\quad -\hat{x}_{i,k|k-1}\}\bar{z}_i^T H_{i,k}^{-1} f_{\bar{z}_{i,k|k-1}}(\bar{z}_i|I_{i,k-1})d\bar{z}_i \\ &= K_{i,k}H_{i,k}^{T,-1}\frac{1}{p_{\theta_i}} \int_\Omega \bar{z}_i\bar{z}_i^T f_{\bar{z}_{i,k|k-1}}(\bar{z}_i|I_{i,k-1})d\bar{z}_i H_{i,k}^{-1} \end{aligned}$$

$$
\begin{aligned}
&= K_{i,k} H_{i,k}^{T,-1} E\{\bar{z}_{i,k|k-1} \bar{z}_{i,k|k-1}^T | \hat{I}_{i,k}\} H_{i,k}^{-1} \\
&= K_{i,k} E\{\tilde{z}_{i,k|k-1} \tilde{z}_{i,k|k-1}^T | \hat{I}_{i,k}\}
\end{aligned} \tag{8.37}
$$

where, the last equality is from Lemma 8.2 and

$$
K_{i,k} = [P_{i,k|k-1} C_{i,k}^T + S_{i,k}^*][C_{i,k} P_{i,k|k-1} C_{i,k}^T + C_{i,k} S_{i,k}^* + S_{i,k}^{*,T} C_{i,k}^T + R_{i,k}]^{-1} \tag{8.38}
$$

then, we have

$$
\begin{aligned}
& E\{[\tilde{x}_{i,k|k-1} - K_{i,k} \tilde{z}_{i,k|k-1}] \tilde{z}_{i,k|k-1}^T | \hat{I}_{i,k}\} \\
&= E\{\tilde{x}_{i,k|k-1} \tilde{z}_{i,k|k-1}^T | \hat{I}_{i,k}\} - K_{i,k} E\{\tilde{z}_{i,k|k-1}^T \tilde{z}_{i,k|k-1}^T | \hat{I}_{i,k}\} \\
&= 0
\end{aligned} \tag{8.39}
$$

and

$$
\begin{aligned}
& E\{[\tilde{x}_{i,k|k-1} - K_{i,k} \tilde{z}_{i,k|k-1}][\tilde{x}_{i,k|k-1} - K_{i,k} \tilde{z}_{i,k|k-1}]^T | \hat{I}_{i,k}\} \\
&= \int_{\Omega} E\{[\tilde{x}_{i,k|k-1} - K_{i,k} \tilde{z}_{i,k|k-1}][\tilde{x}_{i,k|k-1} - K_{i,k} \tilde{z}_{i,k|k-1}]^T | I_{i,k-1}, \\
& ||\bar{z}_{i,k|k-1}||_\infty \le \theta_i, \tilde{z}_{i,k|k-1} = H_{i,k}^{T,-1} \bar{z}_i\} \frac{f_{\bar{z}_{i,k|k-1}}(\bar{z}_i | I_{i,k-1})}{p_{\theta_i}} d\bar{z}_i \\
&= \frac{1}{p_{\theta_i}} [P_{i,k|k-1} - K_{i,k} P_{\tilde{z}_i,k|k-1} K_{i,k}^T] \int_{\Omega} f_{\bar{z}_{i,k|k-1}}(\bar{z}_i | I_{i,k-1}) d\bar{z}_i \\
&= P_{i,k|k-1} - K_{i,k} P_{\tilde{z}_i,k|k-1} K_{i,k}^T
\end{aligned} \tag{8.40}
$$

from Eq. (8.34), we have $\int_{\Omega} f_{\bar{z}_{i,k|k-1}}(\bar{z}_i | I_{i,k-1}) d\bar{z}_i = p_{\theta_i}$.

Now from Eqs. (8.39), (8.40), Lemmas 8.1 and 8.2, the corresponding estimation error covariance matrix $P_{i,k|k}$ can be computed as

$$
\begin{aligned}
P_{i,k|k} &= E\{[x_k - \hat{x}_{i,k|k}][x_k - \hat{x}_{i,k|k}]^T | \hat{I}_{i,k}\} \\
&= E\{[x_k - \hat{x}_{i,k|k-1}][x_k - \hat{x}_{i,k|k-1})]^T | \hat{I}_{i,k}\} \\
&= E\{[(\tilde{x}_{i,k|k-1} - K_{i,k} \tilde{z}_{i,k|k-1}) + K_{i,k} \tilde{z}_{i,k|k-1}] \\
&\quad \cdot [(\tilde{x}_{i,k|k-1} - K_{i,k} \tilde{z}_{i,k|k-1}) + K_{i,k} \tilde{z}_{i,k|k-1}]^T | \hat{I}_{i,k}\} \\
&= E\{[(\tilde{x}_{i,k|k-1} - K_{i,k} \tilde{z}_{i,k|k-1})(\tilde{x}_{i,k|k-1} - K_{i,k} \tilde{z}_{i,k|k-1})^T] \\
&\quad + E[(\tilde{x}_{i,k|k-1} - K_{i,k} \tilde{z}_{i,k|k-1})(K_{i,k} \tilde{z}_{i,k|k-1})^T] \\
&\quad + E[K_{i,k} \tilde{z}_{i,k|k-1}(\tilde{x}_{i,k|k-1} - K_{i,k} \tilde{z}_{i,k|k-1})^T] \\
&\quad + E[(K_{i,k} \tilde{z}_{i,k|k-1})(K_{i,k} \tilde{z}_{i,k|k-1})^T]| \hat{I}_{i,k}\} \\
&= P_{i,k|k-1} - K_{i,k} P_{\tilde{z}_i,k|k-1} K_{i,k}^T + K_{i,k} E[(\tilde{z}_{i,k|k-1} \tilde{z}_{i,k|k-1})^T | \hat{I}_{i,k}] K_{i,k}^T \\
&= P_{i,k|k-1} - K_{i,k} P_{\tilde{z}_i,k|k-1} K_{i,k}^T + [1 - \beta_{\theta_i}] K_{i,k} (H_{i,k} H_{i,k}^T)^{-1} K_{i,k}^T
\end{aligned}
$$

$$
\begin{aligned}
&= P_{i,k|k-1} - K_{i,k} P_{\tilde{z}_i,k|k-1} K_{i,k}^T + [1-\beta_{\theta_i}] K_{i,k} P_{\tilde{z}_i,k|k-1} K_{i,k}^T \\
&= P_{i,k|k-1} - \beta_{\theta_i} K_{i,k} P_{\tilde{z}_i,k|k-1} K_{i,k}^T
\end{aligned} \tag{8.41}
$$

Because

$$
\hat{x}_{i,k|k-1} = A_{k-1}\hat{x}_{i,k-1|k-1} + \hat{w}_{i,k-1|k-1} \tag{8.42}
$$

Next, we have the following one step prediction error equations for the state based on the projection theory

$$
\begin{aligned}
\tilde{x}_{i,k|k-1} &= x_k - \hat{x}_{i,k|k-1} \\
&= A_{k-1}\tilde{x}_{i,k-1|k-1} + \tilde{w}_{i,k-1|k-1}
\end{aligned} \tag{8.43}
$$

The corresponding prediction error covariance matrix $P_{i,k|k-1}$ can be computed as

$$
\begin{aligned}
P_{i,k|k-1} &= E\{\tilde{x}_{i,k|k-1}\tilde{x}_{i,k|k-1}^T | I_{i,k-1}\} \\
&= A_k P_{i,k-1|k-1} A_k^T + A_k E\{\tilde{x}_{i,k-1|k-1}\tilde{w}_{i,k-1|k-1}^T\} \\
&\quad + E\{\tilde{w}_{i,k-1|k-1}\tilde{x}_{i,k-1|k-1}^T\} A_k^T + E\{\tilde{w}_{i,k-1|k-1}\tilde{w}_{i,k-1|k-1}^T\} \\
&= A_k P_{i,k-1|k-1} A_k^T + A_k P_{\tilde{x}\tilde{w},k-1|k-1}^i + P_{\tilde{w}\tilde{x},k-1|k-1}^i A_k^T + P_{\tilde{w},k-1|k-1}^i
\end{aligned} \tag{8.44}
$$

From the projection theory [1], we have the white noise filter

$$
\begin{aligned}
\hat{w}_{i,k|k} &= E[w_{i,k} | \hat{I}_{i,k}] \\
&= \frac{1}{p_{\theta_i}} \int_{\Omega} E\{w_{i,k} | I_{i,k-1}, ||\bar{z}_{i,k|k-1}||_\infty \le \theta_i, \tilde{z}_{i,k|k-1} = H_{i,k}^{T,-1}\bar{z}_i\} \\
&\quad \cdot f_{\bar{z}_{i,k|k-1}}(\bar{z}_i | I_{i,k-1}) d\bar{z}_i \\
&= \frac{1}{p_{\theta_i}} \int_{\Omega} (K_{w,k}^i H_{i,k}^{T,-1}\bar{z}_i) f_{\bar{z}_{i,k|k-1}}(\bar{z}_i | I_{i,k-1}) d\bar{z}_i \\
&= \frac{K_{w,k}^i H_{i,k}^{T,-1}}{p_{\theta_i}} \int_{\Omega} \bar{z}_{i,k|k-1} f_{\bar{z}_{i,k|k-1}}(\bar{z}_i | I_{i,k-1}) d\bar{z}_i \\
&= 0
\end{aligned} \tag{8.45}
$$

where

$$
K_{w,k}^i = S_k [C_{i,k} P_{i,k|k-1} C_{i,k}^T + C_{i,k} S_{i,k}^* + S_{i,k}^{*,T} C_{i,k}^T + R_{i,k}]^{-1} \tag{8.46}
$$

then

$$
\begin{aligned}
\tilde{w}_{i,k|k} &= w_{i,k} - \hat{w}_{i,k|k} \\
&= w_{i,k}
\end{aligned} \tag{8.47}
$$

and

$$\begin{aligned} P^i_{\tilde{w},k|k} &= E\{\tilde{w}_{i,k|k}\tilde{w}^T_{i,k|k}|\hat{I}_{i,k}\} \\ &= E\{w_{i,k}w^T_{i,k}\} \\ &= Q_{i,k} \end{aligned} \tag{8.48}$$

and

$$\begin{aligned} P^i_{\tilde{x}\tilde{w},k|k} &= E\{\tilde{x}_{i,k|k}\tilde{w}^T_{i,k|k}|\hat{I}_{i,k}\} \\ &= E\{\tilde{x}_{i,k|k-1}w^T_{i,k}|\hat{I}_{i,k}\} \\ &= 0 \end{aligned} \tag{8.49}$$

8.3.2 Batch Fusion Algorithm with Correlated Noise

When it comes to data fusion, the most intuitional method is to collect information from all sensors for centralized estimation fusion. The centralized fusion algorithm in cross-correlated noise is given by the following theorem.

Theorem 8.2 (The optimal centralized batch fusion(BF)) *For system (8.1)–(8.2), the estimation of x_k by the centralized batch fusion is given by*

$$\begin{cases} \hat{x}_{b,k|k} &= \hat{x}_{b,k|k-1} + K_{b,k}\gamma_k(z_k - C_k\hat{x}_{b,k|k-1}) \\ P_{b,k|k} &= P_{b,k|k-1} - g_k K_{b,k} P_{\tilde{z}_i,k|k-1} K^T_{b,k} \\ \hat{x}_{b,k|k-1} &= A_{k-1}\hat{x}_{b,k-1|k-1} + \hat{w}_{b,k-1|k-1} \\ P_{b,k|k-1} &= A_{k-1}P_{b,k-1|k-1}A^T_{k-1} + A_{k-1}P^b_{\tilde{x}\tilde{w},k-1|k-1} \\ & \quad + P^b_{\tilde{w}\tilde{x},k-1|k-1}A^T_{k-1} + P^b_{\tilde{w},k-1|k-1} \\ K_{b,k} &= (P_{b,k|k-1}C^T_k + S^*_k)(C_k P_{b,k|k-1}C^T_k + R_k \\ & \quad + C_k S^*_k + S^{*,T}_k C^T_k)^{-1} \\ \gamma_k &= \mathrm{diag}\{\gamma_{1,k}I_{m_1}, \ldots, \gamma_{i,k}I_{m_i}, \ldots, \gamma_{N,k}I_{m_N}\} \\ g_k &= \mathrm{diag}\{g_{1,k}I_{m_1}, \ldots, g_{i,k}I_{m_i}, \ldots, g_{N,k}I_{m_N}\} \end{cases} \tag{8.50}$$

where, the white noise estimator is computed by

$$\begin{cases} \hat{w}_{b,k|k} = K^b_{w,k}\gamma_k(z_k - C_k\hat{x}_{b,k|k-1}) \\ K^b_{w,k} = S_k(C_k P_{b,k|k-1}C^T_k + R_k + C_k S^*_k + S^{*,T}_k C^T_k)^{-1} \\ P^b_{\tilde{w},k|k} = Q_k - K^b_{w,k}\gamma_k S^T_k \end{cases} \tag{8.51}$$

The filtering error cross-covariance matrix between the state and the system noise is computed by

$$\begin{cases} P^b_{\tilde{x}\tilde{w},k|k} = -K_{b,k}\gamma_k S_k^T \\ P^b_{\tilde{w}\tilde{x},k|k} = -S_k K_{b,k}^T \gamma_k \end{cases} \tag{8.52}$$

The subscript b and superscript b in (8.50)–(8.52) stands for the batch fusion and

$$z_k = \left[\, z_{1,k}^T \cdots z_{i,k}^T \cdots z_{N,k}^T \,\right]^T \tag{8.53}$$

$$C_k = \left[\, C_{1,k}^T \cdots C_{i,k}^T \cdots C_{N,k}^T \,\right]^T \tag{8.54}$$

$$v_k = \left[\, v_{1,k}^T \cdots v_{i,k}^T \cdots v_{N,k}^T \,\right]^T \tag{8.55}$$

$$R_k = \begin{bmatrix} R_{1,k} & \cdots & R_{1i,k} & \cdots & R_{1N,k} \\ \vdots & & \vdots & & \vdots \\ R_{i1,k} & \cdots & R_{i,k} & \cdots & R_{iN,k} \\ \vdots & & \vdots & & \vdots \\ R_{N1,k} & \cdots & R_{Ni,k} & \cdots & R_{N,k} \end{bmatrix} \tag{8.56}$$

$$S_k^* = \left[\, S_{1,k}^* \cdots S_{i,k}^* \cdots S_{N,k}^* \,\right] \tag{8.57}$$

$$S_k = \left[\, S_{1,k} \cdots S_{i,k} \cdots S_{N,k} \,\right] \tag{8.58}$$

where, $g_{i,k}$, β_{θ_i}, and Q_{θ_i} are given in Eqs. (8.28)–(8.30). $z_{i,k}$ is the kth measurement of sensor i in time $(k-1, k]$, $i = 1, 2, \ldots, N$, $k = 1, 2, \ldots$, $R_{ij,k} = E[v_{i,k}v_{j,k}^T]$.

8.4 Numerical Example

An example is used to illustrate the effectiveness of the presented algorithm in this section. A tracking system with two sensors can be described by

$$x_{k+1} = \begin{bmatrix} 1 & T_s \\ 0 & 1 \end{bmatrix} x_k + \Gamma_k \xi_k \tag{8.59}$$

$$z_{1,k} = C_1 x_k + v_{1,k},\ k = 1, 2, \ldots, L \tag{8.60}$$

$$z_{2,k} = C_2 x_k + v_{2,k} \tag{8.61}$$

$$v_{1,k} = \eta_{1,k} + \beta_{11}\xi_{k-1} + \beta_{12}\xi_k \tag{8.62}$$

$$v_{2,k} = \eta_{2,k} + \beta_{21}\xi_{k-1} + \beta_{22}\xi_k \tag{8.63}$$

where $T_s = 0.01$ is the sampling period. $L = 300$ is the length of the signal x to be estimated. The state $x_k = [s_k \ \dot{s}_k]^T$, where s_k and $\dot{s}_k$ are the position and velocity of the target at time kT_s, respectively. $\Gamma_k = [T_s \ \ 1]$ is the noise transition matrix. $\xi_k \in R$ is the system noise, assumed to be white and Gaussian distributed with zero mean and variance σ_ξ^2.

$z_{i,k}$ $(i = 1, 2)$ are the observation vector of the two sensors, which observe the position and velocity, respectively, i.e., $C_1 = [1 \ 0]$, $C_2 = [0 \ 1]$. $v_{i,k}$ $(i = 1, 2)$ are the measurement noises of sensor i, and are cross-correlated and coupled with the system noise ξ_{k-1} and ξ_k, which may have originated from the discretization of a continuous time system. The strength of the correlation is determined by β_{i1} and β_{i2}. $\eta_{i,k}$ $(i = 1, 2)$ are zero mean white Gaussian noises with variances $\sigma_{\eta_i}^2$ and are independent of ξ_k, $k = 1, 2, \ldots$. The initial values are $\bar{x}_0 = [1 \ 1]^T$, $\bar{P}_0 = I_2$, which is independent of ξ_k and $v_{i,k}$, $k = 1, 2, \ldots, i = 1, 2$.

From Eq. (8.59), we have $Q_k = \Gamma_k \Gamma_k^T \sigma_\xi^2$, which is the system noise covariance corresponding to $w_k = \Gamma_k \xi_k$. From Eqs. (8.62) and (8.63), the measurement noise covariance is computed by

$$R_k = \begin{bmatrix} (\beta_{11}^2 + \beta_{12}^2)\sigma_\xi^2 + \sigma_{\eta_1}^2 & (\beta_{11}\beta_{21} + \beta_{12}\beta_{22})\sigma_\xi^2 \\ (\beta_{21}\beta_{11} + \beta_{22}\beta_{12})\sigma_\xi^2 & (\beta_{21}^2 + \beta_{22}^2)\sigma_\xi^2 + \sigma_{\eta_2}^2 \end{bmatrix} \tag{8.64}$$

The covariance between w_{k-1} and v_{i,k_i} is

$$S_k^* = [\beta_{11}\sigma_\xi^2 \Gamma_{k-1} \ \ \beta_{21}\sigma_\xi^2 \Gamma_{k-1}] \tag{8.65}$$

and the covariance between w_k and v_{i,k_i} is

$$S_k = [\beta_{12}\sigma_\xi^2 \Gamma_k \ \ \beta_{22}\sigma_\xi^2 \Gamma_k] \tag{8.66}$$

Let the event-triggering threshold $\theta_i = \theta$ for $i = 1, 2$ for simplicity. The value of θ is changing within the set $\theta \in \{0, 0.5, 0.8, 1\}$ to illustrate the impact of θ on the estimation performance, where $\theta = 0$ means that the scheduler is time-triggered, that is always activated, and the estimator can receive all the measurements of the corresponding sensor at each instant.

Next, the state estimation of x_k by fusing information from the two sensors is given, and the differences between the estimated results obtained by different estimation algorithms in case of noise correlation are compared. In the case of noise correlation, the influence on fusion results when noise correlation is ignored is analyzed.

Denote $\sigma_\xi^2 = 0.01$, $\sigma_{\eta_1}^2 = 0.25$ and $\sigma_{\eta_2}^2 = 0.16$. For β_{i1} and β_{i2}, $i = 1, 2$, we denote $\beta_{11} = 3, \beta_{12} = 2, \beta_{21} = 2, \beta_{22} = 3$, thence the measurement noises are cross-correlated and coupled with the system noise. We set 300 sampling time for 100 Monte Carlo simulations, and observe the effectiveness of the presented algorithm. The simulation results are shown in Figs. 8.1, 8.2 and 8.3 and Tables 8.1 and 8.2.

The statistical simulation curves of Root Mean Square Errors (RMSEs) of proposed event-triggered centralized batched estimation fusion algorithm (ETBF) with

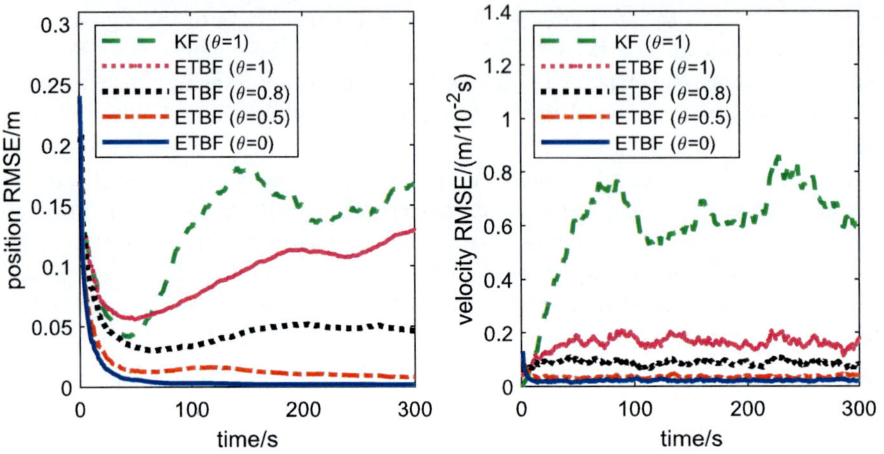

Fig. 8.1 RMSE curves of proposed batched algorithm with different thresholds

Fig. 8.2 Sensor communication rate γ versus scheduling parameter δ

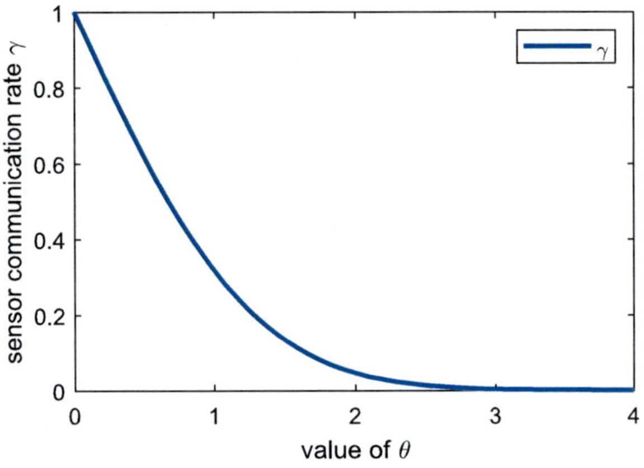

different triggering thresholds are presented in Fig. 8.1, and are contrasted with the Kalman filter. The blue solid line, the red dash-dotted line and black dotted lines denote ETBF algorithm with $\theta = 0$, $\theta = 0.5$, $\theta = 0.8$, respectively. The purple dotted lines denote ETBF algorithm with $\theta = 1.0$ and the green dashed lines denote Kalman filter (KF) with $\theta = 1.0$. It is shown in Fig. 8.1 that the RMSE curves of the proposed batched algorithm with $\theta = 1.0$ are much lower than the corresponding ones of the Kalman filtering with the same θ, which illustrates that the proposed algorithm has better estimation performance than the classical Kalman filter. From Fig. 8.1, we can also see that RMSE curves of proposed algorithm with smaller triggering thresholds are always lower than those with larger triggering thresholds. This is consistent with our theoretical analysis since a lower triggering threshold means a higher data transmission rate and thus has outperform. The average sensor com-

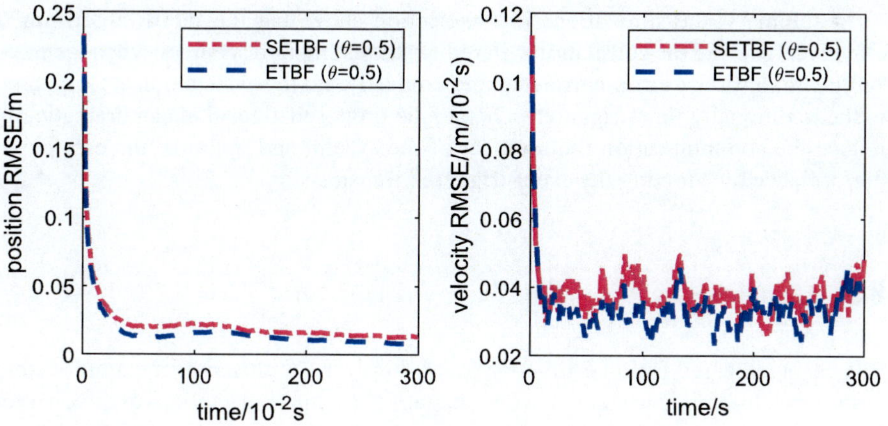

Fig. 8.3 RMSE curves of SETBF and ETBF with $\theta = 0.5$

Table 8.1 Communication rate γ for ETBF with different θ

θ	0	0.5	0.8	1
γ	1	0.6171	0.4237	0.3173

Table 8.2 Time-averaged RMSEs for algorithms with different θ

θ	0	0.5	0.8	1
RMSE of ETBF	0.0070	0.0162	0.0468	0.0940
RMSE of KF	0.0185	0.0265	0.0642	0.1288
RMSE of SETBF	0.0128	0.0216	0.0803	0.1782

munication rate γ from Lemma 8.3 under different values of δ is shown in Fig. 8.2 and Table 8.1. We note that $\theta = 0$ means that all sensor measurements have been sent to the estimator and the system degenerates into time triggered. Consequently, the proposed algorithms have the best estimation performance with $\theta = 0$. Time-averaged RMSEs of ETBF algorithm and KF algorithm are shown in the first and second columns of Table 8.2. We can come to the same conclusion.

The statistical simulation curves of RMSEs of propose ETBF algorithm and the suboptimal event-triggered batched estimation fusion algorithm (SETBF) with thresholds $\theta = 0.5$ is shown in Fig. 8.3. The SETBF algorithm means suboptimal event-triggered batched fusion algorithm neglecting noise correlation. It shows that the RMSE curves of the proposed ETBF algorithm with $\theta = 0.5$ are much lower than the corresponding ones of the SETBF algorithm, which illustrates that ETBF algorithm with consideration of noise correlation is effective while neglecting noise correlation can reduce estimation precision. Time-averaged RMSEs of ETBF algorithm and SETBF algorithm are shown in first and third columns of Table 8.2. We can draw the same conclusion.

In summary, the simulations in this section show that the ETBF algorithm is effective. Ignoring the correlation will reduce the estimation accuracy when the measurement noises are cross-correlated and coupled with the system noise. Compared to the traditionally time-triggered scheme, the proposed algorithm can dramatically reduce the communication requirement of the system and optimize the estimation performance by adopting the event-triggered transmission.

8.5 Conclusions

The event-triggered fusion estimation algorithms for a multisensor dynamic system with correlated noise are studied in this chapter. The main contributions of this chapter are: (1) In the case of sensor noise is cross-correlated and coupled with the system noise of the previous time step and the same time step, an optimal state estimation algorithm based on iterative estimation of white noise estimator for an multisensor dynamic system is generated. (2) Compared with the traditionally time-triggered scheme, by adopting event-triggered transmission, the proposed algorithms can dramatically reduce the communication requirement of the system only with a slightly deterioration of the estimation performance. The performance of the algorithm is compared through an example in terms of root mean square error. From theoretical analysis and simulation results, it can be concluded that the proposed algorithm is effective and has potential value in many application fields related to multisensor networked control or networked estimation systems.

References

1. Anderson, B.D., and J.B. Moore. 1979. *Optimal Filtering*. Englewood Cliffs, New Jersey: Prentice-Hall.
2. Årzén, K.E. 1999. A simple event-based PID controller. In *Proceedings of the14th IFAC World Congress*, pages 423–428.
3. Åström, K.J., and B. Bernhardsson. 1999. Comparison of periodic and event based sampling for first-order stochastic systems. In *Proceedings of the14th IFAC World Congress*, pages 1–7.
4. Dong, Hongli, Zidong Wang, Fuad E Alsaadi, and Bashir Ahmad. 2015. Event-triggered robust distributed state estimation for sensor networks with state-dependent noises. *International Journal of General Systems* 44 (2): 254–266.
5. Ho, Y.C., Xiren Cao, and Christos Cassandras. 1983. Infinitesimal and finite perturbation analysis for queuing networks. *Automatica* 19 (4): 439–445.
6. Jin, Zengwang, Yanyan Hu, Changyin Sun, and Lan Zhang. 2015. Event-triggered state fusion estimation for wireless sensor networks with feedback. In *34th Chinese Control Conference*, pages 4610–4614, Hangzhou, China.
7. Liu, Qinyuan, Zidong Wang, Xiao He, and Donghua Zhou. 2015. Event-based distributed filtering with stochastic measurement fading. *IEEE Transactions on Industrial Informatics* 11 (6): 1643–1652.

8. Liu, Qinyuan, Zidong Wang, Xiao He, and Donghua Zhou. 2015. Event-based recursive distributed filtering over wireless sensor networks. *IEEE Transactions on Automatic Control* 60 (9): 2470–2475.
9. Liu, Qinyuan, Zidong Wang, Weibo Liu, and Wenshuo Li. 2016. Event-triggered resilient filtering with missing measurements. In *IEEE 22nd International Conference on Automation and Computing*, pages 162–167.
10. Rabi, M., G. Moustakides, and J. Baras. 2012. Adaptive sampling for linear state estimation. *SIAM Journal on Control and Optimization* 50 (2): 672–702.
11. Shi, Dawei, Tongwen Chen, and Ling Shi. 2014. An event-triggered approach to state estimation with multiple point- and set-valued measurements. *Automatica* 50 (6): 1641–1648.
12. Shi, Dawei, Tongwen Chen, and Ling Shi. 2014. Event-triggered maximum likelihood state estimation. *Automatica* 50 (1): 247–254.
13. Shi, Dawei, Tongwen Chen, and Ling Shi. 2015. On set-valued Kalman filtering and its application to event-based state estimation. *IEEE Transactions on Automatic Control* 60 (5): 1275–1290.
14. Sijs, J., and M. Lazar. 2012. Event based state estimation with time synchronous updates. *IEEE Transactions on Automatic Control* 57 (10): 2650–2655.
15. Sijs, Joris, Benjamin Noack, and Uwe D. Hanebeck. 2013. Event-based state estimation with negative information. In *Proceedings of the 16th International Conference on Information Fusion*, pages 2192–2199, Istanbul, Turkey.
16. Trimpe, S., and R. D'Andrea. 2014. Event-based state estimation with variance-based triggering. *IEEE Transactions on Automatic Control* 59 (12): 3266–3281.
17. Wang, Licheng, Zidong Wang, Tingwen Huang, and Guoliang Wei. 2016. An event-triggered approach to state estimation for a class of complex networks with mixed time delays and nonlinearities. *IEEE Transactions on Cybernetics* 46 (11): 2497–2508.
18. Weerakkody, Sean, Yilin Mo, Bruno Sinopoli, Duo Han, and Ling Shi. 2016. Multi-sensor scheduling for state estimationwith event-based, stochastic triggers. *IEEE Transactions on Automatic Control* 61 (9): 2695–2701.
19. Wu, Junfeng, Qingshan Jia, Karl Henrik Johansson, and Ling Shi. 2013. Event-based sensor data scheduling: Trade-off between communication rate and estimation quality. *IEEE Transactions on Automatic Control* 58 (4): 1041–1046.
20. Yan, Lei, Xiaomei Zhang, Zhenjuan Zhang, and Yongjie Yang. 2014. Distributed state estimation in sensor networks with event-triggered communication. *Nonlinear Dynamics* 76 (1): 169–181.
21. Yan, Liping, X. Rong Li, and Yuanqing Xia. 2013. Optimal sequential and distributed fusion for state estimation in cross-correlated noise. *Automatica* 49 (12): 3607–3612.

Chapter 9
Event-Triggered Distributed Fusion Estimation for WSN Systems

9.1 Introduction

Owing to the limitation of network resources and bandwidth in the wireless sensor networks (WSNs), in order to reduce unnecessary waste, it is necessary to design a better system communication mechanism. The event-triggered mechanism can save the network bandwidth, as well as the sensor and state estimator energy consumption to some extent, because it is determined by a preset event-triggered constraint to determine whether the information shall be transmitted. The optimization design of the event-triggered sampling/transmission strategy has gradually become a hotspot in domestic and foreign research.

In the early 1980s, an event-triggered sampling strategy for discrete systems was first proposed by Ho et al. in [6]. In 1999, Åström and Bernhardsson given a comparison between periodic and event-based sampling for first order stochastic systems in [2]. A variable sampling event-triggered system based on the deviation signal and apply it to water tank level control was proposed by Årzén et al. in [1]. By mining the information behind the event, Han and Wu et al. deduced the event-triggered state estimator based on the residual triggering condition [4, 16]. The state estimation based on the send-on-delta (SOD) triggered strategy is addressed in [10]. With an event-based communication mechanism, Liu et al. investigated the distributed filtering problem for a class of discrete time-varying systems in wireless sensor networks [9]. Shi et al. take advantage of the "set-valued" filtering architecture to implement event-triggered state estimation in [11]. The event-triggered state estimation problem of the hidden Markov model is studied in [12], and the reliable channel and the unreliable channel with packet loss is compared and analyzed. Han et al. proposed an open-loop and a closed-loop stochastic event-triggered sensor schedule for remote state estimation in [5]. The problem of event-triggered state estimation under mixed delay and nonlinear interference is studied in [15]. To reduce the communication burden and overcomes the network constraints, Wu et al. studied the consensus problem of multi-agent system with external interference through com-

L. Yan et al., *Multisensor Fusion Estimation Theory and Application*,
https://doi.org/10.1007/978-981-15-9426-7_9

bining quantized control with event-triggered control [18]. The problem of event-triggered synchronization of heterogeneous complex networks is investigated, and the influence of transmission delays on event-triggered synchronization is considered in [19]. With quantized communication based on a directed graph, the problem of distributed event-triggered pinning control for practical consensus of multi-agent systems is investigated by Wu et al. in order to decrease communication load of integrant in [17]. The event-triggered strategies mentioned in the above literatures are effective, but couldn't provide the actual communication rate directly.

In the WSNs, data fusion could be accomplished by centralized distributed or distributed fusion architecture. Centralized fusion is not suitable for WSNs because there are large quantities of sensors in the WSNs, so it takes lots of calculation to augment all the sensors, and it is highly possible to be a failure at the center node. In contrast, the distributed fusion architecture shows great advantages, because each sensor performs not only as a sensor but also as an estimator with both sensing and computational capabilities, so each sensor may be used as a fusion center. In addition, the distributed fusion architecture has higher robustness and less communication burden, because the estimation is generated by the measurements which is collected only from its neighborhood. Trimpe et al. proposed an event-triggered distributed data fusion algorithm in which the event-triggered mechanism was based on the estimated error covariance, and each sensor was triggered individually [13]. And in the next year, he studied the problem of distributed event-triggered state estimation in the background of multi-agent coordination, and the triggering conditions of the event were designed according to the value of the residual [14]. For a class of discrete nonlinear stochastic systems with randomly occurring uncertainties, time-varying delays and randomly occurring nonlinearities, the event-triggered distributed state estimation problem was concerned in [7]. Considering the random measurement fading phenomenon, an adaptive algorithm for determining the triggering threshold was given In [8], and the desired average transmission rate could be obtained under limited wireless channel resources. A set of event-triggered time-varying distributed state estimators that allow dynamic state estimation errors to satisfy the average $H\infty$ performance constraints was designed by Dong et al., which to some extent could save constrained computational resources and network bandwidth [3]. In [20], the problem of distributed event-triggered $H\infty$ filtering over sensor networks for a class of discrete-time system modeled by a set of linear Takagi-Sugeno (T-S) fuzzy models was studied. With event-triggered communication schedules on both sensor-to-estimator channel and estimator-to-estimator channel, the distributed estimation problem for networked sensing system was studied in [22].

In summary, there are many event-triggered distributed fusion algorithms. These existed event-triggered transmission/sampling strategies can be divided into the following categories: (1) triggering an event based on the observations, including SOD or measurement-based triggering (MBT); (2) based on the estimation error covariance; (3) based on the observation residual. The first category based on the measurements only, it is therefore more efficient; The last two strategies need to compute the state estimation at the sensor side, which require large calculation. As we all know, event-triggered strategy is different from time-triggered strategy, in that not all the measurements shall be sent to the fusion center, therefore its communication rate is related to the event-triggered threshold. However, the conventional event-triggered methods cannot give the specific communication rate intuitively and directly. In response to this problem, we propose a new event triggering strategy, and on this basis, a novel event-triggered distributed data fusion algorithm for wireless sensor networks is proposed in this chapter. We improve SOD method by setting the threshold according to the chi-square distribution table, which considers the difference between the measurement of the current time and measurement of the last sampled moment. The event will be triggered and the observation will be sampled for estimation, when the chi-square distribution value exceeds the preset threshold. According to the chi-square distribution table, the sampling/communication rate of the proposed algorithm under a certain threshold can be obtained directly, that is, the threshold of the novel dynamic event-triggered strategy is corresponded to the relevant value in the chi-square distribution table. Since the energy of nodes in wireless sensor networks is limited, based on the network topology of sensors, in the distributed fusion algorithm, we only fuse the neighbor nodes and the neighbor's neighbor nodes of the fusion center, which is more practical and energy efficient.

This chapter is organized as follows. Section 9.2 formulates the problem. Section 9.3 presents the event-triggered multisensor state estimation algorithm. In Sect. 9.4, the simulation results is given and Sect. 9.5 draws the conclusion.

9.2 Problem Formulation

9.2.1 System Model Characterization

Consider the following linear dynamic system

$$x_{k+1} = A_k x_k + w_k, \quad k = 0, 1, \ldots \tag{9.1}$$

$$z_{i,k} = C_{i,k} x_k + v_{i,k}, \quad i = 1, 2, \ldots, N \tag{9.2}$$

where $x_k \in R^n$ is the system state, $A_k \in R^{n \times n}$ is the state transition matrix, w_k is the system noise assumed to be white Gaussian distributed and satisfies

$$E\{w_k\} = 0 \tag{9.3}$$

$$E\{w_k w_l^T\} = Q_k \delta_{kl} \tag{9.4}$$

where δ_{kl} is the Kronecker delta function. $z_{i,k} \in R^{m_i}$ is the measurement of sensor i at time k, and $C_{i,k} \in R^{m_i \times n}$ is the measurement matrix. The measurement noise $v_{i,k}$ is assumed to be white Gaussian and satisfies

$$E\{v_{i,k}\} = 0 \tag{9.5}$$

$$E\{v_{i,k} v_{j,l}^T\} = R_{i,k} \delta_{kl} \delta_{ij} \tag{9.6}$$

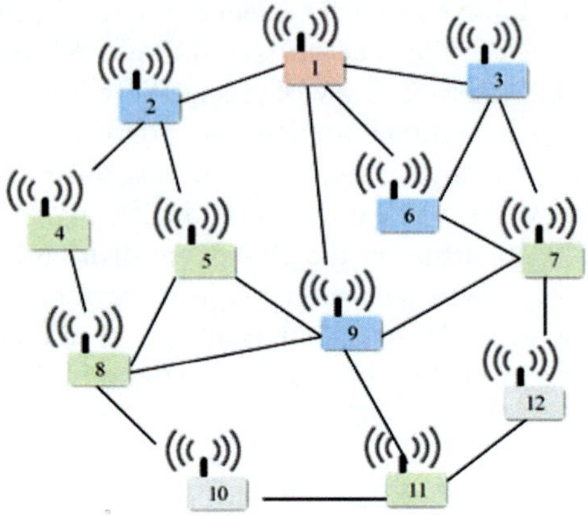

Fig. 9.1 The wireless sensor networks with 12 sensors

The initial state x_0 is independent of w_k and $v_{i,k}$ for $k = 1, 2, \ldots$ and $i = 1, 2, \ldots, N$, and is assumed to be Gaussian distributed with mean $\bar{x}_0$ and variance $\bar{P}_0$.

As shown in Fig. 9.1, an example is used to illustrate the distributed fusion for the WSNs. The WSNs consist of 12 sensors in this scenario, every sensor in the network is equal acting also as an estimator. The sensor only communicates with its neighbors that connected by undirected lines within one hop. Notice that a sensor is always connected to itself. The distributed algorithm includes two steps. (1) Based on the event-triggered sampling mechanism, every sensor i, $i \in \{1, 2, \ldots, N\}$ generates a local estimation by use of information of its own. (2) By use of the local estimations of its neighbors and the neighbor's neighbors which are called the neighborhood and are denoted by $\mathcal{N}_r$, the distributed fusion estimation of the target sensor r is generated, where we use $N_r = |\mathcal{N}_r|$ to denote the number of the sensors in $\mathcal{N}_r$. Only measurements of sensors in the neighborhood are fused to generate the distributed fusion estimation. For example, sensor 1 uses the information of sensors 1, 2, 3, 4, 7 (neighbors of sensor 1) and sensors 5, 6, 9, 10, 11 (neighbors of sensors 2, 3, 4, 7) to generate the fusion estimation, $\mathcal{N}_1 = \{1, 2, 3, 4, 5, 6, 7, 9, 10, 11\}$.

Remark 9.1 Since the energy of the wireless sensor network is limited, in the distributed fusion algorithm, we fuse the neighbor nodes and the neighbor's neighbor nodes of the fusion center based on the network topology of sensors, which has higher survivability and less communicational burden and is more flexible and reliable. The algorithm can be extended to the fusion of sensors with other topologies easily depending on the amount of energy in the WSNs, for example, by changing "neighbors' neighbors' information" to $i \in \mathcal{N}_r$, where $\mathcal{N}_r$ is the set of the sensors that to be fused in the fusion center.

The purpose of this chapter is to obtain the optimal estimation of state x_k by effectively fuse of the observations under limited communication resources.

9.2.2 Event-Triggered Mechanism of Sensors

An event-triggered mechanism is equipped with each sensor to reduce the sensor-to-estimator communication and prolong the lifetime of sensors.

x_k is assumed to be Gaussian distributed with mean

$$E\{x_k\} = A_0 A_1 \cdots A_{k-1}\bar{x}_0 \tag{9.7}$$

and the corresponding estimation error covariance P_k is given by

$$\begin{aligned} P_k &= E\{(x_k - E\{x_k\})(x_k - E\{x_k\})^T\} \\ &= A_{k-1}P_{k-1}A_{k-1}^T + Q_{k-1} \end{aligned} \tag{9.8}$$

and the state second-order moment matrix Ξ_k is given by

$$\begin{aligned} \Xi_k &= E\{x_k x_k^T\} \\ &= A_{k-1}\Xi_{k-1}A_{k-1}^T + Q_{k-1} \end{aligned} \tag{9.9}$$

where, $\Xi_0 = \bar{P}_0 + \bar{x}_0\bar{x}_0^T$. $z_{i,k}$ is Gaussian distributed with mean

$$E\{z_{i,k}\} = C_{i,k}A_0 A_1 \cdots A_{k-1}\bar{x}_0 \tag{9.10}$$

and the covariance of the residual is

$$E\{(z_{i,k} - E\{z_{i,k}\})(z_{i,k} - E\{z_{i,k}\})^T\} = C_{i,k}P_k C_{i,k}^T + R_{i,k} \tag{9.11}$$

For subsequent derivation, the measurement $\bar{z}_{i,k}$ of the current sampled time is defined by

$$\bar{z}_{i,k} = \lambda_{i,k}z_{i,k} + (1 - \lambda_{i,k})\bar{z}_{i,k-1} \tag{9.12}$$

where $\bar{z}_{i,k-1}$ represents the measurement previously sampled, $z_{i,k}$ stands for the current measurement and $\lambda_{i,k}$ is a given positive constant.

For convenience, it is assumed that the last sampled moment of sensor i is time l_i. Therefore, the difference between the measurement of the current time and the measurement of the last sampled moment is

$$\begin{aligned} z_{i,k} - \bar{z}_{i,k-1} &= z_{i,k} - z_{i,l_i} \\ &= (C_{i,k} - C_{i,l_i}\prod_{p=l_i}^{k-1} A_p^{-1})x_k + C_{i,l_i}(\sum_{m=l_i}^{k-1}\prod_{p=l_i}^{m} A_p^{-1}w_m) + v_{i,k} - v_{i,l_i} \end{aligned} \tag{9.13}$$

which satisfies

$$E\{z_{i,k} - \bar{z}_{i,k-1}\} = (C_{i,k} - C_{i,l_i} \prod_{p=l_i}^{k-1} A_p^{-1}) A_0 A_1 \cdots A_{k-1} \bar{x}_0 \tag{9.14}$$

$$\begin{aligned} & E\{[(z_{i,k} - \bar{z}_{i,k-1}) - E\{z_{i,k} - \bar{z}_{i,k-1}\}][(z_{i,k} - \bar{z}_{i,k-1}) - E\{z_{i,k} - \bar{z}_{i,k-1}\}]^T\} \\ & = [C_{i,k} - C_{i,l_i} \prod_{p=l_i}^{k-1} A_p^{-1}] P_k [C_{i,k} - C_{i,l_i} \prod_{p=l_i}^{k-1} A_p^{-1}]^T + R_{i,k} + R_{i,l_i} \\ & + C_{i,l_i} (\sum_{m=l_i}^{k-1} \prod_{p=l_i}^{m} A_p^{-1} Q_m \prod_{p=m+1}^{k-1} A_p^T)[C_{i,k} - C_{i,l_i} \prod_{p=l_i}^{k-1} A_p^{-1}]^T \\ & + C_{i,k} (\sum_{m=l_i}^{k-1} \prod_{p=m+1}^{k-1} A_p Q_m \prod_{p=l_i}^{m} A_p^{-T}) C_{i,l_i}^T \end{aligned} \tag{9.15}$$

For $m = l_i = k - 1$, $\prod_{p=m+1}^{k-1} A_p = \prod_{k}^{k-1} A_p = I$, and

$$\begin{aligned} & E\{(z_{i,k} - \bar{z}_{i,k-1})(z_{i,k} - \bar{z}_{i,k-1})^T\} \\ & = (C_{i,k} - C_{i,l_i} \prod_{p=l_i}^{k-1} A_p^{-1}) \Xi_k (C_{i,k} - C_{i,l_i} \prod_{p=l_i}^{k-1} A_p^{-1})^T + R_{i,k} + R_{i,l_i} \\ & + C_{i,l_i} (\sum_{m=l_i}^{k-1} \prod_{p=l_i}^{m} A_p^{-1} Q_m \prod_{p=m+1}^{k-1} A_p^T)(C_{i,k} - C_{i,l_i} \prod_{p=l_i}^{k-1} A_p^{-1})^T \\ & + C_{i,k} (\sum_{m=l_i}^{k-1} \prod_{p=m+1}^{k-1} A_p Q_m \prod_{p=l_i}^{m} A_p^{-T}) C_{i,l_i}^T \end{aligned} \tag{9.16}$$

Denote

$$y_{i,k} = z_{i,k} - \bar{z}_{i,k-1} - (C_{i,k} - C_{i,l_i} \prod_{p=l_i}^{k-1} A_p^{-1}) A_0 A_1 \cdots A_{k-1} \bar{x}_0 \tag{9.17}$$

then $E\{y_{i,k}\} = 0$, and

$$\begin{aligned} S_{i,k} &= E\{(y_{i,k} - E\{y_{i,k}\})(y_{i,k} - E\{y_{i,k}\})^T\} \\ &= (C_{i,k} - C_{i,l_i} \prod_{p=l_i}^{k-1} A_p^{-1}) P_k (C_{i,k} - C_{i,l_i} \prod_{p=l_i}^{k-1} A_p^{-1})^T + R_{i,k} + R_{i,l_i} \\ &+ C_{i,l_i} (\sum_{m=l_i}^{k-1} \prod_{p=l_i}^{m} A_p^{-1} Q_m \prod_{p=m+1}^{k-1} A_p^T)(C_{i,k} - C_{i,l_i} \prod_{p=l_i}^{k-1} A_p^{-1})^T \\ &+ C_{i,k} (\sum_{m=l_i}^{k-1} \prod_{p=m+1}^{k-1} A_p Q_m \prod_{p=l_i}^{m} A_p^{-T}) C_{i,l_i}^T \end{aligned} \tag{9.18}$$

Constructing a chi-square distribution quantity:

$$\rho_{i,k} = y_{i,k}^T S_{i,k}^{-1} y_{i,k} \tag{9.19}$$

The event-triggered condition of the sensor scheduler is defined as

$$\lambda_{i,k} = \begin{cases} 0, & \text{if } \rho_{i,k} \le \chi_\alpha^2(m_i) \\ 1, & \text{otherwise} \end{cases} \tag{9.20}$$

Remark 9.2 $\rho_{i,k} \sim \chi_\alpha^2(m_i)$ is standard chi-square distribution with m_i degrees of freedom, and its mean and variance are m_i and $2m_i$, respectively, where m_i is the dimension of $z_{i,k}$. Therefore, $\rho_{i,k}$ can be used as a metric to evaluate whether $z_{i,k}$ shall be sampled for an estimator for further processing. That is, in this event-triggered problem, the original hypothesis is $H_0 : \lambda_{i,k} = 1$, and the antithetic hypothesis is $H_1 : \lambda_{i,k} \neq 1$. The rejection interval is $(0, \chi_\alpha^2(m_i))$, where $\chi_\alpha^2(m_i)$ is the one-side χ^2 value with confidence α, $1 \le i \le N_k$. Basically, if $\rho_{i,k} \le \chi_\alpha^2(m_i)$, the original sensor measurement will not be sampled and transmitted to the fusion center, where one has $\lambda_{i,k} = 0$. Therefore, if $\rho_{i,k} > \chi_\alpha^2(m_i)$, the fusion center can obtain the accurate measurement $z_{i,k}$, So we set $\lambda_{i,k} = 1$.

9.3 Fusion Algorithm with Event-Triggered Mechanism

9.3.1 Kalman Filter with Event-Triggered Mechanism

Theorem 9.1 (The Kalman filter (KF) with event-triggered mechanism) *For system (9.1)–(9.2), the estimation of x_k by the Kalman filter is given by*

$$\begin{cases} \hat{x}_{i,k|k-1} &= A_{k-1}\hat{x}_{i,k-1|k-1} \\ P_{i,k|k-1} &= A_{k-1}P_{i,k-1|k-1}A_{k-1}^T + Q_{k-1} \\ \hat{x}_{i,k|k} &= \hat{x}_{i,k|k-1} + K_{i,k}\tilde{z}_{i,k|k-1} \end{cases} \tag{9.21}$$

$$P_{i,k|k} = \begin{cases} (I - K_{i,k}C_{i,k})P_{i,k|k-1}(I - K_{i,k}C_{i,k})^T + K_{i,k}R_{i,l_i}K_{i,k}^T + K_{i,k} \\ \cdot \{(C_{i,k} - C_{i,l_i}\prod_{p=l_i}^{k-1} A_p^{-1})\Xi_k(C_{i,k} - C_{i,l_i}\prod_{p=l_i}^{k-1} A_p^{-1})^T + C_{i,l_i} \\ \cdot (\sum_{m=l_i}^{k-1}\prod_{p=l_i}^{m} A_p^{-1}Q_m \prod_{p=m+1}^{k-1} A_p^T)(C_{i,k} - C_{i,l_i}\prod_{p=l_i}^{k-1} A_p^{-1})^T + C_{i,k} \\ \cdot (\sum_{m=l_i}^{k-1}\prod_{p=m+1}^{k-1} A_p Q_m \prod_{p=l_i}^{m} A_p^{-T})C_{i,l_i}^T\}K_{i,k}^T - (I - K_{i,k}C_{i,k}) \\ \cdot \{P_{i,k|k-1}(C_{i,l_i}\prod_{p=l_i}^{k-1} A_p^{-1} - C_{i,k})^T - \sum_{m=l_i}^{k-1}\prod_{p=m+1}^{k-1} A_p Q_m \prod_{p=l_i}^{m} A_p^{-T} \\ \cdot C_{i,l_i}^T\}K_{i,k}^T - K_{i,k}\{(C_{i,l_i}\prod_{p=l_i}^{k-1} A_p^{-1} - C_{i,k})P_{i,k|k-1} \\ - C_{i,l_i}\sum_{m=l_i}^{k-1}\prod_{p=l_i}^{m} A_p^{-1}Q_m \prod_{p=m+1}^{k-1} A_p^T\}(I - K_{i,k}C_{i,k})^T \\ \qquad\qquad , \text{if } \lambda_{i,k} = 0 \\ P_{i,k|k-1} - K_{i,k}P_{\tilde{z}_i,k|k-1}K_{i,k}^T \qquad , \text{if } \lambda_{i,k} = 1 \end{cases} \tag{9.22}$$

$$K_{i,k} = \begin{cases} [P_{i,k|k-1}\prod_{p=l_i}^{k-1} A_p^{-T} - \sum_{m=l_i}^{k-1}\prod_{p=m+1}^{k-1} A_p Q_m \prod_{p=l_i}^{m} A_p^{-T}C_{i,l_i}^T]P_{\tilde{z}_i,k|k-1}^{-1} \\ \qquad\qquad , \text{if } \lambda_{i,k} = 0 \\ P_{i,k|k-1}C_{i,k}^T[C_{i,k}P_{i,k|k-1}C_{i,k}^T + R_{i,k}]^{-1} \qquad , \text{if } \lambda_{i,k} = 1 \end{cases} \tag{9.23}$$

where

$$\tilde{z}_{i,k|k-1} = \begin{cases} C_{i,k}\tilde{x}_{i,k|k-1} + v_{i,k} + \bar{z}_{i,k-1} - z_{i,k}, & \text{if } \lambda_{i,k} = 0 \\ C_{i,k}\tilde{x}_{i,k|k-1} + v_{i,k}, & \text{if } \lambda_{i,k} = 1 \end{cases} \tag{9.24}$$

and

$$P_{\tilde{z}_i,k|k-1} = \begin{cases} -C_{i,k}P_{i,k|k-1}C_{i,k}^T + R_{i,l_i} + (C_{i,k} - C_{i,l_i}\prod_{p=l_i}^{k-1} A_p^{-1})\Xi_k[C_{i,k} \\ - C_{i,l_i}\prod_{p=l_i}^{k-1} A_p^{-1}]^T - C_{i,l_i}(\sum_{m=l_i}^{k-1}\prod_{p=l_i}^{m} A_p^{-1}Q_m \prod_{p=l_i}^{m} A_p^{-T})C_{i,l_i}^T \\ + C_{i,k}P_{i,k|k-1}\prod_{p=l_i}^{k-1} A_p^{-T}C_{i,l_i}^T + C_{i,l_i}\prod_{p=l_i}^{k-1} A_p^{-1}P_{i,k|k-1}C_{i,k}^T, \\ \qquad\qquad \text{if } \lambda_{i,k} = 0 \\ C_{i,k}P_{i,k|k-1}C_{i,k}^T + R_{i,k}, \qquad \text{if } \lambda_{i,k} = 1 \end{cases} \tag{9.25}$$

Proof. After $z_{i,k}$ is obtained, the sensor decides whether it shall be sampled for an estimator for further processing. Suppose $\lambda_{i,k} = 1$ or 0 is the event-triggered decision variable to decide whether to send $z_{i,k}$.

(1) For $\lambda_{i,k} = 1$, the filter estimation is the same with the time-driven schedule, and we have

$$\begin{cases} \hat{x}_{i,k|k-1} & = A_{k-1}\hat{x}_{i,k-1|k-1} \\ P_{i,k|k-1} & = A_{k-1}P_{i,k-1|k-1}A_{k-1}^T + Q_{k-1} \\ \hat{x}_{i,k|k} & = \hat{x}_{i,k|k-1} + K_{i,k}\tilde{z}_{i,k|k-1} \\ P_{i,k|k} & = P_{i,k|k-1} - K_{i,k}P_{\tilde{z}_i,k|k-1}K_{i,k}^T \\ K_{i,k} & = P_{i,k|k-1}C_{i,k}^T(C_{i,k}P_{i,k|k-1}C_{i,k}^T + R_{i,k})^{-1} \\ \tilde{z}_{i,k|k-1} & = C_{i,k}\tilde{x}_{i,k|k-1} + v_{i,k} \\ P_{\tilde{z}_i,k|k-1} & = C_{i,k}P_{i,k|k-1}C_{i,k}^T + R_{i,k} \end{cases} \tag{9.26}$$

(2) For $\lambda_{i,k} = 0$, from formula (9.12), we have

$$\begin{aligned} \bar{z}_{i,k} &= \bar{z}_{i,k-1} \\ &= \bar{z}_{i,k-1} - z_{i,k} + z_{i,k} \end{aligned} \tag{9.27}$$

The state prediction of x_k is

$$\hat{x}_{i,k|k-1} = A_{k-1}\hat{x}_{i,k-1|k-1} \tag{9.28}$$

and

$$\hat{z}_{i,k|k-1} = C_{i,k}\hat{x}_{i,k|k-1} \tag{9.29}$$

The residual is

$$\begin{aligned} \tilde{z}_{i,k|k-1} &= \bar{z}_{i,k} - \hat{z}_{i,k|k-1} \\ &= z_{i,k} - \hat{z}_{i,k|k-1} + \bar{z}_{i,k-1} - z_{i,k} \end{aligned} \tag{9.30}$$

The state estimation of x_k is

$$\hat{x}_{i,k|k} = \hat{x}_{i,k|k-1} + K_{i,k}\tilde{z}_{i,k|k-1} \tag{9.31}$$

The state estimation error is computed by

$$\begin{aligned} \tilde{x}_{i,k|k} &= x_k - \hat{x}_{i,k|k} \\ &= \tilde{x}_{i,k|k-1} - K_{i,k}\tilde{z}_{i,k|k-1} \end{aligned} \tag{9.32}$$

The state prediction error is defined by

$$\begin{aligned} \tilde{x}_{i,k|k-1} &= x_k - \hat{x}_{i,k|k-1} \\ &= A_{k-1}\tilde{x}_{i,k-1|k-1} + w_{k-1} \end{aligned} \tag{9.33}$$

The state prediction error covariance is computed by

$$\begin{aligned} P_{i,k|k-1} &= E\{\tilde{x}_{i,k|k-1}(\tilde{x}_{i,k|k-1})^T\} \\ &= A_{k-1}P_{i,k-1|k-1}A_{k-1}^T + Q_{k-1} \end{aligned} \tag{9.34}$$

Substituting Eq. (9.29) into Eq. (9.30), we have

$$\begin{aligned} \tilde{z}_{i,k|k-1} &= z_{i,k} - C_{i,k}\hat{x}_{i,k|k-1} + \bar{z}_{i,k-1} - z_{i,k} \\ &= C_{i,k}\tilde{x}_{i,k|k-1} + v_{i,k} + \bar{z}_{i,k-1} - z_{i,k} \end{aligned} \tag{9.35}$$

and the state estimation error is rewritten as

$$\begin{aligned} \tilde{x}_{i,k|k} &= \tilde{x}_{i,k|k-1} - K_{i,k}(C_{i,k}\tilde{x}_{i,k|k-1} + v_{i,k} + \bar{z}_{i,k-1} - z_{i,k}) \\ &= (I - K_{i,k}C_{i,k})\tilde{x}_{i,k|k-1} - K_{i,k}v_{i,k} - K_{i,k}(\bar{z}_{i,k-1} - z_{i,k}) \end{aligned} \tag{9.36}$$

Because

$$E\{\tilde{x}_{i,k|k-1}v_{i,k}^T\} = 0 \tag{9.37}$$

$$E\{v_{i,k}(\bar{z}_{i,k-1} - z_{i,k})^T\} = -R_{i,k} \tag{9.38}$$

and

$$\begin{aligned} & E\{\tilde{x}_{i,k|k-1}(\bar{z}_{i,k-1} - z_{i,k})^T\} \\ &= E\{(A_{k-1}\tilde{x}_{i,k-1|k-1} + w_{k-1}) \\ &\quad \cdot [(C_{i,l_i}\prod_{p=l_i}^{k-1} A_p^{-1} - C_{i,k})x_k - C_{i,l_i}(\sum_{m=l_i}^{k-1}\prod_{p=l_i}^{m} A_p^{-1}w_m) - v_{i,k} + v_{i,l_i}]^T\} \\ &\approx P_{i,k|k-1}(C_{i,l_i}\prod_{p=l_i}^{k-1} A_p^{-1} - C_{i,k})^T - \sum_{m=l_i}^{k-1}\prod_{p=m+1}^{k-1} A_p Q_m \prod_{p=l_i}^{m} A_p^{-T}C_{i,l_i}^T \end{aligned} \tag{9.39}$$

We obtain the state estimation error covariance matrix as

$$\begin{aligned} & P_{i,k|k} = E\{\tilde{x}_{i,k|k}\tilde{x}_{i,k|k}^T\} \\ &= E\{[(I - K_{i,k}C_{i,k})\tilde{x}_{i,k|k-1} - K_{i,k}v_{i,k} - K_{i,k}(\bar{z}_{i,k-1} - z_{i,k})] \\ &\quad \cdot [(I - K_{i,k}C_{i,k})\tilde{x}_{i,k|k-1} - K_{i,k}v_{i,k} - K_{i,k}(\bar{z}_{i,k-1} - z_{i,k})]^T\} \\ &= (I - K_{i,k}C_{i,k})P_{i,k|k-1}(I - K_{i,k}C_{i,k})^T + K_{i,k}R_{i,k}K_{i,k}^T - 2K_{i,k}R_{i,k}K_{i,k}^T \\ &\quad + K_{i,k}\{(C_{i,k} - C_{i,l_i}\prod_{p=l_i}^{k-1} A_p^{-1})\Xi_k(C_{i,k} - C_{i,l_i}\prod_{p=l_i}^{k-1} A_p^{-1})^T + C_{i,l_i} \end{aligned}$$

$$
\begin{aligned}
&\cdot(\sum_{m=l_i}^{k-1}\prod_{p=l_i}^{m}A_p^{-1}Q_m\prod_{p=m+1}^{k-1}A_p^T)(C_{i,k}-C_{i,l_i}\prod_{p=l_i}^{k-1}A_p^{-1})^T+C_{i,k}\\
&\cdot(\sum_{m=l_i}^{k-1}\prod_{p=m+1}^{k-1}A_pQ_m\prod_{p=l_i}^{m}A_p^{-T})C_{i,l_i}^T+R_{i,k}+R_{i,l_i}\}K_{i,k}^T-(I-K_{i,k}C_{i,k})\\
&\cdot\{P_{i,k|k-1}(C_{i,l_i}\prod_{p=l_i}^{k-1}A_p^{-1}-C_{i,k})^T-\sum_{m=l_i}^{k-1}\prod_{p=m+1}^{k-1}A_pQ_m\prod_{p=l_i}^{m}A_p^{-T}C_{i,l_i}^T\}K_{i,k}^T\\
&-K_{i,k}\{(C_{i,l_i}\prod_{p=l_i}^{k-1}A_p^{-1}-C_{i,k})P_{i,k|k-1}-C_{i,l_i}\sum_{m=l_i}^{k-1}\prod_{p=l_i}^{m}A_p^{-1}Q_m\prod_{p=m+1}^{k-1}A_p^T\}\\
&\cdot(I-K_{i,k}C_{i,k})^T\\
=\;&(I-K_{i,k}C_{i,k})P_{i,k|k-1}(I-K_{i,k}C_{i,k})^T+K_{i,k}R_{i,l_i}K_{i,k}^T+K_{i,k}\{(C_{i,k}\\
&-C_{i,l_i}\prod_{p=l_i}^{k-1}A_p^{-1})\Xi_k(C_{i,k}-C_{i,l_i}\prod_{p=l_i}^{k-1}A_p^{-1})^T+C_{i,l_i}(\sum_{m=l_i}^{k-1}\prod_{p=l_i}^{m}A_p^{-1}Q_m\\
&\cdot\prod_{p=m+1}^{k-1}A_p^T)(C_{i,k}-C_{i,l_i}\prod_{p=l_i}^{k-1}A_p^{-1})^T+C_{i,k}(\sum_{m=l_i}^{k-1}\prod_{p=m+1}^{k-1}A_pQ_m\prod_{p=l_i}^{m}A_p^{-T})\\
&\cdot C_{i,l_i}^T\}K_{i,k}^T-(I-K_{i,k}C_{i,k})\{P_{i,k|k-1}(C_{i,l_i}\prod_{p=l_i}^{k-1}A_p^{-1}-C_{i,k})^T\\
&-\sum_{m=l_i}^{k-1}\prod_{p=m+1}^{k-1}A_pQ_m\prod_{p=l_i}^{m}A_p^{-T}C_{i,l_i}^T\}K_{i,k}^T-K_{i,k}\{(C_{i,l_i}\prod_{p=l_i}^{k-1}A_p^{-1}-C_{i,k})\\
&\cdot P_{i,k|k-1}-C_{i,l_i}\sum_{m=l_i}^{k-1}\prod_{p=l_i}^{m}A_p^{-1}Q_m\prod_{p=m+1}^{k-1}A_p^T\}(I-K_{i,k}C_{i,k})^T
\end{aligned}
\tag{9.40}
$$

where

$$
\begin{aligned}
K_{i,k}&=E\{\tilde{x}_{i,k|k-1}\tilde{z}_{i,k|k-1}^T\}P_{\tilde{z}_i,k|k-1}^{-1}\\
&=E\{\tilde{x}_{i,k|k-1}(C_{i,k}\tilde{x}_{i,k|k-1}+v_{i,k}+\bar{z}_{i,k-1}-z_{i,k})^T\}P_{\tilde{z}_i,k|k-1}^{-1}\\
&=(P_{i,k|k-1}\prod_{p=l_i}^{k-1}A_p^{-T}C_{i,l_i}^T-\sum_{m=l_i}^{k-1}\prod_{p=m+1}^{k-1}A_pQ_m\prod_{p=l_i}^{m}A_p^{-T}C_{i,l_i}^T)P_{\tilde{z}_i,k|k-1}^{-1}
\end{aligned}
\tag{9.41}
$$

and

$$
\begin{aligned}
&P_{\tilde{z}_i,k|k-1}=E\{\tilde{z}_{i,k|k-1}\tilde{z}_{i,k|k-1}^T\}\\
=\;&E\{(C_{i,k}\tilde{x}_{i,k|k-1}+v_{i,k}+\bar{z}_{i,k-1}-z_{i,k})(C_{i,k}\tilde{x}_{i,k|k-1}+v_{i,k}+\bar{z}_{i,k-1}-z_{i,k})^T\}
\end{aligned}
$$

$$
\begin{aligned}
&= C_{i,k}P_{i,k|k-1}C_{i,k}^T + R_{i,k} + (C_{i,k} - C_{i,l_i}\prod_{p=l_i}^{k-1} A_p^{-1})\Xi_k(C_{i,k} - C_{i,l_i}\prod_{p=l_i}^{k-1} A_p^{-1})^T \\
&+ C_{i,l_i}(\sum_{m=l_i}^{k-1}\prod_{p=l_i}^{m} A_p^{-1}Q_m \prod_{p=m+1}^{k-1} A_p^T)(C_{i,k} - C_{i,l_i}\prod_{p=l_i}^{k-1} A_p^{-1})^T \\
&+ C_{i,k}(\sum_{m=l_i}^{k-1}\prod_{p=m+1}^{k-1} A_p Q_m \prod_{p=l_i}^{m} A_p^{-T})C_{i,l_i}^T + R_{i,k} + R_{i,l_i} \\
&+ C_{i,k}P_{i,k|k-1}(C_{i,l_i}\prod_{p=l_i}^{k-1} A_p^{-1})^T - C_{i,k}\sum_{m=l_i}^{k-1}\prod_{p=m+1}^{k-1} A_p Q_m \prod_{p=l_i}^{m} A_p^{-T}C_{i,l_i}^T - R_{i,k} \\
&+ (C_{i,l_i}\prod_{p=l_i}^{k-1} A_p^{-1})P_{i,k|k-1}C_{i,k}^T - C_{i,l_i}\sum_{m=l_i}^{k-1}\prod_{p=l_i}^{m} A_p^{-1}Q_m \prod_{p=m+1}^{k-1} A_p^T C_{i,k}^T - R_{i,k} \\
&= -C_{i,k}P_{i,k|k-1}C_{i,k}^T + R_{i,l_i} + (C_{i,k} - C_{i,l_i}\prod_{p=l_i}^{k-1} A_p^{-1})\Xi_k(C_{i,k} - C_{i,l_i}\prod_{p=l_i}^{k-1} A_p^{-1})^T \\
&- C_{i,l_i}(\sum_{m=l_i}^{k-1}\prod_{p=l_i}^{m} A_p^{-1}Q_m \prod_{p=l_i}^{m} A_p^{-T})C_{i,l_i}^T + C_{i,l_i}\prod_{p=l_i}^{k-1} A_p^{-1}P_{i,k|k-1}C_{i,k}^T \\
&+ C_{i,k}P_{i,k|k-1}\prod_{p=l_i}^{k-1} A_p^{-T}C_{i,l_i}^T
\end{aligned} \tag{9.42}
$$

9.3.2 Distributed Fusion Algorithm in WSNs

For $i = 1, 2, \ldots, N_r$, denote

$$
Y_k^{(i)} = \{y_{j,k}, j = 1, 2, \ldots, i\} \tag{9.43}
$$

$$
Y_i^k = \{y_{i,t}, 0 < t \le k\} \tag{9.44}
$$

$$
\begin{aligned}
Y^{k,i} &= \{y_{j,t}, 0 < t \le k, j = 1, 2, \ldots, i\} \\
&= \{Y_j^k\}_{j=1}^i = \{Y_l^{(i)}\}_{l=1}^k
\end{aligned} \tag{9.45}
$$

then $Y^{k,i}$ denotes the measurements observed by sensors $1, 2, \ldots, i$ up to and including time k. Y_i^k stands for the measurements observed by sensor i during time $(0, k]$. $Y_k^{(i)}$ denotes the measurements sampled by sensors $1, 2, \ldots, i$ at time k. Then, Y^{k,N_r} denotes the measurements observed by all the sensors in $\mathcal{N}_r$ up to and including time k, where N_r denotes the number of the sensors in $\mathcal{N}_r$. The distributed fusion algorithm is given by the following theorem.

Theorem 9.2 (The distributed fusion algorithm (DF) with event-triggered mechanism) *Denote $\hat{x}_{i,k|k}$, $i \in \mathcal{N}_r$ be the local estimations, $\hat{x}_{i,k|k}$ is the state estimation of*

x_k based on data Y_i^k and $P_{i,k|k}$ is the corresponding estimation error covariance. Then, the optimal fusion estimation of $\hat{x}_{d,k|k}^r$ at sensor r is given by

$$\hat{x}_{d,k|k}^r = \sum_{i \in \mathcal{N}_r} \alpha_{i,k} \hat{x}_{i,k|k} \tag{9.46}$$

$$P_{d,k|k}^r = (e^T \Sigma_k^{-1} e)^{-1} \tag{9.47}$$

where the local state estimation of x_k based on data Y_i^k is given by Theorem 9.1, and the optimal matrix weights $\alpha_{i,k}$ are computed by

$$\alpha_k \equiv [\alpha_k^1 \ \alpha_k^2 \ \cdots \ \alpha_k^{N_r}] = (e^T \Sigma_k^{-1} e)^{-1} e^T \Sigma_k^{-1} \tag{9.48}$$

where $\Sigma_k = [P_{ij,k|k}]$, $i, j \in \mathcal{N}_r$ is an $nN_r \times nN_r$ symmetric positive definite matrix, the covariance and cross-covariance matrices as $P_{i,k|k} = P_{ii,k|k} = E\{\tilde{x}_{i,k|k}\tilde{x}_{i,k|k}^T\}$ and $P_{ij,k|k} = E\{\tilde{x}_{i,k|k}\tilde{x}_{j,k|k}^T\}(j \in \mathcal{N}_r, i \neq j)$, respectively. $e = [\underbrace{I_n \ \cdots \ I_n}_{N_r}]^T$, $I_n \in R^{n \times n}$ is an identity matrix. $l = \max\{l_i, l_j\}$, represents the larger value of l_i and l_j. $P_{ij,k|k}, i, j \in \mathcal{N}_r$ are computed by

$$P_{ij,k|k} = \begin{cases} (I - K_{i,k}C_{i,k})P_{ij,k|k-1}(I - K_{j,k}C_{j,k})^T + K_{i,k}\{(C_{i,k} - C_{i,l_i}\prod_{p=l_i}^{k-1} A_p^{-1}) \\ \cdot \Xi_k (C_{j,k} - C_{j,l_j}\prod_{p=l_j}^{k-1} A_p^{-1})^T + C_{i,l_i}(\sum_{m=l}^{k-1}\prod_{p=l}^{m} A_p^{-1} Q_m \prod_{p=m+1}^{k-1} A_p^T) \\ \cdot (C_{j,k} - C_{j,l_j}\prod_{p=l_j}^{k-1} A_p^{-1})^T + C_{i,k}(\sum_{m=l}^{k-1}\prod_{p=m+1}^{k-1} A_p Q_m \prod_{p=l}^{m} A_p^{-T})C_{j,l_j}^T\}K_{j,k}^T \\ - (I - K_{i,k}C_{i,k})\{P_{ij,k|k-1}(C_{j,l_j}\prod_{p=l_j}^{k-1} A_p^{-1} - C_{j,k})^T \\ - \sum_{m=l}^{k-1}\prod_{p=m+1}^{k-1} A_p Q_m \prod_{p=l}^{m} A_p^{-T} C_{j,l_j}^T\}K_{j,k}^T - K_{i,k}\{(C_{i,l_i}\prod_{p=l_i}^{k-1} A_p^{-1} \\ - C_{i,k})P_{ij,k|k-1} - C_{i,l_i}\sum_{m=l}^{k-1}\prod_{p=l}^{m} A_p^{-1} Q_m \prod_{p=m+1}^{k-1} A_p^T\}(I - K_{j,k}C_{j,k})^T \\ \qquad , \text{if } \gamma_{i,k} = 0, \gamma_{j,k} = 0 \\ (I - K_{i,k}C_{i,k})P_{ij,k|k-1}(I - K_{j,k}C_{j,k})^T - (I - K_{i,k}C_{i,k})\{P_{ij,k|k-1} \\ \cdot (C_{j,l_j}\prod_{p=l_j}^{k-1} A_p^{-1} - C_{j,k})^T - \sum_{m=l}^{k-1}\prod_{p=m+1}^{k-1} A_p Q_m \prod_{p=l}^{m} A_p^{-T} C_{j,l_j}^T\}K_{j,k}^T \\ \qquad , \text{if } \gamma_{i,k} = 1, \gamma_{j,k} = 0 \\ (I - K_{i,k}C_{i,k})P_{ij,k|k-1}(I - K_{j,k}C_{j,k})^T - K_{i,k}\{(C_{i,l_i}\prod_{p=l_i}^{k-1} A_p^{-1} \\ - C_{i,k})P_{i,k|k-1} - C_{i,l_i}\sum_{m=l}^{k-1}\prod_{p=l}^{m} A_p^{-1} Q_m \prod_{p=m+1}^{k-1} A_p^T\}(I - K_{j,k}C_{j,k})^T \\ \qquad , \text{if } \gamma_{i,k} = 0, \gamma_{j,k} = 1 \\ (I - K_{i,k}C_{i,k})P_{ij,k|k-1}(I - K_{j,k}C_{j,k})^T \quad , \text{if } \gamma_{i,k} = 1, \gamma_{j,k} = 1 \end{cases} \tag{9.49}$$

where

$$P_{ij,k|k-1} = A_{k-1} P_{ij,k-1|k-1} A_{k-1}^T + Q_{k-1} \tag{9.50}$$

and

$$\Xi_k = A_{k-1} \Xi_{k-1} A_{k-1}^T + Q_{k-1} \tag{9.51}$$

the triggering condition is

$$\lambda_{i,k} = \begin{cases} 0, & \text{if } \rho_{i,k} \le \chi_\alpha^2(m_i) \\ 1, & \text{otherwise} \end{cases} \tag{9.52}$$

It can be proven that the covariance matrix of the fusion estimation error $P_{d,k|k}^r$ *meets*

$$P_{d,k|k}^r \le P_{i,k|k},\ i \in \mathcal{N}_r. \tag{9.53}$$

Which means the fusion estimation is effective in the sense of minimizing the estimation error covariance.

Proof. In the event-triggered mechanism, if sensors i and j both obtain observations at time k, that is, $\lambda_{i,k} = \lambda_{j,k} = 1$, then the estimation error covariance between local estimations $\hat{x}_{i,k|k}$ and $\hat{x}_{j,k|k}$ could be calculated by

$$\begin{aligned} & P_{ij,k|k} \\ &= E\{(x_k - \hat{x}_{i,k|k})(x_k - \hat{x}_{j,k|k})^T\} \\ &= E\{[(I - K_{i,k}C_{i,k})\tilde{x}_{i,k|k-1} - K_{i,k}v_{i,k}][(I - K_{j,k}C_{j,k})\tilde{x}_{j,k|k-1} - K_{j,k}v_{j,k}]^T\} \\ &= (I - K_{i,k}C_{i,k})P_{ij,k|k-1}(I - K_{j,k}C_{j,k})^T \end{aligned} \tag{9.54}$$

If the measurement of sensor i is observed at time k, but sensor j does not, that is, $\lambda_{i,k} = 1$, $\lambda_{j,k} = 0$, one can calculate the estimation error covariance between local estimations $\hat{x}_{i,k|k}$ and

$$\begin{aligned} P_{ij,k|k} &= E\{(x_k - \hat{x}_{i,k|k})(x_k - \hat{x}_{j,k|k})^T\} \\ &= E\{[(I - K_{i,k}C_{i,k})\tilde{x}_{i,k|k-1} - K_{i,k}v_{i,k}] \\ &\quad \cdot [(I - K_{j,k}C_{j,k})\tilde{x}_{j,k|k-1} - K_{j,k}v_{j,k} - K_{j,k}(z_{j,k-1} - z_{j,k})]^T\} \\ &= (I - K_{i,k}C_{i,k})P_{ij,k|k-1}(I - K_{j,k}C_{j,k})^T - (I - K_{i,k}C_{i,k})\{P_{ij,k|k-1} \\ &\quad \cdot (C_{j,l_j}\prod_{p=l_j}^{k-1} A_p^{-1} - C_{j,k})^T - \sum_{m=l}^{k-1}\prod_{p=m+1}^{k-1} A_p Q_m \prod_{p=l}^{m} A_p^{-T} C_{j,l_j}^T\} K_{j,k}^T \end{aligned} \tag{9.55}$$

If the measurement of sensor j is observed at time k, but sensor i does not, that is, $\lambda_{i,k} = 0$, $\lambda_{j,k} = 1$, one can calculate the estimation error covariance between local estimations $\hat{x}_{i,k|k}$ and $\hat{x}_{j,k|k}$ by

$$\begin{aligned}
P_{ij,k|k} &= E\{(x_k - \hat{x}_{i,k|k})(x_k - \hat{x}_{j,k|k})^T\} \\
&= E\{[(I - K_{i,k}C_{i,k})\tilde{x}_{i,k|k-1} - K_{i,k}v_{i,k} - K_{i,k}(\bar{z}_{i,k-1} - z_{i,k})] \\
&\quad \cdot [(I - K_{j,k}C_{j,k})\tilde{x}_{j,k|k-1} - K_{j,k}v_{j,k}]^T\} \\
&= (I - K_{i,k}C_{i,k})P_{ij,k|k-1}(I - K_{j,k}C_{j,k})^T - K_{i,k}\{(C_{i,l_i}\prod_{p=l_i}^{k-1} A_p^{-1} - C_{i,k}) \\
&\quad \cdot P_{ij,k|k-1} - C_{i,l_i}\sum_{m=l}^{k-1}\prod_{p=l}^{m} A_p^{-1}Q_m \prod_{p=m+1}^{k-1} A_p^T\}(I - K_{j,k}C_{j,k})^T
\end{aligned} \tag{9.56}$$

If the measurements of sensor i and j are not obtained at time k, that is, $\lambda_{i,k} = \lambda_{j,k} = 0$, the estimation error covariance between local estimations $\hat{x}_{i,k|k}$ and $\hat{x}_{j,k|k}$ could be calculated by

$$\begin{aligned}
P_{ij,k|k} &= E\{(x_k - \hat{x}_{i,k|k})(x_k - \hat{x}_{j,k|k})^T\} \\
&= E\{[(I - K_{i,k}C_{i,k})\tilde{x}_{i,k|k-1} - K_{i,k}v_{i,k} - K_{i,k}(\bar{z}_{i,k-1} - z_{i,k})] \\
&\quad \cdot [(I - K_{j,k}C_{j,k})\tilde{x}_{j,k|k-1} - K_{j,k}v_{j,k} - K_{j,k}(z_{j,k-1} - z_{j,k})]^T\} \\
&= (I - K_{i,k}C_{i,k})P_{ij,k|k-1}(I - K_{j,k}C_{j,k})^T + K_{i,k}\{(C_{i,k} - C_{i,l_i}\prod_{p=l_i}^{k-1} A_p^{-1}) \\
&\quad \cdot \Xi_k(C_{j,k} - C_{j,l_j}\prod_{p=l_j}^{k-1} A_p^{-1})^T + C_{i,l_i}(\sum_{m=l}^{k-1}\prod_{p=l}^{m} A_p^{-1}Q_m \prod_{p=m+1}^{k-1} A_p^T)(C_{j,k} \\
&\quad - C_{j,l_j}\prod_{p=l_j}^{k-1} A_p^{-1})^T + C_{i,k}(\sum_{m=l}^{k-1}\prod_{p=m+1}^{k-1} A_p Q_m \prod_{p=l}^{m} A_p^{-T})C_{j,l_j}^T\}K_{j,k}^T \\
&\quad - (I - K_{i,k}C_{i,k})\{P_{ij,k|k-1}(C_{j,l_j}\prod_{p=l_j}^{k-1} A_p^{-1} - C_{j,k})^T - \sum_{m=l}^{k-1}\prod_{p=m+1}^{k-1} A_p Q_m \\
&\quad \cdot \prod_{p=l}^{m} A_p^{-T}C_{j,l_j}^T\}K_{j,k}^T - K_{i,k}\{(C_{i,l_i}\prod_{p=l_i}^{k-1} A_p^{-1} - C_{i,k})P_{ij,k|k-1} \\
&\quad - C_{i,l_i}\sum_{m=l}^{k-1}\prod_{p=l}^{m} A_p^{-1}Q_m \prod_{p=m+1}^{k-1} A_p^T\}(I - K_{j,k}C_{j,k})^T
\end{aligned}$$

Let

$$F_x = (y_k - ex_k)^T \Sigma_k^{-1}(y_k - ex_k) \tag{9.57}$$

where $y_k = [\hat{x}^1_{k|k}\ \hat{x}^2_{k|k}\ \cdots\ \hat{x}^{N_r}_{k|k}]^T$, $e = [\underbrace{I_n\ \cdots\ I_n}_{N_r}]^T$, and $\Sigma = [P_{ij,k|k}]$ is an $nN_r \times nN_r$ symmetrical positive definite matrix.

Evidently, F_x is a quadratic form of x_k and thus a convex function of x_k for each $k = 1, 2, \ldots$. So the minimum of F_x exists and can be found by $\frac{dF_x}{dx} = 0$, from which we obtain

$$x_k = (e^T \Sigma_k^{-1} e)^{-1} e^T \Sigma_k^{-1} y_k \tag{9.58}$$

We have the optimal distributed estimation from the unbiasedness property

$$\hat{x}^r_{d,k|k} = \alpha_k y_k = \sum_{i=1}^{N_r} \alpha_{i,k} \hat{x}_{i,k|k} \tag{9.59}$$

where $\alpha_k = [\alpha^1_k\ \alpha^2_k\ \cdots\ \alpha^{N_r}_k] = (e^T \Sigma_k^{-1} e)^{-1} e^T \Sigma_k^{-1}$ is a $bn \times bnN_r$ matrix.

The error covariance of the optimal distributed estimation is given by

$$\begin{aligned} P^r_{d,k|k} &= \sum_{i \in \mathcal{N}_r} \sum_{j \in \mathcal{N}_r} \alpha_{i,k} P_{ij,k|k} (\alpha_{j,k})^T \\ &= \alpha_k \Sigma_k \alpha_k^T = (e^T \Sigma_k^{-1} e)^{-1} \end{aligned} \tag{9.60}$$

Notice that by setting $\alpha_{i,k} = I_n$ and $\alpha_{j,k} = 0\ \ (j \in \mathcal{N}_r,\ j \neq i)$, we have $P^r_{d,k|k} = P_{i,k|k}$. Thence, $P^r_{d,k|k} \leq P_{i,k|k}$, $i \in \mathcal{N}_r$ is obtained.

9.4 Numerical Example

An example is used to illustrate the effectiveness of the presented algorithm in this section. A wireless sensor network with 12 sensor nodes is deployed to monitor the target, and the topology of the WSNs is shown in Fig. 9.1 [21, 23].

$$x_{k+1} = \begin{bmatrix} 1 & T_s \\ 0 & 1 \end{bmatrix} x_k + \Gamma_k \xi_k \tag{9.61}$$

$$z_{i,k} = C_i x_k + v_{i,k},\ k = 1, 2, \ldots, L \tag{9.62}$$

where $L = 1000$ is the length of the signal x to be estimated. $T_s = 0.01$ is the sampling period. The state $x_k = [s_k\ \dot{s}_k]^T$, where s_k and $\dot{s}_k$ are the position and velocity of the target at time kT_s, respectively. $\Gamma_k = [T_s\ \ 1]$ is the noise transition matrix. $\xi_k \in R$ is the system noise, assumed to be white and Gaussian distributed with zero mean and variance σ^2_ξ. $z_{i,k}$ $(i = 1, 2, \ldots 12)$ are the observation vector of the twelve sensors, which observe the position and velocity, respectively, i.e.,

$$\begin{cases} C_1 = [1\ 0], & C_2 = [0\ 0.8], & C_3 = [0.7\ 0] \\ C_4 = [0.6\ 0], & C_5 = [0\ 0.5], & C_6 = [0\ 0.4] \\ C_7 = [0.3\ 0], & C_8 = [0.2\ 0], & C_9 = [0\ 1] \\ C_{10} = [0.8\ 0], & C_{11} = [0\ 0.6], & C_{12} = [0.7\ 0] \end{cases} \tag{9.63}$$

$v_{i,k}$ $(i = 1, 2, \ldots, 12)$ are the measurement noises of sensor i. The initial values are $\bar{x}_0 = [1\ 1]^T$, $\bar{P}_0 = I_2$.

From Eq. (9.61), we have $Q_k = \Gamma_k \sigma_\xi^2 \Gamma_k^T$, which is the system noise covariance corresponding to $w_k = \Gamma_k \xi_k$, Denote $\sigma_\xi^2 = 0.01$ and the measurement noise covariance is computed by

$$\begin{cases} R_1 = 0.4, & R_2 = 0.7, & R_3 = 0.4 \\ R_4 = 0.4, & R_5 = 0.3, & R_6 = 0.2 \\ R_7 = 0.3, & R_8 = 0.3, & R_9 = 0.5 \\ R_{10} = 0.4, & R_{11} = 0.3, & R_{12} = 0.1 \end{cases} \tag{9.64}$$

For simplicity, let the threshold of event-triggered $\chi_\alpha^2(m_i) = \chi^2$ for $i = 1, 2, \ldots, 12$. In order to prove the impact of threshold χ^2 on the estimation performance, the value of χ^2 is changing within the set $\chi^2 \in \{0, 0.45, 0.6\}$, where $\chi^2 = 0$ means that the transmission event is always in the triggered state, and the estimator can receive all the measurements of the corresponding sensors at each instant, that is time-triggered. Therefore, the larger the χ^2, the less the information received by the estimator.

Next, we will fuse the information from the 12 sensors to give the state estimation of x_k and compare the differences between the estimation results obtained by different estimation algorithms.

We select 1000 sampling time for 100 Monte Carlo simulations and the effectiveness of the presented algorithm is observed. The simulation results are shown in Figs. 9.2, 9.3, 9.4 and 9.5 and Table 9.1.

According to the chi-square distribution table, when $\chi^2 = 0, 0.45, 0.6$, the corresponding communication rates are 100%, 50% and 44%, which is shown in the first row and the second row of Table 9.1. Figure 9.2 and the third row of Table 9.1 show numbers of sensor transmission/sampling under different χ^2 in simulation, which is basically consistent with the theoretical results. The results show that our event-triggered mechanism is reasonable and can give the communication rate under different triggering conditions directly and accurately.

Figures 9.3 and 9.4 show the statistical simulation curves of Root Mean Square Errors (RMSEs) of Kalman filter (KF) and the proposed event-triggered distributed fusion algorithm (DF). The black dashed lines denote different algorithms with $\chi^2 = 0.6$, purple dotted lines denote different algorithms with $\chi^2 = 0.45$ and blue solid lines denote different algorithms with $\chi^2 = 0$. It can be seen that the higher the communication/sampling rate, the lower the estimation errors, so the better the estimation results.

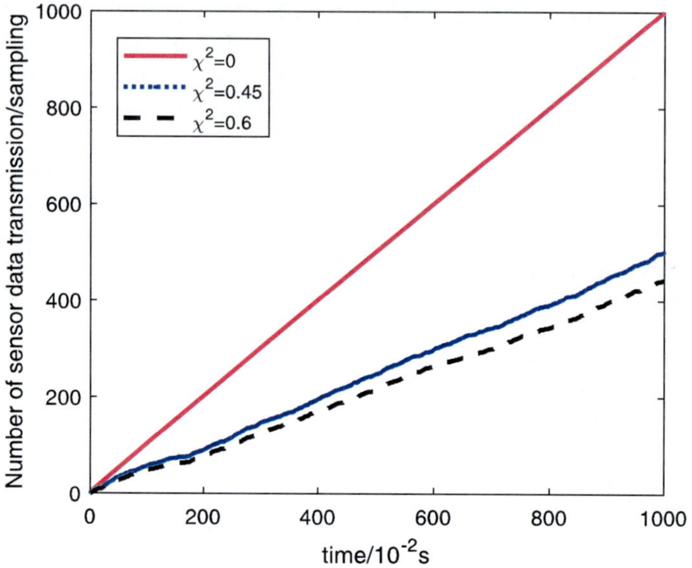

Fig. 9.2 Numbers of transmission/sampling under different χ^2 in 1000 sampling time

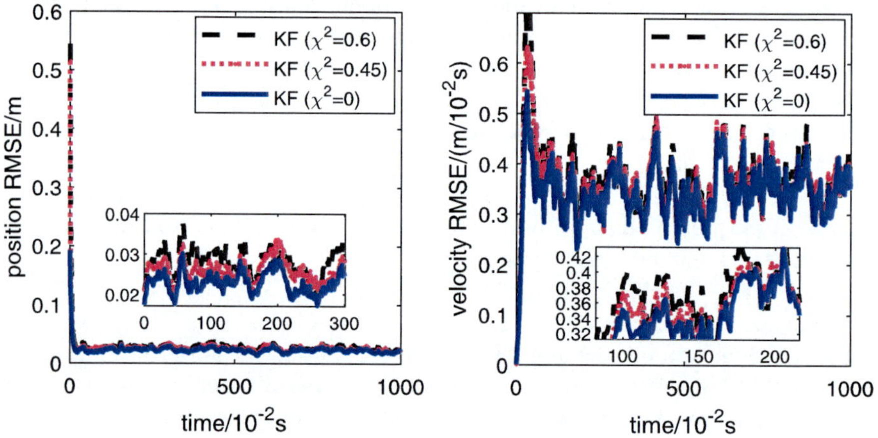

Fig. 9.3 Statistical simulation curves of RMSEs of Kalman filter

The statistical simulation curves of RMSEs of the Kalman filter algorithm (purple dashed line) and the proposed DF algorithm (blue solid line) with thresholds $\chi^2 = 0$ is shown in Fig. 9.5. It can be seen that the RMSE curves of the DF algorithm are much lower than the KF algorithm, which shows that DF algorithm is effective. Time-averaged RMSEs of DF algorithm and KF algorithm under different χ^2 are shown in the third and the fourth rows of Table 9.1. We can draw the same conclusion.

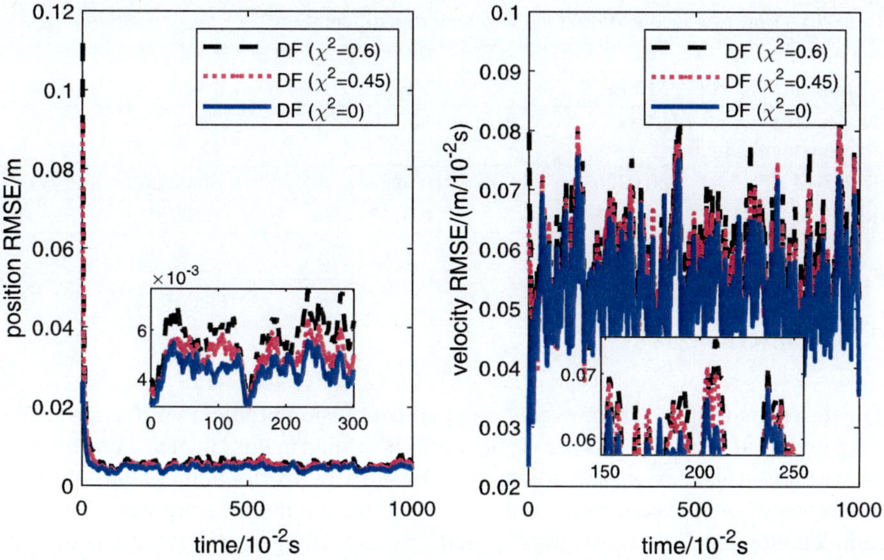

Fig. 9.4 Statistical simulation curves of RMSEs of distributed fusion algorithm

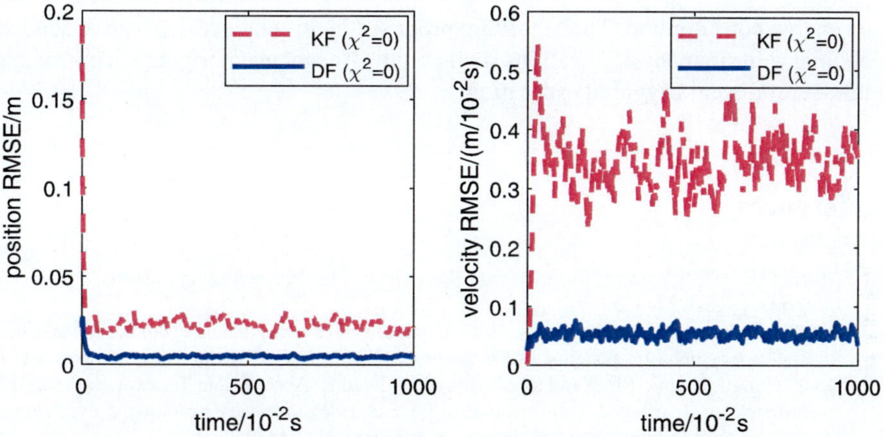

Fig. 9.5 Statistical simulation curves of RMSEs of Kalman filter and distributed fusion

In summary, the simulations in this section show that our event-triggered mechanism is reasonable and the DF algorithm is effective. By adopting the event-triggered transmission/sampling and only fuse the measurements in the neighborhood of the fusion center. Compared with the traditionally time-triggered data fusion algorithms, the proposed algorithm can significantly reduce the communication requirement of the system and optimize the estimation performance.

Table 9.1 Time averaged RMSEs for algorithms with different χ^2

χ^2	0	0.45	0.6
Communication rate	100%	50%	44%
Numbers of transmission/sampling	1000	501	444
RMSE of KF	0.0226	0.0263	0.0293
RMSE of DF	0.0045	0.0055	0.0065

9.5 Conclusions

The distributed multi-sensor event-triggered data fusion problem with a new event-triggered strategy in wireless sensor networks is studied in this chapter. The threshold of the event-trigger is set according to the chi-square distribution composed by the difference of the measurement of the current time and the measurement of the last sampled moment. This novel trigger method can easily give the specific communication rate according to the chi-square distribution table. Only the measurements in the neighborhood of the fusion center are fused in the distributed fusion algorithm for wireless sensor networks, so that it can achieve the required state estimation with less energy consumption. The algorithm proposed in this chapter is effective and has potential value in many applications such as multi-sensor event triggered control and multi-sensor event triggered estimation.

References

1. Årzén, K.E. 1999. A simple event-based PID controller. In *Proceedings of the14th IFAC World Congress*, pages 423–428.
2. Åström, K. J., and B. Bernhardsson. 1999. Comparison of periodic and event based sampling for first-order stochastic systems. In *Proceedings of the14th IFAC World Congress*, pages 1–7.
3. Dong, Hongli, Xianye Bu, Nan Hou, Yurong Liu, Fuad E. Alsaadi, and Tasawar Hayat. 2017. Event-triggered distributed state estimation for a class of time-varying systems over sensor networks with redundant channels. *Information Fusion* 36: 243–250.
4. Han, Duo, Yilin Mo, Junfeng Wu, Bruno Sinopoli, and Ling Shi. 2013. Stochastic event-triggered sensor scheduling for remote state estimation. In *in Proc. 52nd IEEE Conf. Decision and Control*, pages 6079–6084.
5. Han, Duo, Yilin Mo, Junfeng Wu, Sean Weerakkody, Bruno Sinopoli, and Ling Shi. 2015. Stochastic event-triggered sensor schedule for remote state estimation. *IEEE Transactions on Automatic Control* 60 (10): 2662–2675.
6. Ho, Y.C., Xiren Cao, and Christos Cassandras. 1983. Infinitesimal and finite perturbation analysis for queuing networks. *Automatica* 19 (4): 439–445.
7. Jun, Hu, Zidong Wang, Jinling Liang, and Hongli Dong. 2015. Event-triggered distributed state estimation with randomly occurring uncertainties and nonlinearities over sensor networks: a delay-fractioning approach. *Journal of the Franklin Institute* 352 (9): 3750–3763.

8. Liu, Qinyuan, Zidong Wang, Xiao He, and Donghua Zhou. 2015. Event-based distributed filtering with stochastic measurement fading. *IEEE Transactions on Industrial Informatics* 11 (6): 1643–1652.
9. Liu, Qinyuan, Zidong Wang, Xiao He, and Donghua Zhou. 2015. Event-based recursive distributed filtering over wireless sensor networks. *IEEE Transactions on Automatic Control* 60 (9): 2470–2475.
10. Miskowicz, M. 2006. Send-on-delta concept: an event-based data reporting strategy. *sensors* 6 (1): 49–63.
11. Shi, Dawei, Tongwen Chen, and Ling Shi. 2015. On set-valued Kalman filtering and its application to event-based state estimation. *IEEE Transactions on Automatic Control* 60 (5): 1275–1290.
12. Shi, Dawei, Robert J Elliott, and Tongwen Chen. 2016. Event-based state estimation of discrete-state hidden Markov models. *Automatica* 65: 12–26.
13. Trimpe, S., and R. D'Andrea. 2014. Event-based state estimation with variance-based triggering. *IEEE Transactions on Automatic Control* 59 (12): 3266–3281.
14. Trimpe, Sebastian. 2015. Distributed event-based state estimation. *IET Control Theory and Application* 59 (12): 3266–3281.
15. Wang, Licheng, Zidong Wang, Tingwen Huang, and Guoliang Wei. 2016. An event-triggered approach to state estimation for a class of complex networks with mixed time delays and nonlinearities. *IEEE Transactions on Cybernetics* 46 (11): 2497–2508.
16. Wu, Junfeng, Qingshan Jia, Karl Henrik Johansson, and Ling Shi. 2013. Event-based sensor data scheduling: Trade-off between communication rate and estimation quality. *IEEE Transactions on Automatic Control* 58 (4): 1041–1046.
17. Wu, Zhengguang, Yong Xu, Yajun Pan, Peng Shi, and Qian Wang. 2017. Event-triggered pinning control for consensus of multi-agent systems with quantized information. *IEEE Transactions on Systems, Man, and Cybernetic: Systems* 48 (11): 1929–1938.
18. Wu, Zhengguang, Yong Xu, Yajun Pan, Housheng Su, and Yang Tang. 2017. Event-triggered control for consensus problem in multi-agent systems with quantized relative state measurement and external disturbance. *IEEE Transactions on Circuit and Systems I: Regular paper* 65 (7): 2232–2242.
19. Wu, Zongze, Yuanqing Wu, Zhengguang Wu, and Jianquan Lu. 2017. Event-based synchronization of heterogeneous complex networks subject to transmission delays. *IEEE Transactions on Systems, Man and Cybernetics: Systems* 48 (12): 2126–2134.
20. Yan, Huaicheng, Xu Xiaoli, Hao Zhang, and Fuwen Yang. 2017. Distributed event-triggered H-infinity state estimation for T-S fuzzy systems over filtering networks. *Journal of the Franklin Institute* 354 (9): 3760–3779.
21. Yan, Liping, Lu Jiang, Yuanqing Xia, and Mengyin Fu. 2016. State estimation and data fusion for multirate sensor networks. *International Journal of Adaptive Control and Signal Processing* 30 (1): 3–15.
22. Zhang, Cui, and Yingmin Jia. 2017. Distributed Kalman consensus filter with event-triggered communication: Formulation and stability analysis. *Journal of the Franklin Institute* 354 (13): 5486–5502.
23. Zhang, Wen-An, Gang Feng, and Li Yu. 2012. Multi-rate distributed fusion estimation for sensor networks with packet losses. *Automatica* 58 (9): 2016–2028.

Chapter 10
Event-Triggered Sequential Fusion for Systems with Correlated Noises

10.1 Introduction

The event–triggered centralized fusion and the event–triggered distributed fusion algorithms are introduced in the above two chapters. In this chapter, we aim to introduce the event–triggered sequential fusion estimation for dynamic systems with correlated noises.

In the past few decades, owing to the advantages of reducing the costs of installation as well as implementation and decreasing the need for hardwiring, research interest in Wireless Sensor Network Systems (WSNs) increases with each passing day. WSNs nodes are powered by batteries and generally work in an unattended environment. They carry less energy and are difficult to charge. It is an important issue to effectively reduce the energy consumption of the WSNs to extend their life cycle. Thus, how to design an event–triggered fusion estimation algorithm to reduce unnecessary waste has great significance. The event–triggered control and data transmission strategies have found wide applications including earthquake monitoring, forest fire prevention, intrusion detection, etc.

After more than 40 years of development, the event–triggered control method has made considerable progress, and a variety of event–triggered strategies have emerged. The send-on-delta (SOD) data collecting strategy to capture information from the environment was first put forward by Miskowicz et al. [14]. Nguyen et al. improved the SOD strategy and reduced the estimation error [15]. The SOD transmission method to the problem of recursive state estimation for discrete–time nonlinear systems was applied by Zheng et al. [23]. Liu et al. investigated the distributed filtering problem with the SOD communication mechanism [13]. A new event–triggered state estimator based on the observation residual triggering condition was deduced by mining the information behind the event in [20]. Furthermore, Shi et al. studied an event–triggered scheme quantifying the magnitude of the innovation of the estimator at each time instant in [16]. With event-based communication, random packet dropouts and the additive filter gain variations, Zhang et al. studied the distributed

L. Yan et al., *Multisensor Fusion Estimation Theory and Application*,
https://doi.org/10.1007/978-981-15-9426-7_10

non-fragile filtering problem for a class of discrete–time T–S systems [22]. Bian et al. proposed an online data-driven communication scheme based on cumulative innovation, and derived the corresponding minimum mean square error (MMSE) estimator [1]. Based on the estimation error covariance, Trimpe et al. presented a state estimation method which allowed the designer to trade off estimator performance for communication bandwidth in a networked control system [19]. The above three categories of event–triggered strategies were compared in [18]. The event–triggered model predictive control (MPC) problem was studied for continuous–time nonlinear systems subject to bounded disturbances in [9]. Li et al. made the first attempt to introduce a dynamic event–triggered strategy into the design of synchronization controllers for complex dynamical networks [12].

Since the WSNs are multi–sensor systems, multiple sensors configured with multiple event schedulers are assembled to monitor the same system. This naturally leads to the multi–sensor fusion problem under event–triggered mechanism. Jin et al. proposed two event–triggered fusion estimation algorithms under sequential and parallel fusion structures, respectively [7]. The problem of event–triggered reliable H_∞ filtering for networked systems with multiple sensor distortions was investigated in [3]. Sun et al. concerned the problem of event–triggered Kalman-consensus filter for two- target tracking sensor networks [17]. The event–triggered distributed multi–sensor data fusion algorithm based on a new event–triggered strategy was presented in [4]. An event–triggered distributed fusion estimation problem was investigated for a multi–sensor nonlinear networked system with random transmission delays in [10].

The aforementioned researches on event–triggered sampling systems rarely consider the correlation of noise. However, when the dynamic process is observed in a common noisy environment, many multi–sensor systems have correlated noises in practical applications. The problem of event–triggered centralized fusion estimation with correlated noises was studied in [5]. Li et al. designed an event–triggered nonlinear filtering algorithm for networked systems with correlated noises in [11]. Gao et al. proposed an optimal communication scheduling and remote estimation over an additive noise channel [2]. The problem of measurements missing and noise correlation in event–triggered information fusion of networked systems was considered in [6]. The difference in between this chapter and them is that this chapter proposes a sequential fusion estimation algorithm, which is desirable for applications that require less computation complexity. The main contributions of this chapter are summarized as follows:

(1) An event–triggered sequential fusion estimation algorithm is generated for a multi–sensor dynamic system whose sensor noises are cross–correlated and coupled with the system noise of the previous time step, and the relation between the transmission rate and the estimation performance is analyzed.

(2) Compared with the traditionally time–triggered scheme, an event–triggered data transmission mechanism is adopted to reduce the redundant measurements transmission only with a slight deterioration of the state estimation performance.

(3) The convergence of the designed fusion algorithm is ensured by defining standard values of the correlation parameters, and an upper bound of the estimation error covariance is given.

This chapter is organized as follows. The problem is formulated in Sect. 10.2. Section 10.3 presents the event–triggered sequential fusion estimation algorithm. Sufficient conditions for boundness of the proposed fusion estimation error covariance are given in Sect. 10.4. Simulation results are given in Sects. 10.5 and 10.6 draws the conclusions.

10.2 Problem Formulation

10.2.1 System Modeling

Consider the following linear dynamic system

$$x_{k+1} = A_k x_k + w_k, \quad k = 0, 1, \ldots \tag{10.1}$$

$$z_{i,k} = C_{i,k} x_k + v_{i,k}, \quad i = 1, 2, \ldots, N \tag{10.2}$$

where $x_k \in R^n$ is the system state, $A_k \in R^{n\times n}$ is the state transition matrix, w_k is the system noise assumed to be white Gaussian distributed and satisfies

$$E\{w_k\} = 0 \tag{10.3}$$

$$E\{w_k w_l^T\} = Q_k \delta_{kl} \tag{10.4}$$

where δ_{kl} is the Kronecker delta function. $z_{i,k} \in R^{m_i}$ is the measurement of sensor i at time k, and $C_{i,k} \in R^{m_i\times n}$ is the measurement matrix. The measurement noise $v_{i,k}$ is white Gaussian and satisfies

$$E\{v_{i,k}\} = 0 \tag{10.5}$$

$$E\{v_{i,k} v_{j,l}^T\} = R_{ij,k} \delta_{kl} \delta_{ij} \tag{10.6}$$

$$E\{w_{k-1} v_{i,k}^T\} = S_{i,k}^* \tag{10.7}$$

Note that the measurement noise of different sensors $v_{i,k}$ and $v_{j,k}$ are cross–correlated at time k with $E[v_{i,k} v_{j,k}^T] = R_{ij,k} \neq 0$ for $i, j = 1, 2, \ldots, N$. For simplicity, suppose $R_{i,k} := R_{ii,k} > 0$ in the sequel for $i = 1, 2, \ldots, N$; the measurement noise is correlated with the system noise—$v_{i,k}$ are dependent on w_{k-1} for $k = 1, 2, \ldots$ and $i = 1, 2, \ldots, N$.

The initial state x_0 is Gaussian distributed with mean $\bar{x}_0$ and variance $\bar{P}_0$, and is supposed to be independent of w_k and $v_{i,k}$ for $k = 1, 2, \ldots$ and $i = 1, 2, \ldots, N$.

The purpose of this chapter is to obtain the fusion estimation of state x_k by effectively fusing observations under limited communication resources.

10.2.2 Event–Triggered Mechanism of Sensors

The event–triggered mechanism of sensors in the Kalman filter algorithm is similar to Sect. 8.2.2. For event–triggered mechanism of sensors in the sequential fusion algorithm, it can be designed as:

The event–triggered condition of the ith sensor scheduler is associated with the Kalman filtering innovation

$$\tilde{z}_{i,k}^{s} = z_{i,k} - C_{i,k}\hat{x}_{i-1,k|k}^{s} \tag{10.8}$$

where $\hat{x}_{i-1,k|k}^{s}$ is the estimation of x_k associated with sensor $i-1, i-2, \ldots, 1$.

Similar to [21], the covariance of $\tilde{z}_{i,k}^{s}$ is

$$P_{\tilde{z}_i,k}^{s} = C_{i,k}P_{i-1,k|k}^{s}C_{i,k}^{T} + R_{i,k} + C_{i,k}\Delta_{i-1,k} + \Delta_{i-1,k}^{T}C_{i,k}^{T} \tag{10.9}$$

where $\Delta_{0,k} = S_{1,k}^{*}$, and for $i = 2, 3, \ldots, N$,

$$\Delta_{i-1,k} = \prod_{1}^{p=i-1}(I - K_{p,k}^{s}C_{p,k})S_{i,k}^{*} - \sum_{q=2}^{i-1}\prod_{q}^{p=i-1}(I - K_{p,k}^{s}C_{p,k}) \cdot K_{q-1,k}^{s}R_{q-1,i,k} - K_{i-1,k}^{s}R_{i-1,i,k} \tag{10.10}$$

where $\prod_{q}^{p=i-1} D_p = D_{i-1}D_{i-2} \cdots D_q$, $P_{i-1,k|k}^{s}$ is the state estimation covariance of $\hat{x}_{i-1,k|k}^{s}$, and $K_{i,k}^{s}$ is the gain matrix.

Since $P_{\tilde{z}_i,k}^{s}$ is a positive semidefinite matrix, we can obtain a unitary matrix $U_{i,k}^{s} \in R^{m_i \times m_i}$ to get the diagonalization matrix:

$$U_{i,k}^{(s),T}P_{\tilde{z}_i,k}^{s}U_{i,k}^{s} = \Lambda_{i,k}^{s} \tag{10.11}$$

where the matrix

$$\Lambda_{i,k}^{s} = \mathrm{diag}\{\lambda_{i1,k}^{s}, \ldots, \lambda_{im_i,k}^{s}\} \in R^{m_i \times m_i} \tag{10.12}$$

and these diagonal scalar elements $\lambda_{i1,k}^{s}, \ldots, \lambda_{im_i,k}^{s} \in R^{m_i \times m_i}$ are the eigenvalues of $P_{\tilde{z}_i,k}^{s}$. Define $H_{i,k}^{s} \in R^{m_i \times m_i}$ by

$$H_{i,k}^{s} = U_{i,k}^{s}\Lambda_{i,k}^{(s),-1/2} \tag{10.13}$$

Evidently, $H_{i,k}^{s}H_{i,k}^{s,T} = P_{\tilde{z}_i,k}^{(s),-1}$. Then, $\tilde{z}_{i,k}^{s}$ is normalized and decorrelated by

$$\bar{z}_{i,k}^{s} = H_{i,k}^{s,T}\tilde{z}_{i,k}^{s} \tag{10.14}$$

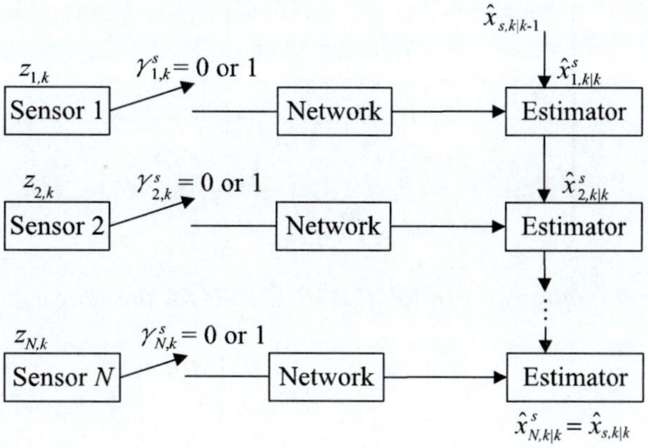

Fig. 10.1 Flow diagram of the event–triggered sequential fusion estimation algorithm

Self-evidently, elements of the normalized innovation $\bar{z}^s_{i,k}$ are standard Gaussian distributed and independent with each other. Similar to [20], we define the event-triggering condition of the ith sensor scheduler as

$$\gamma^s_{i,k} = \begin{cases} 0, & \text{if } \;||\bar{z}^s_{i,k}||_\infty \le \theta_i \\ 1, & \text{otherwise} \end{cases} \tag{10.15}$$

where $||\cdot||_\infty$ represents the infinity-norm of a vector, $|\bar{z}^s_{im_1,k}|, \ldots, |\bar{z}^s_{im_i,k}|$ are the absolute values of the first to the m_ith dimensional elements of $\bar{z}^s_{i,k}$, respectively. Consequently, we know that when $\gamma^s_{i,k} = 0$, the fusion center can not get the exact measurement, and we only have the knowledge that $\max(|\bar{z}^s_{im_1,k}|, \ldots, |\bar{z}^s_{im_i,k}|) \le \theta_i$; Otherwise, when $\gamma^s_{i,k} = 1$, the raw sensor measurement $z_{i,k}$ is transmitted to the fusion center. The flow diagram of the event–triggered sequential fusion estimation algorithm is shown in Fig. 10.1, where $z_{i,k}, i = 1, 2, \ldots, N$ are the measurements detected during $(k-1, k]$ and $\hat{x}_{s,k|k-1}$ is the state prediction of x_k by the use of all the measurements up to time $k-1$. The event–triggered mechanism is used to decide whether $z_{i,k}$ should be transmitted to the fusion center. At the fusion center, the observations are processed according to the sequence of the order received to update $\hat{x}_{s,k|k-1}$, so the fusion estimation of x_k at time k is obtained.

10.3 Fusion Algorithm with Event–Triggered Mechanism

10.3.1 Event–Triggered Kalman Filter with Correlated Noises

Similar to the derivation of [5, 20], we can get the following results.

Proposition 10.1 (The Event–triggered Kalman filter (ETKF)) *For system (10.1)–(10.2), the estimation of* x_k *by the event–triggered Kalman filter is given by*

$$\begin{cases} \hat{x}_{i,k|k} &= \hat{x}_{i,k|k-1} + K_{i,k}\gamma_{i,k}(z_{i,k} - C_{i,k}\hat{x}_{i,k|k-1}) \\ P_{i,k|k} &= P_{i,k|k-1} - g_{i,k}K_{i,k}P_{\tilde{z}_i,k}K_{i,k}^T \\ \hat{x}_{i,k|k-1} &= A_{k-1}\hat{x}_{i,k-1|k-1} \\ P_{i,k|k-1} &= A_{k-1}P_{i,k-1|k-1}A_{k-1}^T + Q_{i,k} \\ P_{\tilde{z}_i,k} &= C_{i,k}P_{i,k|k-1}C_{i,k}^T + C_{i,k}S_{i,k}^* + S_{i,k}^{*,T}C_{i,k}^T + R_{i,k} \\ K_{i,k} &= (P_{i,k|k-1}C_{i,k}^T + S_{i,k}^*)(C_{i,k}P_{i,k|k-1}C_{i,k}^T + R_{i,k} + C_{i,k}S_{i,k}^* \\ &\quad + S_{i,k}^{*,T}C_{i,k}^T)^{-1} \end{cases} \tag{10.16}$$

The subscript i in Eq. (10.16) stands for the sensor i and

$$g_{i,k} = \begin{cases} \beta(\theta_i), & \text{if } \gamma_{i,k} = 0 \\ 1, & \text{if } \gamma_{i,k} = 1 \end{cases} \tag{10.17}$$

$$\beta(\theta_i) = \frac{2}{\sqrt{2\pi}}\theta_i e^{-\frac{\theta_i^2}{2}}[1 - 2Q(\theta_i)]^{-1} \tag{10.18}$$

$$Q(\theta_i) = \int_{\theta_i}^{\infty} \frac{1}{\sqrt{2\pi}} e^{-\frac{x^2}{2}} dx \tag{10.19}$$

where, $z_{i,k}$, $i = 1, 2, \ldots, N$, $k = 1, 2, \ldots$ *is the kth measurement of sensor i in time* $(k-1, k]$.

10.3.2 Event–Triggered Sequential Fusion Estimation Algorithm with Correlated Noises

By event–triggered mechanism, the sensor decides whether $z_{i,k}$ shall be sent to a remote estimator for further processing. Let $\gamma_{i,k}^s = 1$ or 0 be the decision variable whether $z_{i,k}$ shall be sent or not. Define $I_{1,k}^N \triangleq \{\gamma_{i,0}^s z_{i,0}, \ldots, \gamma_{i,k}^s z_{i,k}, \ldots, \gamma_{N,k}^s z_{N,k}\} \cup \{\gamma_{i,0}^s, \ldots, \gamma_{i,k}^s, \ldots \gamma_{N,k}^s\}$ with $I_{1,-1}^N = \emptyset$, $\hat{I}_{1,k}^i = \{I_{1,k-1}^N, I_{1,k}^{i-1}\} \cup \{\gamma_{i,k}^s = 0\}$.

Lemma 10.1 ([20]) *Let* $x \in R$ *be a Gaussian r.v. with zero mean and variance* $E\{x^2\} = \sigma^2$. *Denoting* $\Delta = \theta\sigma$, *then* $E\{x^2 | |x| \le \Delta\} = \sigma^2(1 - \beta(\theta))$, *where*

$$\beta(\theta) = \frac{2}{\sqrt{2\pi}}\theta e^{-\frac{\theta^2}{2}}[1 - 2Q(\theta)]^{-1} \tag{10.20}$$

$$Q(\theta) = \int_{\theta}^{\infty} \frac{1}{\sqrt{2\pi}} e^{-\frac{x^2}{2}} dx \tag{10.21}$$

Lemma 10.2 *A preliminary result is obtained*

$$\begin{aligned}&E\{\bar{z}^{s}_{i,k}\bar{z}^{s,T}_{i,k}|\hat{I}^{i}_{1,k}\}\\&=H^{s,T}_{i,k}E\{\bar{z}^{s}_{i,k}\bar{z}^{s,T}_{i,k}|\hat{I}^{i}_{1,k}\}H^{s}_{i,k}\\&=[1-\beta^{s}(\theta_i)]I_{m_i}\end{aligned} \tag{10.22}$$

Proof Similar to the results of [20], according to the design process of event–triggered mechanism, the following equations are established

$$E\{\bar{z}^{s}_{i,k}|I^{N}_{1,k-1}\}=H^{s,T}_{i,k}E\{\bar{z}^{s}_{i,k}|I^{N}_{1,k-1}\}H^{s}_{i,k}=0 \tag{10.23}$$

$$E\{\bar{z}^{s}_{i,k}\bar{z}^{s,T}_{i,k}|I^{N}_{1,k-1}\}=H^{s,T}_{i,k}E\{\bar{z}^{s}_{i,k}\bar{z}^{s,T}_{i,k}|I^{N}_{1,k-1}\}H^{s}_{i,k}=I_{m_i} \tag{10.24}$$

$$E\{\bar{z}^{s}_{i,k}\bar{z}^{s,T}_{i,k}|\hat{I}^{i}_{1,k}\}=H^{s,T}_{i,k}E\{\bar{z}^{s}_{i,k}\bar{z}^{s,T}_{i,k}|\hat{I}^{i}_{1,k}\}H^{s}_{i,k} \tag{10.25}$$

where $\bar{z}^{s}_{i,k}$ given $I^{N}_{1,k-1}$ is a zero-mean Gaussian variable with unit covariance. Denoting $\bar{z}^{s,m}_{i,k}$ as the mth element of $\bar{z}^{s}_{i,k}$, we can see that $\bar{z}^{s,m}_{i,k}$ and $\bar{z}^{s,n}_{i,k}$ are mutually independent when $m \neq n$. Therefore

$$\begin{aligned}&E\{\bar{z}^{s,m}_{i,k}\bar{z}^{(s,n),T}_{i,k}|\hat{I}^{i}_{1,k}\}\\&=E\{\bar{z}^{s,m}_{i,k}\bar{z}^{(s,n),T}_{i,k}|I^{N}_{1,k-1},I^{i-1}_{1,k},||\bar{z}^{s}_{i,k}||_\infty\le\theta_i\}\\&=E\{\bar{z}^{s,m}_{i,k}\bar{z}^{(s,n),T}_{i,k}|I^{N}_{1,k-1},I^{i-1}_{1,k},||\bar{z}^{s,m}_{i,k}||_\infty\le\theta_i,||\bar{z}^{s,n}_{i,k}||_\infty\le\theta_i\}\\&=0\end{aligned} \tag{10.26}$$

On the basis of Lemma 10.1, it is derived that

$$\begin{aligned}E\{(\bar{z}^{s}_{i,k})^2|\hat{I}^{i}_{1,k}\}&=E\{(\bar{z}^{s}_{i,k})^2|I^{N}_{1,k-1},I^{i-1}_{1,k},||\bar{z}^{s}_{i,k}||_\infty\le\theta_i\}\\&=1-\beta^{s}(\theta_i)\end{aligned} \tag{10.27}$$

As a result, we have

$$E\{\bar{z}^{s}_{i,k}\bar{z}^{s,T}_{i,k}|\hat{I}^{i}_{1,k}\}=(1-\beta^{s}(\theta_i))I_{m_i} \tag{10.28}$$

Then, we have

$$\begin{aligned}E\{\bar{z}^{s}_{i,k}\bar{z}^{s,T}_{i,k}|\hat{I}^{i}_{1,k}\}&=H^{s,T}_{i,k}E\{\bar{z}^{s}_{i,k}\bar{z}^{s,T}_{i,k}|\hat{I}^{i}_{1,k}\}H^{s}_{i,k}\\&=[1-\beta^{s}(\theta_i)]I_{m_i}\end{aligned} \tag{10.29}$$

Proof completed.

Theorem 10.1 (The Event–triggered Sequential Fusion(ETSF)) *For system (10.1)–(10.2), given the state estimation* $\hat{x}^{s}_{N,k-1|k-1}$ *and* $P^{s}_{N,k-1|k-1}$ *based on* $I^{N}_{1,k-1}$, *if* $z_{i,k}$, $i=1,2,\ldots,N$ *are detected during* $(k-1,k]$, *then the state estimation of* x_k *can be* *computed by*

$$\begin{cases} \hat{x}^s_{i,k|k} &= \hat{x}^s_{i-1,k|k} + K^s_{i,k}\gamma^s_{i,k}(z_{i,k} - C_{i,k}\hat{x}^s_{i-1,k|k}) \\ P^s_{i,k|k} &= P^s_{i-1,k|k} - g^s_{i,k}K^s_{i,k}P^s_{\tilde{z}_i,k}K^{s,T}_{i,k} \\ K^s_{i,k} &= (P^s_{i-1,k|k}C^T_{i,k} + \Delta_{i-1,k})(C_{i,k}P^s_{i-1,k|k}C^T_{i,k} + R_{i,k} \\ &\quad + C_{i,k}\Delta_{i-1,k} + \Delta^T_{i-1,k}C^T_{i,k})^{-1} \\ P^s_{\tilde{z}_i,k} &= C_{i,k}P^s_{i-1,k|k}C^T_{i,k} + R_{i,k} + C_{i,k}\Delta_{i-1,k} + \Delta^T_{i-1,k}C^T_{i,k} \end{cases} \tag{10.30}$$

where, the subscript i in Eq. (10.30) stands for the sensor i and

$$g^s_{i,k} = \begin{cases} \beta^s(\theta_i), & \text{if } \gamma^s_{i,k} = 0 \\ 1, & \text{if } \gamma^s_{i,k} = 1 \end{cases} \tag{10.31}$$

$$\beta^s(\theta_i) = \frac{2}{\sqrt{2\pi}}\theta_i e^{-\frac{\theta_i^2}{2}}[1 - 2Q^s(\theta_i)]^{-1} \tag{10.32}$$

$$Q^s(\theta_i) = \int_{\theta_i}^{\infty} \frac{1}{\sqrt{2\pi}} e^{-\frac{x^2}{2}} dx \tag{10.33}$$

where, $z_{i,k}, i = 1, 2, \ldots, N, k = 1, 2, \ldots$ is the kth measurement of sensor i in time $(k-1, k]$.

$$\Delta_{i-1,k} = \prod_{1}^{p=i-1}(I - K^s_{p,k}C_{p,k})S_{i,k} - \sum_{q=2}^{i-1}\prod_{q}^{p=i-1}(I - K^s_{p,k}C_{p,k}) \cdot K^s_{q-1,k}R_{q-1,i,k} - K^s_{i-1,k}R_{i-1,i,k} \tag{10.34}$$

where $\hat{x}_{0,k|k}$, $P_{0,k|k}$ are given by

$$\begin{cases} \hat{x}_{0,k|k} = A_{k-1}\hat{x}^s_{N,k-1|k-1} \\ P_{0,k|k} = A_{k-1}P^s_{N,k-1|k-1}A^T_{k-1} + Q_{k-1} \end{cases} \tag{10.35}$$

Denote

$$\begin{cases} \hat{x}_{s,k|k} = \hat{x}^s_{N,k|k} \\ P_{s,k|k} = P^s_{N,k|k} \end{cases} \tag{10.36}$$

then they are the fused state estimation.

Proof We use inductive method and projection theorem to prove this theorem.

For whether $z_{i,k}$ shall be sent or not, we have the following two cases:

(1) $\gamma^s_{i,k} = 1$: For $\gamma^s_{i,k} = 1$, $z_{i,k}$ is sent to the estimator. In this case, the filter estimation is the same as the time-driven schedule, and we have [21]

$$\begin{cases} \hat{x}^s_{i,k|k} &= \hat{x}^s_{i-1,k|k} + K^s_{i,k}(z_{i,k} - C_{i,k}\hat{x}^s_{i-1,k|k}) \\ P^s_{i,k|k} &= P^s_{i-1,k|k} - K^s_{i,k}P^s_{\tilde{z}_i,k}K^{s,T}_{i,k} \\ K^s_{i,k} &= (P^s_{i-1,k|k}C^T_{i,k} + \Delta_{i-1,k})(C_{i,k}P^s_{i-1,k|k}C^T_{i,k} \\ &\quad + R_{i,k} + C_{i,k}\Delta_{i-1,k} + \Delta^T_{i-1,k}C^T_{i,k})^{-1} \\ P^s_{\tilde{z}_i,k} &= C_{i,k}P^s_{i-1,k|k}C^T_{i,k} + R_{i,k} + C_{i,k}\Delta_{i-1,k} + \Delta^T_{i-1,k}C^T_{i,k} \end{cases} \tag{10.37}$$

where $\hat{x}_{0,k|k}$, $P_{0,k|k}$ is given by

$$\begin{cases} \hat{x}_{0,k|k} = A_{k-1}\hat{x}^s_{N,k-1|k-1} \\ P_{0,k|k} = A_{k-1}P^s_{N,k-1|k-1}A^T_{k-1} + Q_{k-1} \end{cases} \tag{10.38}$$

(2) $\gamma^s_{i,k} = 0$: The remote estimator can not get the exact measurement $z_{i,k}$ and we only know that $||\bar{z}^s_{i,k}||_\infty \le \theta_i$.

Define the set $\Omega_i \subset R^{m_i}$ as

$$\Omega_i = \{\bar{z}^s_{i,k} \in R^{m_i} : ||\bar{z}^s_{i,k}||_\infty \le \theta_i\} \tag{10.39}$$

Given $I^N_{1,k-1}$ and $I^{i-1}_{1,k}$, $\bar{z}^s_{i,k}$ is Gaussian with zero mean and unit covariance, so we can define $p(\theta_i) = \text{Prob}(||\bar{z}^s_{i,k}||_\infty \le \theta_i | I^N_{1,k-1}, I^{i-1}_{1,k})$. Using the conditional probability density function (pdf)

$$f_{\bar{z}^s_{i,k}}(\bar{z}^s_i|\hat{I}^i_{1,k}) = \begin{cases} \frac{f_{\bar{z}^s_{i,k}}(\bar{z}^s_i|I^N_{1,k-1}, I^{i-1}_{1,k})}{p(\theta_i)}, & \text{if } ||\bar{z}^s_{i,k}||_\infty \le \theta_i \\ 0, & \text{otherwise} \end{cases} \tag{10.40}$$

We have

$$\begin{aligned} \hat{x}^s_{i,k|k} &= E[x_k|\hat{I}^i_{1,k}] \\ &= \frac{1}{p(\theta_i)}\int_{\Omega_i} E\{x_k|I^N_{1,k-1}, I^{i-1}_{1,k}, ||\bar{z}^s_{i,k}||_\infty \le \theta_i, \\ &\quad \bar{z}^s_{i,k} = H^{s,-T}_{i,k}\bar{z}^s_i\} f_{\bar{z}^s_{i,k}}(\bar{z}^s_i|I^N_{1,k-1}, I^{i-1}_{1,k})d\bar{z}^s_i \\ &= \frac{1}{p(\theta_i)}\int_{\Omega_i} (\hat{x}^s_{i-1,k|k} + K^s_{i,k}H^{s,-T}_{i,k}\bar{z}^s_i) f_{\bar{z}^s_{i,k}}(\bar{z}^s_i|I^N_{1,k-1}, I^{i-1}_{1,k})d\bar{z}^s_i \\ &= \hat{x}^s_{i-1,k|k} + \frac{K^s_{i,k}H^{s,-T}_{i,k}}{p(\theta_i)} \cdot \int_{\Omega_i} \bar{z}^s_i f_{\bar{z}^s_{i,k}}(\bar{z}^s_i|I^N_{1,k-1}, I^{i-1}_{1,k})d\bar{z}^s_i \\ &= \hat{x}^s_{i-1,k|k} \end{aligned} \tag{10.41}$$

where, the last equality is due to

$$\int_{\Omega_i} \bar{z}^s_i f_{\bar{z}^s_{i,k}}(\bar{z}^s_i|I^N_{1,k-1}, I^{i-1}_{1,k})d\bar{z}^s_i = 0 \tag{10.42}$$

Denote

$$\tilde{x}^s_{i-1,k|k} = x_k - \hat{x}^s_{i-1,k|k} \tag{10.43}$$

then, we have

$$\begin{aligned}
&E\{\tilde{x}^s_{i-1,k|k}\bar{z}^{s,T}_{i,k}|\hat{I}^i_{1,k}\} \\
&= \frac{1}{p(\theta_i)}\int_{\Omega_i} E\{\tilde{x}^s_{i-1,k|k}|I^N_{1,k-1}, I^{i-1}_{1,k}, ||\bar{z}^s_{i,k}||_\infty \le \theta_i, \bar{z}^s_{i,k} = H^{s,-T}_{i,k}\bar{z}^s_i\} \\
&\cdot \bar{z}^{s,T}_i H^{s,-1}_{i,k} f_{\bar{z}^s_{i,k}}(\bar{z}^s_i|I^N_{1,k-1}, I^{i-1}_{1,k})d\bar{z}^s_i \\
&= \frac{1}{p(\theta_i)}\int_{\Omega_i} E\{x_k - \hat{x}^s_{i-1,k|k}|I^N_{1,k-1}, I^{i-1}_{1,k}, ||\bar{z}^s_{i,k}||_\infty \le \theta_i, \bar{z}^s_{i,k} = H^{s,-T}_{i,k}\bar{z}^s_i\} \\
&\cdot \bar{z}^{s,T}_i H^{s,-1}_{i,k} f_{\bar{z}^s_{i,k}}(\bar{z}^s_i|I^N_{1,k-1}, I^{i-1}_{1,k})d\bar{z}^s_i \\
&= \frac{1}{p(\theta_i)}\int_{\Omega_i} \{E\{x_k|I^N_{1,k-1}, I^{i-1}_{1,k}, ||\bar{z}^s_{i,k}||_\infty \le \theta_i, \bar{z}^s_{i,k} = H^{s,-T}_{i,k}\bar{z}^s_i\} \\
&-\hat{x}^s_{i-1,k|k}\}\bar{z}^{s,T}_i H^{s,-1}_{i,k} f_{\bar{z}^s_{i,k}}(\bar{z}^s_i|I^N_{1,k-1}, I^{i-1}_{1,k})d\bar{z}^s_i \\
&= K^s_{i,k}H^{s,-T}_{i,k}\frac{1}{p(\theta_i)}\cdot\int_{\Omega_i} \bar{z}^s_i\bar{z}^{s,T}_i f_{\bar{z}^s_{i,k}}(\bar{z}^s_i|I^N_{1,k-1}, I^{i-1}_{1,k})d\bar{z}^s_i H^{s,-1}_{i,k} \\
&= K^s_{i,k}H^{s,-T}_{i,k}E\{\bar{z}_{i,k}\bar{z}^{s,T}_{i,k}|\hat{I}^i_{1,k}\}H^{s,-1}_{i,k} \\
&= K^s_{i,k}E\{\bar{z}^s_{i,k}\bar{z}^{s,T}_{i,k}|\hat{I}^i_{1,k}\}
\end{aligned} \tag{10.44}$$

and

$$\begin{aligned}
K^s_{i,k} &= (P^s_{i-1,k|k}C^T_{i,k} + \Delta_{i-1,k})(C_{i,k}P^s_{i-1,k|k}C^T_{i,k} + R_{i,k} \\
&\quad + C_{i,k}\Delta_{i-1,k} + \Delta^T_{i-1,k}C^T_{i,k})^{-1}
\end{aligned} \tag{10.45}$$

where

$$\begin{aligned}
\Delta_{i-1,k} &= E\{\tilde{x}^s_{i-1,k|k}v^T_{i,k}|\hat{I}^i_{1,k}\} \\
&= E\{(x_k - \hat{x}^s_{i-1,k|k})v^T_{i,k}|\hat{I}^i_{1,k}\} \\
&= (I - K^s_{i-1,k}C_{i-1,k})E\{\tilde{x}^s_{i-2,k|k}v^T_{i,k}\} - K^s_{i-1,k}E\{v_{i-1,k}v^T_{i,k}\} \\
&= \prod_{1}^{p=i-1}(I - K^s_{p,k}C_{p,k})S_{i,k} - \sum_{q=2}^{i-1}\prod_{q}^{p=i-1}(I - K^s_{p,k}C_{p,k}) \\
&\quad \cdot K^s_{q-1,k}R_{q-1,i,k} - K^s_{i-1,k}R_{i-1,i,k}
\end{aligned} \tag{10.46}$$

where, $\prod_{q}^{p=i-1} D_p = D_{i-1}D_{i-2}\cdots D_q$. Then, we have

$$
\begin{aligned}
& E\{(\tilde{x}^s_{i-1,k|k} - K^s_{i,k}\tilde{z}^s_{i,k})\tilde{z}^{s,T}_{i,k} | \hat{I}^i_{1,k}\} \\
& = E\{\tilde{x}^s_{i-1,k|k}\tilde{z}^{s,T}_{i,k} | \hat{I}^i_{1,k}\} - K^s_{i,k}E\{\tilde{z}^s_{i,k}\tilde{z}^{s,T}_{i,k} | \hat{I}^i_{1,k}\} \\
& = 0
\end{aligned} \tag{10.47}
$$

and

$$
\begin{aligned}
& E\{(\tilde{x}^s_{i-1,k|k} - K^s_{i,k}\tilde{z}^s_{i,k})(\tilde{x}^s_{i-1,k|k} - K^s_{i,k}\tilde{z}^s_{i,k})^T | \hat{I}^i_{1,k}\} \\
& = \int_{\Omega_i} E\{(\tilde{x}^s_{i-1,k|k} - K^s_{i,k}\tilde{z}^s_{i,k})\cdot(\tilde{x}^s_{i-1,k|k} - K^s_{i,k}\tilde{z}^s_{i,k})^T | I^N_{1,k-1}, I^{i-1}_{1,k} \\
& , \|\tilde{z}^s_{i,k}\|_\infty \le \theta_i, \tilde{z}^s_{i,k} = H^{s,-T}_{i,k}\tilde{z}^s_i\} \frac{f_{\tilde{z}^s_{i,k}}(\tilde{z}^s_i | I^N_{1,k-1}, I^{i-1}_{1,k})}{p(\theta_i)} d\tilde{z}^s_i \\
& = \frac{1}{p(\theta_i)}(P^s_{i-1,k|k} - K^s_{i,k}P^s_{\tilde{z}_i,k}K^{s,T}_{i,k}) \int_{\Omega_i} f_{\tilde{z}^s_{i,k}}(\tilde{z}^s_i | I^N_{1,k-1}, I^{i-1}_{1,k}) d\tilde{z}^s_i \\
& = P^s_{i-1,k|k} - K^s_{i,k}P^s_{\tilde{z}_i,k}K^{s,T}_{i,k}
\end{aligned} \tag{10.48}
$$

from Eq. (10.40), we have $\int_{\Omega_i} f_{\tilde{z}^s_{i,k}}(\tilde{z}^s_i | I^N_{1,k-1}, I^{i-1}_{1,k}) d\tilde{z}^s_i = p(\theta_i)$.

Now from Eqs. (10.47), (10.48), Lemmas 10.1 and 10.2, the corresponding estimation error covariance matrix $P^s_{i,k|k}$ is computed by

$$
\begin{aligned}
& P^s_{i,k|k} \\
& = E\{(x_k - \hat{x}^s_{i,k|k})(x_k - \hat{x}^s_{i,k|k})^T | \hat{I}^i_{1,k}\} \\
& = E\{(x_k - \hat{x}^s_{i-1,k|k})(x_k - \hat{x}^s_{i-1,k|k})^T | \hat{I}^i_{1,k}\} \\
& = E\{[(\tilde{x}^s_{i-1,k|k} - K^s_{i,k}\tilde{z}^s_{i,k}) + K^s_{i,k}\tilde{z}^s_{i,k}][(\tilde{x}^s_{i-1,k|k} - K^s_{i,k}\tilde{z}^s_{i,k}) + K^s_{i,k}\tilde{z}^s_{i,k}]^T | \hat{I}^i_{1,k}\} \\
& = E\{(\tilde{x}^s_{i-1,k|k} - K^s_{i,k}\tilde{z}^s_{i,k})(\tilde{x}^s_{i-1,k|k} - K^s_{i,k}\tilde{z}^s_{i,k})^T + (\tilde{x}^s_{i-1,k|k} - K^s_{i,k}\tilde{z}^s_{i,k})(K^s_{i,k}\tilde{z}^s_{i,k})^T \\
& \quad + K^s_{i,k}\tilde{z}^s_{i,k}(\tilde{x}^s_{i-1,k|k} - K^s_{i,k}\tilde{z}^s_{i,k})^T + (K^s_{i,k}\tilde{z}^s_{i,k})(K^s_{i,k}\tilde{z}^s_{i,k})^T | \hat{I}^i_{1,k}\} \\
& = P^s_{i-1,k|k} - K^s_{i,k}P^s_{\tilde{z}_i,k}K^{s,T}_{i,k} + K^s_{i,k}E\{\tilde{z}^s_{i,k}\tilde{z}^{s,T}_{i,k} | \hat{I}^i_{1,k}\}K^{(s),T}_{i,k} \\
& = P^s_{i-1,k|k} - K^s_{i,k}P^s_{\tilde{z}_i,k}K^{s,T}_{i,k} + [1 - \beta^s(\theta_i)]K^s_{i,k}(H^s_{i,k}H^{s,T}_{i,k})^{-1}K^{(s),T}_{i,k} \\
& = P^s_{i-1,k|k} - K^s_{i,k}P^s_{\tilde{z}_i,k}K^{s,T}_{i,k} + [1 - \beta^s(\theta_i)]K^s_{i,k}P^s_{\tilde{z}_i,k}K^{s,T}_{i,k} \\
& = P^s_{i-1,k|k} - \beta^s(\theta_i)K^s_{i,k}P^s_{\tilde{z}_i,k}K^{s,T}_{i,k}
\end{aligned} \tag{10.49}
$$

when $i = N$, we have $\hat{x}_{s,k|k} = \hat{x}^s_{N,k|k}$, $P_{s,k|k} = P^s_{N,k|k}$, $k = 1, 2, \ldots$.

Because

$$
\hat{x}_{0,k|k} = A_{k-1}\hat{x}^s_{N,k-1|k-1} \tag{10.50}
$$

Next, we have the following one step prediction error equations for the state based on the projection theory

$$
\begin{aligned}
\tilde{x}_{0,k|k} &= x_k - \hat{x}_{0,k|k} \\
&= A_{k-1}x_{k-1} + w_{k-1} - A_{k-1}\hat{x}^s_{N,k-1|k-1} \\
&= A_{k-1}\tilde{x}^s_{N,k-1|k-1} + w_{k-1}
\end{aligned}
\tag{10.51}
$$

The corresponding prediction error covariance matrix $P^s_{i-1,k|k}$ is given by

$$
\begin{aligned}
P_{0,k|k} &= E\{\tilde{x}^s_{N,k-1|k-1}\tilde{x}^{s,T}_{N,k-1|k-1}|I^N_{1,k-1}, I^{i-1}_{1,k}\} \\
&= A_{k-1}P^s_{N,k-1|k-1}A^T_{k-1} + Q_{k-1}
\end{aligned}
\tag{10.52}
$$

Remark 10.1 The event–triggered sequential fusion (ETSF) algorithm is deduced in Theorem 10.1. As stated in [20], $P^s_{i,k|k}$ is a function of $\gamma^s_{i,k}$ and $\beta^s(\theta_i)$, both of which depend on event–triggered threshold θ_i. By appropriately adjusting θ_i, a desired tradeoff between estimation performance and communication rate can be achieved. In addition, the estimation performance can also be influenced by the correlation parameter $S^*_{i,k}$ between measurement noise and the system noise.

10.4 Boundness of the Fusion Estimation Error Covariance

In this section, the convergence of the designed event–triggered sequential fusion estimation algorithm is analyzed.

Lemma 10.3 ([8]) *For two matrices $Z \in R^{m\times n}$, $G \in R^{m\times n}$, if there exists a symmetric positive definite matrix $P \in R^{m\times m}$, then*

$$
-G^T P^{-1} G \le Z^T P Z - G^T Z - Z^T G \tag{10.53}
$$

Equality holds if and only if $Z = P^{-1}G$.

Theorem 10.2 *Suppose that the linearization of multi–sensor systems (10.1)–(10.2) satisfies the uniform observability condition, $C_{i,k}$ is nonsingular, and real constants $\bar{a}, \bar{c}_i, \bar{q}, \bar{r}_i, \bar{\theta}_i$ is existed, such that*

$$
\| A_k \| \le \bar{a}, \quad \| C_{i,k} \| \le \bar{c}_i, \quad Q_k \le \bar{q}I, \quad R_{i,k} \le \bar{r}_i I, \quad \theta_i \le \bar{\theta}_i \tag{10.54}
$$

where, $\|\cdot\|$ denotes the Euclidean norm of corresponding vectors or the spectral norm of corresponding matrices. If $\Delta_{i-1,k} \le S^*_{i,k} \le \bar{s}_i I < \{[\frac{1-\bar{a}^2}{1-\beta^s(\bar{\theta}_i)\gamma^s_i+\beta^s(\bar{\theta}_i)} - \bar{a}^2\bar{c}_i][\bar{c}^2_i(1+\bar{c}_i)]^{-1}\}^{\frac{1}{2}}$,then

$$
E\{P^s_{i,k|k}\} \le E\{P_{0,k|k}\} < \bar{p}I_n \tag{10.55}
$$

Proof Based on the property of Kalman filter, we have

$$
E\{P^s_{i,k|k}\} \le E\{P^s_{i-1,k|k}\} \le \cdots \le E\{P^s_{1,k|k}\} \le E\{P_{0,k|k}\} \tag{10.56}
$$

So, we only need to prove that $P_{0,k|k}$ is bounded. Following Eqs. (10.30) and (10.35), the prediction error covariance can be computed by

$$\begin{aligned}
P_{0,k|k} &= A_{k-1}P^s_{N,k-1|k-1}A^T_{k-1} + Q_{k-1} \\
&= A_{k-1}P^s_{N-1,k-1|k-1}A^T_{k-1} - \{[\gamma^s_{N-1,k-1} + (1-\gamma^s_{N-1,k-1}\beta^s(\theta_{N-1}))] \\
&\quad \cdot A_{k-1}(P^s_{N-1,k-1|k-1}C^T_{N,k-1} + \Delta_{N-1,k-1})(C_{N,k-1}P^s_{N-1,k-1|k-1} \\
&\quad \cdot C^T_{N,k-1} + R_{N,k-1} + C_{N,k-1}\Delta_{N-1,k-1} + \Delta^T_{N-1,k-1}C^T_{N,k-1})^{-1} \\
&\quad \cdot (C_{N,k-1}P^s_{N-1,k-1|k-1} + \Delta^T_{N-1,k-1})\}A^T_{k-1} + Q_{k-1}
\end{aligned} \tag{10.57}$$

Define

$$G^T = A_{k-1}(P^s_{N-1,k-1|k-1}C^T_{N,k-1} + \Delta_{N-1,k-1}) \tag{10.58}$$

$$\begin{aligned}
P &= C_{N,k-1}P^s_{N-1,k-1|k-1}C^T_{N,k-1} + C_{N,k-1}\Delta_{N-1,k-1} \\
&\quad + \Delta^T_{N-1,k-1}C^T_{N,k-1} + R_{N,k-1}
\end{aligned} \tag{10.59}$$

$$Z = \Delta_{N-1,k-1} \tag{10.60}$$

From Lemma 10.3, we have

$$\begin{aligned}
&P_{0,k|k} \\
&< A_{k-1}P^s_{N-1,k-1|k-1}A^T_{k-1} - [(1-\beta^s(\theta_{N-1}))\gamma^s_{N-1,k-1} + \beta^s(\theta_{N-1})]A_{k-1} \\
&\quad \cdot [P^s_{N-1,k-1|k-1}C^T_{N,k-1} + \Delta_{N-1,k-1}]\Delta_{N-1,k-1} \\
&\quad - [(1-\beta^s(\theta_{N-1}))\gamma^s_{N-1,k-1} + \beta^s(\theta_{N-1})]\Delta^T_{N-1,k-1}[C_{N,k-1}P^s_{N-1,k-1|k-1} \\
&\quad + \Delta^T_{N-1,k-1}]A^T_{k-1} + [(1-\beta^s(\theta_{N-1}))\gamma^s_{N-1,k-1} + \beta^s(\theta_{N-1})][\Delta^T_{N-1,k-1} \\
&\quad \cdot C_{N,k-1}P^s_{N-1,k-1|k-1}C^T_{N,k-1}\Delta_{N-1,k-1} + \Delta^T_{N-1,k-1}C_{N,k-1} \\
&\quad \cdot \Delta_{N-1,k-1}\Delta_{N-1,k-1} + \Delta^T_{N-1,k-1}\Delta^T_{N-1,k-1}C^T_{N,k-1}\Delta_{N-1,k-1} \\
&\quad + \Delta_{N-1,k-1}R_{N,k-1}\Delta_{N-1,k-1}] + Q_{k-1} \\
&< A_{k-1}P^s_{N-1,k-1|k-1}A^T_{k-1} + [(1-\beta^s(\theta_{N-1}))\gamma^s_{N-1,k-1} + \beta^s(\theta_{N-1})]A_{k-1} \\
&\quad \cdot [P^s_{N-1,k-1|k-1}C^T_{N,k-1} + \Delta_{N-1,k-1}]A^T_{k-1} \\
&\quad + [(1-\beta^s(\theta_{N-1}))\gamma^s_{N-1,k-1} + \beta^s(\theta_{N-1})]\Delta^T_{N-1,k-1}[C_{N,k-1}P^s_{N-1,k-1|k-1} \\
&\quad + \Delta^T_{N-1,k-1}]\Delta_{N-1,k-1} + [(1-\beta^s(\theta_{N-1}))\gamma^s_{N-1,k-1} + \beta^s(\theta_{N-1})][\Delta^T_{N-1,k-1} \\
&\quad \cdot C_{N,k-1}P^s_{N-1,k-1|k-1}C^T_{N,k-1}\Delta_{N-1,k-1} + \Delta^T_{N-1,k-1}C_{N,k-1}\Delta_{N-1,k-1} \\
&\quad \cdot \Delta_{N-1,k-1} + \Delta^T_{N-1,k-1}\Delta^T_{N-1,k-1}C^T_{N,k-1}\Delta_{N-1,k-1} \\
&\quad + \Delta_{N-1,k-1}R_{N,k-1}\Delta_{N-1,k-1}] + Q_{k-1}
\end{aligned} \tag{10.61}$$

Note that $\gamma^s_{N-1,k-1}$ is independent of $P_{0,k-1|k-1}$, from Eq. (10.54), we have

$$\begin{aligned} &E\{P_{0,k|k}\} \\ &< \{\bar{a}^2 + [(1-\beta^s(\bar{\theta}_{N-1}))\gamma^s_{N-1} + \beta^s(\bar{\theta}_{N-1})](\bar{a}^2\bar{c}_N + \bar{s}^2_{N-1}\bar{c}_N + \bar{s}^2_{N-1}\bar{c}^2_N)\} \\ &\cdot E\{P^s_{N-1,k-1|k-1}\} + [(1-\beta^s(\bar{\theta}_{N-1}))\gamma^s_{N-1} + \beta^s(\bar{\theta}_{N-1})] \\ &\cdot (\bar{a}^2\bar{s}_{N-1} + \bar{s}^3_{N-1} + 2\bar{s}^3_{N-1}\bar{c}_N + \bar{s}^2_{N-1}\bar{r}_N)I_n + \bar{q}I_n \end{aligned} \tag{10.62}$$

where $\gamma^s_{N-1} = E[\gamma^s_{N-1,k}]$. Because

$$E\{P^s_{N,k|k}\} \leq E\{P^s_{N-1,k|k}\} \leq \cdots \leq E\{P^s_{1,k|k}\} \leq E\{P_{0,k|k}\} \tag{10.63}$$

then,

$$\begin{aligned} E\{P_{0,k|k}\} &< \{\bar{a}^2 + [(1-\beta^s(\bar{\theta}_{N-1}))\gamma^s_{N-1} + \beta^s(\bar{\theta}_{N-1})](\bar{a}^2\bar{c}_N + \bar{s}^2_{N-1}\bar{c}_N \\ &\quad + \bar{s}^2_{N-1}\bar{c}^2_N)\}E\{P_{0,k-1|k-1}\} + [(1-\beta^s(\bar{\theta}_{N-1}))\gamma^s_{N-1} + \beta^s(\bar{\theta}_{N-1})] \\ &\quad \cdot (\bar{a}^2\bar{s}_{N-1} + \bar{s}^3_{N-1} + 2\bar{s}^3_{N-1}\bar{c}_N + \bar{s}^2_{N-1}\bar{r}_N)I_n + \bar{q}I_n \end{aligned} \tag{10.64}$$

An inductive method is introduced to derive the boundness of $E\{P_{0,k|k}\}$. First, assume that $E\{\bar{P}_0(1|1)\} > 0$, then we have

$$\begin{aligned} E\{\bar{P}_0(2|2)\} &< \{\bar{a}^2 + [(1-\beta^s(\bar{\theta}_{N-1}))\gamma^s_{N-1} + \beta^s(\bar{\theta}_{N-1})](\bar{a}^2\bar{c}_N + \bar{s}^2_{N-1}\bar{c}_N \\ &\quad + \bar{s}^2_{N-1}\bar{c}^2_N)\}E\{\bar{P}_0(1|1)\} + [(1-\beta^s(\bar{\theta}_{N-1}))\gamma^s_{N-1} + \beta^s(\bar{\theta}_{N-1})] \\ &\quad \cdot (\bar{a}^2\bar{s}_{N-1} + \bar{s}^3_{N-1} + 2\bar{s}^3_{N-1}\bar{c}_N + \bar{s}^2_{N-1}\bar{r}_N)I_n + \bar{q}I_n \\ &< \{\bar{a}^2 + [(1-\beta^s(\bar{\theta}_{N-1}))\gamma^s_{N-1} + \beta^s(\bar{\theta}_{N-1})](\bar{a}^2\bar{c}_N \\ &\quad + \bar{s}^2_{N-1}\bar{c}_N + \bar{s}^2_{N-1}\bar{c}^2_N)\}\hat{p}I_n + \hat{p}I_n \end{aligned} \tag{10.65}$$

where $\hat{p}$ =max{$\|\ \bar{P}_0(1|1)\ \|$, $[(1-\beta^s(\bar{\theta}_{N-1}))\gamma^s_{N-1} + \beta^s(\bar{\theta}_{N-1})](\bar{a}^2\bar{s}_{N-1} + \bar{s}^3_{N-1} + 2\bar{s}^3_{N-1}\bar{c}_N + \bar{s}^2_{N-1}\bar{r}_N) + \bar{q}$}.

Next, assume that

$$\begin{aligned} E\{P_{0,k-1|k-1}\} < \hat{p}I_n \sum_{i=0}^{k-2}\{\bar{a}^2 + [(1-\beta^s(\bar{\theta}_{N-1}))\gamma^s_{N-1} + \beta^s(\bar{\theta}_{N-1})] \\ \cdot (\bar{a}^2\bar{c}_N + \bar{s}^2_{N-1}\bar{c}_N + \bar{s}^2_{N-1}\bar{c}^2_N)\}^i \end{aligned} \tag{10.66}$$

Then, we have

$$
\begin{aligned}
&E\{P_{0,k|k}\}\\
&<\{\bar{a}^2+[(1-\beta^s(\bar{\theta}_{N-1}))\gamma^s_{N-1}+\beta^s(\bar{\theta}_{N-1})](\bar{a}^2\bar{c}_N+\bar{s}^2_{N-1}\bar{c}_N+\bar{s}^2_{N-1}\bar{c}^2_N)\}\\
&\cdot E\{P_{0,k-1|k-1}\}+[(1-\beta^s(\bar{\theta}_{N-1}))\gamma^s_{N-1}+\beta^s(\bar{\theta}_{N-1})](\bar{a}^2\bar{s}_{N-1}+\bar{s}^3_{N-1}\\
&+2\bar{s}^3_{N-1}\bar{c}_N+\bar{s}^2_{N-1}\bar{r}_N)I_n+\bar{q}I_n\\
&<\{\bar{a}^2+[(1-\beta^s(\bar{\theta}_{N-1}))\gamma^s_{N-1}+\beta^s(\bar{\theta}_{N-1})](\bar{a}^2\bar{c}_N+\bar{s}^2_{N-1}\bar{c}_N+\bar{s}^2_{N-1}\bar{c}^2_N)\}\\
&\cdot\hat{p}\sum_{i=0}^{k-2}\{\bar{a}^2+[(1-\beta^s(\bar{\theta}_{N-1}))\gamma^s_{N-1}+\beta^s(\bar{\theta}_{N-1})](\bar{a}^2\bar{c}_N+\bar{s}^2_{N-1}\bar{c}_N\\
&+\bar{s}^2_{N-1}\bar{c}^2_N)\}^i I_n+[(1-\beta^s(\bar{\theta}_{N-1}))\gamma^s_{N-1}+\beta^s(\bar{\theta}_{N-1})](\bar{a}^2\bar{s}_{N-1}+\bar{s}^3_{N-1}\\
&+2\bar{s}^3_{N-1}\bar{c}_N+\bar{s}^2_{N-1}\bar{r}_N)I_n+\bar{q}I_n\\
&<\{\bar{a}^2+[(1-\beta^s(\bar{\theta}_{N-1}))\gamma^s_{N-1}+\beta^s(\bar{\theta}_{N-1})](\bar{a}^2\bar{c}_N+\bar{s}^2_{N-1}\bar{c}_N+\bar{s}^2_{N-1}\bar{c}^2_N)\}\\
&\cdot\hat{p}\sum_{i=0}^{k-2}\{\bar{a}^2+[(1-\beta^s(\bar{\theta}_{N-1}))\gamma^s_{N-1}+\beta^s(\bar{\theta}_{N-1})](\bar{a}^2\bar{c}_N+\bar{s}^2_{N-1}\bar{c}_N\\
&+\bar{s}^2_{N-1}\bar{c}^2_N)\}^i I_n+\hat{p}I_n\\
&<\hat{p}\sum_{i=0}^{k-1}\{\bar{a}^2+[(1-\beta^s(\bar{\theta}_{N-1}))\gamma^s_{N-1}+\beta^s(\bar{\theta}_{N-1})](\bar{a}^2\bar{c}_N+\bar{s}^2_{N-1}\bar{c}_N\\
&+\bar{s}^2_{N-1}\bar{c}^2_N)\}^i I_n
\end{aligned}
\tag{10.67}
$$

Setting

$$
\begin{aligned}
&\hat{p}\sum_{i=0}^{k-1}\{\bar{a}^2+[(1-\beta^s(\bar{\theta}_{N-1}))\gamma^s_{N-1}+\beta^s(\bar{\theta}_{N-1})]\\
&\cdot(\bar{a}^2\bar{c}_N+\bar{s}^2_{N-1}\bar{c}_N+\bar{s}^2_{N-1}\bar{c}^2_N)\}^i I_n=\bar{p}I_n
\end{aligned}
\tag{10.68}
$$

It is concluded that the proposed filter converges when

$$
\bar{s}_{N-1}<\{[\frac{1-\bar{a}^2}{1-\beta^s(\bar{\theta}_{N-1})\gamma^s_{N-1}+\beta^s(\bar{\theta}_{N-1})}-\bar{a}^2\bar{c}_N][\bar{c}^2_N(1+\bar{c}_N)]^{-1}\}^{\frac{1}{2}} \tag{10.69}
$$

Then, it is obtained that

$$
E\{P^s_{i,k|k}\}\leq E\{P_{0,k|k}\}<\bar{p}I_n \tag{10.70}
$$

This completes the proof.

Remark 10.2 From Theorem 10.2, we can see that the event–triggered sequential fusion (ETSF) algorithm presented in this chapter is effective. The ETSF is convergent under mild conditions, and the expectation of the associated estimation error covariance is bounded.

10.5 Numerical Example

An example is used to illustrate the effectiveness of the presented algorithm in this section. A tracking system with two sensors can be described by

$$x_{k+1} = \begin{bmatrix} 1 & T_s \\ 0 & 1 \end{bmatrix} x_k + \Gamma_k \xi_k \tag{10.71}$$

$$z_{1,k} = C_1 x_k + v_{1,k} \tag{10.72}$$

$$z_{2,k} = C_2 x_k + v_{2,k} \tag{10.73}$$

$$v_{1,k} = \eta_{1,k} + \beta_1 \xi_{k-1} \tag{10.74}$$

$$v_{2,k} = \eta_{2,k} + \beta_2 \xi_{k-1} \tag{10.75}$$

where $T_s = 0.1$ is the sampling period. $L = 300$ is the length of the signal x to be estimated. The state $x_k = [s_k \ \dot{s}_k]^T$, where s_k and $\dot{s}_k$ are the position and velocity of the target at time kT_s, respectively. $\Gamma_k = [T_s \ \ 1]$ is the noise transition matrix. $\xi_k \in R$ is the system noise, assumed to be white and Gaussian distributed with zero mean and variance σ_ξ^2.

$z_{i,k}$ $(i = 1, 2)$ are the observations of the two sensors, which observe the position and the velocity, respectively, i.e., $C_1 = [1 \ 0]$, $C_2 = [0 \ 1]$. $v_{i,k}$ $(i = 1, 2)$ are the measurement noise of sensor i, and are cross–correlated and coupled with the system noise ξ_{k-1}. The strength of the correlation is determined by β_1 and β_2. $\eta_{i,k}$ $(i = 1, 2)$ are zero mean white Gaussian noise with variances $\sigma_{\eta_i}^2$ and are independent of ξ_k, $k = 1, 2, \ldots$. The initial values are $\bar{x}_0 = [1 \ 1]^T$ and $\bar{P}_0 = I_2$, which is independent of ξ_k and $v_{i,k}$, $k = 1, 2, \ldots, i = 1, 2$.

We have $Q_k = \Gamma_k \Gamma_k^T \sigma_\xi^2$ from Eq. (10.71), which is the system noise covariance corresponding to $w_k = \Gamma_k \xi_k$. From Eqs. (10.74) and (10.75), the measurement noise covariance is given by

$$R_k = \begin{bmatrix} \beta_1^2 \sigma_\xi^2 + \sigma_{\eta_1}^2 & \beta_1 \beta_2 \sigma_\xi^2 \\ \beta_2 \beta_1 \sigma_\xi^2 & \beta_2^2 \sigma_\xi^2 + \sigma_{\eta_2}^2 \end{bmatrix} \tag{10.76}$$

The covariance between w_{k-1} and v_{i,k_i} is

$$S_k^* = [\beta_1 \sigma_\xi^2 \Gamma_{k-1} \ \ \beta_2 \sigma_\xi^2 \Gamma_{k-1}] \tag{10.77}$$

Let the event-triggering threshold $\theta_i = \theta$ for $i = 1, 2$ for simplicity. The value of θ is changing within the set $\theta \in \{0, 0.45, 0.6, 0.8\}$ to illustrate the impact of θ on the estimation performance, where $\theta = 0$ means that the scheduler is time-triggered, that is always activated, and the estimator can receive all the measurements of the corresponding sensor at each instant.

Next, the state estimation of x_k by fusing information from the two sensors is given, and the differences between the estimated results obtained by different estimation

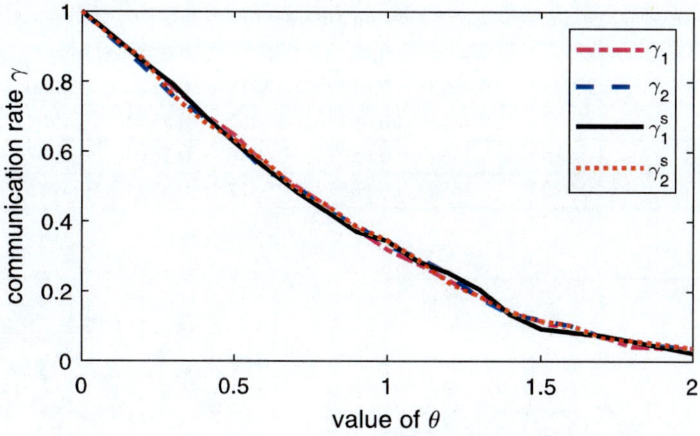

Fig. 10.2 Sensor communication rate γ versus scheduling parameter θ

algorithms in case of noise correlation are compared. In the case of noise correlation, the influence on fusion results when noise correlation is ignored is analyzed.

Denote $\sigma_\xi^2 = 0.3$, $\sigma_{\eta_1}^2 = 25$, $\sigma_{\eta_2}^2 = 36$, $\beta_1 = 6$ and $\beta_2 = 5$, thence the measurement noises are cross-correlated and coupled with the system noise. We set 300 sampling time for 100 Monte Carlo simulations, and observe the effectiveness of the presented algorithm. The simulation results are shown in Figs. 10.3, 10.4, 10.5 and Tables 10.2, 10.3.

The average communication rate of sensor i $(i=1,2)$ for the ETKF algorithm and the ETSF algorithm is defined as follows, respectively [16]

$$\gamma_i = \frac{1}{M}\sum_{k=1}^{M}\gamma_{i,k}, \qquad \gamma_i^s = \frac{1}{M}\sum_{k=1}^{M}\gamma_{i,k}^s \tag{10.78}$$

Figure 10.2 and Table 10.1 show the relationship between event–triggered threshold θ and average sensor communication rate γ. γ_i, $i = 1, 2$ denote the sensor communication rate of the ETKF algorithm, γ_i^s, $i = 1, 2$ denote the sensor communication rate of the ETSF algorithm. It is shown in Fig. 10.2 and Table 10.1 that as the event–triggered threshold increases, the communication rate decreases, and the communication rate of the ETSF algorithm is similar to the ETKF algorithm.

The statistical simulation curves of the Root Mean Square Errors (RMSEs) of the event–triggered Kalman filter (ETKF) and the proposed event–triggered sequential fusion estimation algorithm (ETSF) with different triggering thresholds are shown in Fig. 10.3. The pink solid lines denote $\theta = 0$, the green dash-dotted lines denote $\theta = 0.45$, and the red dashed lines, the blue dotted line denote $\theta = 0.6$ and $\theta = 0.8$, respectively. From which, one can see that the estimation curves of the proposed sequential algorithm have a better estimation than the ETKF with the same θ, which illustrates that the proposed algorithm is superior than the classical Kalman filter. It

Table 10.1 Average communication rate γ with different θ

θ	0	0.1	0.3	0.45	0.5	0.6	0.7	0.8
γ_1	1	0.9170	0.7633	0.6550	0.6213	0.5517	0.4880	0.4327
γ_1^s	1	0.9227	0.7623	0.6577	0.6267	0.5550	0.4963	0.4383
γ_2	1	0.9153	0.7667	0.6593	0.6270	0.5610	0.4907	0.4303
γ_2^s	1	0.9223	0.7650	0.6657	0.6317	0.5593	0.4933	0.4247

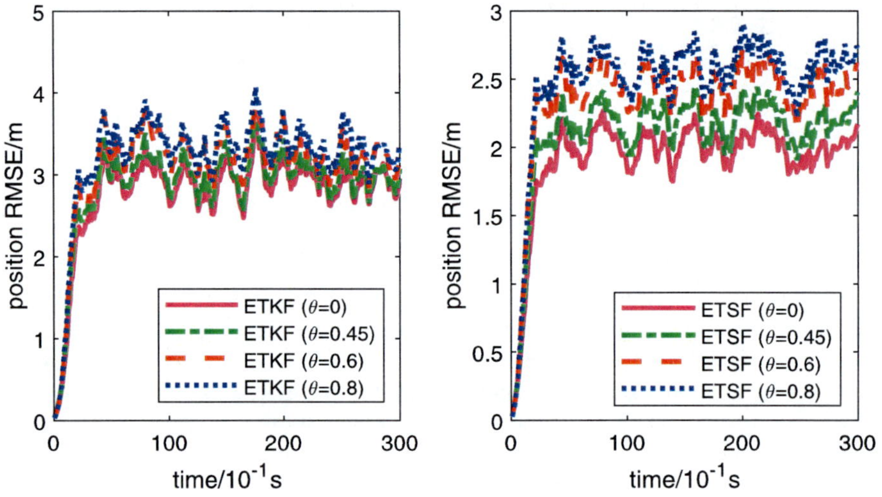

Fig. 10.3 RMSE curves of ETKF and ETSF with different thresholds

Table 10.2 Time-Averaged RMSEs for algorithms with different θ

θ	0	0.45	0.6	0.8
ETSF	1.9056	2.0728	2.3128	2.4341
ETKF	2.8041	2.8929	3.0975	3.2214

is also shown in Fig. 10.3 that state estimation effect of two algorithms with smaller triggering thresholds is always better than those with larger triggering thresholds.

The time-averaged RMSEs of the ETSF algorithm and the ETKF algorithm are shown in Table 10.2. It can be seen from Table 10.2 that the time-averaged RMSEs of the ETSF algorithm are less than those of ETKF algorithm for any given θ, which means the ETSF algorithm outperforms the ETKF algorithm. Note that $\theta = 0$ means all raw sensor measurements have been transmitted, and the system reduces to be time-triggered. Therefore, the proposed algorithm with $\theta = 0$ has the best estimation performance.

Figure 10.4 shows the statistical simulation curves of RMSEs of the proposed ETSF algorithm, ignore noise correlation event–triggered sequential fusion esti-

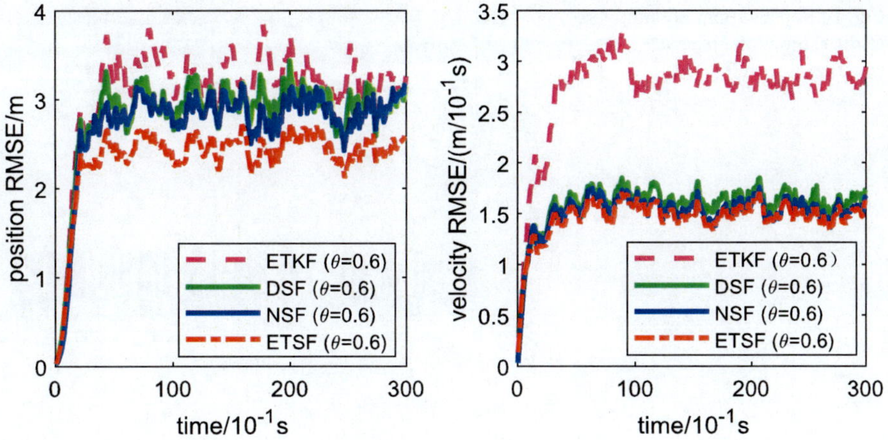

Fig. 10.4 RMSE curves for algorithms with $\theta = 0.6$

Table 10.3 Time-averaged RMSEs for algorithms with different θ

θ	0	0.45	0.6	0.8
ETSF	1.9109	2.0936	2.3461	2.4812
DSF	1.9109	2.2377	2.8180	3.1858
NSF	2.5322	2.6362	2.7245	2.7847
ETKF	2.7775	2.8777	3.0908	3.2096

mation algorithm (NSF), the data dropout sequential fusion estimation algorithm (DSF) and the ETKF algorithm with thresholds $\theta = 0.6$. The NSF algorithm means an event–triggered sequential fusion algorithm neglecting noise correlation. The DSF algorithm means the sequential fusion algorithm that considers the correlation between noises but treats the untriggered measurements as package loss. From which, one can see that the RMSE curves of the proposed ETSF algorithm with $\theta = 0.6$ are much lower than the corresponding ones of the other algorithms, which means that the ETSF algorithm with consideration of noise correlation has better estimation performance while neglecting noise correlation can reduce state estimation precision. Compared with the DSF algorithm, the ETSF algorithm can ensure estimation performance and is more energy efficient.

Time-averaged position RMSEs of the ETSF algorithm, the DSF algorithm, the NSF algorithm and the ETKF algorithm with different triggering thresholds are shown in Table 10.3. From which, one can see that the ETSF algorithm is better than DSF and NSF, while ETKF is the worst when θ takes the same value. As θ increases, the amount of communication data decreases, and the estimation accuracy of each algorithm deteriorates. But regardless of the conditions, the ETSF algorithm proposed in this chapter is optimal.

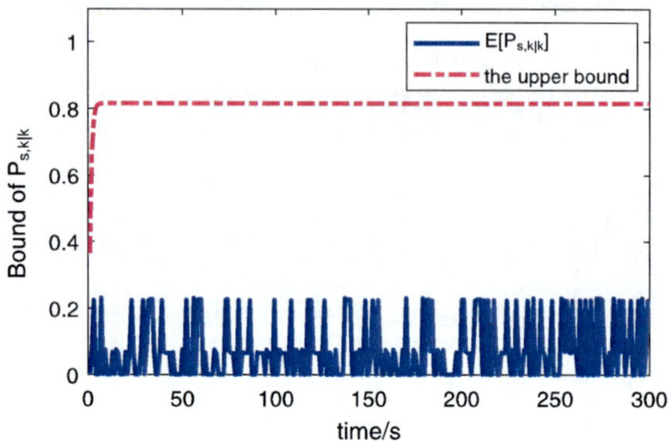

Fig. 10.5 The upper bound of the estimation error covariance with $\theta = 0.6$

The upper bound of the estimation error covariance of the proposed ETSF algorithm with $\theta = 0.6$ is shown in Fig. 10.5, and the conclusion of Theorem 10.2 is verified.

10.6 Conclusions

The event–triggered sequential fusion (ETSF) algorithm for a class of multi–sensor dynamic systems with correlated noises is presented, where the noise of different sensors is cross–correlated and coupled with the system noise of the previous time step. The shown theoretical deductions and conducted simulations lead to the following conclusions:

(1) The proposed event–triggered sequential fusion algorithm is effective, which is superior to the event–triggered Kalman filter of single sensor, the event–triggered sequential fusion algorithm that neglect the correlation of the noises and the sequential fusion algorithm with intermittent observations;

(2) The relation between the estimation performance and the averaged communication rate of the event–triggered algorithm is analyzed, which shows that the estimation performance will exacerbate with the decease of the communication rate, while, a little bit of exacerbation of estimation performance will reduce a large amount of data transmission for example, as shown in Tables 10.2 and 10.3, when the transmission rate reduce 37%, the averaged RMSEs by the ETSF deteriorate from 1.9056 to 2.0728;

(3) The stability of the designed fusion algorithm is proved and an upper bound of the estimation error covariance is given.

The algorithm put forward in this chapter has potential value in many applications including multi–sensor networked control systems and cyber physical systems.

References

1. Bian, Xiaolei, X. Rong Li, and Vesselin P Jilkov. 2018. Remote state estimation with data-driven communication and guaranteed stability. In *21st International Conference on Information Fusion*, pages 846–853, July10–13 Cambridge, UK.
2. Gao, Xiaobin, Emrah Akyol, and Tamer Başar. 2018. Optimal communication scheduling and remote estimation over an additive noise channel. *Automatica* 88: 57–69.
3. Gu, Zhou, Linghui Yang, Engang Tian, and Huan Zhao. 2017. Event-triggered reliable H_∞ filter design for networked systems with multiple sensor distortions: a probabilistic partition approach. *ISA Transactions* 66 (1): 2–9.
4. Jiang, Lu, Liping Yan, and Yuanqing Xia. 2018. Distributed fusion in wireless sensor networks based on a novel event-triggered strategy. *Journal of the Franklin Institute* 365 (17): 1–20.
5. Jiang, Lu, Liping Yan, and Yuanqing Xia. 2017. Event-triggered multisensor data fusion with correlated noise. In *20st International conference on Information Fusion*, pages 1–8. Xi'an, China, July 10–13 Xi'an, China.
6. Jin, Zengwang, Yanyan Hu, and Changyin Sun. 2019. Event-triggered information fusion for networked systems with missing measurements and correlated noises. *Neurocomputing* 332: 15–28.
7. Jin, Zengwang, Yanyan Hu, Changyin Sun, and Lan Zhang. 2015. Event-triggered state fusion estimation for wireless sensor networks with feedback. In *34th Chinese Control Conference*, pages 4610–4614, Hangzhou, China.
8. Julier, S., and J. Uhlmann. 2004. Unscented filtering and nonlinear estimation. *Proceedings of the IEEE* 92 (3): 401–422.
9. Li, Huiping, and Yang Shi. 2014. Event-triggered robust model predictive control of continuous-time nonlinear systems. *Automatica* 50 (5): 1507–1513.
10. Li, Li, Mengfei Niu, Yuanqing Xia, et al. 2018. Event-triggered distributed fusion estimation with random transmission delays. *Information Sciences* 475 (2019): 2–9.
11. Li, Li, Mengfei Niu, Hongjiu Yang, and Zhixin Liu. 2018. Event-triggered nonlinear filtering for networked systems with correlated noises. *Journal of the Franklin Institute* 355 (13): 5811–5829.
12. Li, Qi, Bo Shen, and Zidong Wang. 2018. Synchronization control for a class of discrete time-delay complex dynamical networks: A dynamic event-triggered approach. *IEEE Transactions on Cybernetics* 49 (5): 1979–1986.
13. Liu, Qinyuan, Zidong Wang, Xiao He, and Donghua Zhou. 2015. Event-based recursive distributed filtering over wireless sensor networks. *IEEE Transactions on Automatic Control* 60 (9): 2470–2475.
14. Miskowicz, M. 2006. Send-on-delta concept: an event-based data reporting strategy. *sensors* 6 (1): 49–63.
15. Nguyen, V.H., and Y.S. Suh. 2007. Improving estimation performance in networked control systems applying the send-on-delta transmission method. *Sensors* 7 (10): 2128–2138.
16. Shi, Dawei, Tongwen Chen, and Ling Shi. 2014. Event-triggered maximum likelihood state estimation. *Automatica* 50 (1): 247–254.
17. Su, Housheng, Zhenghao Li, and Yanyan Ye. 2017. Event-triggered Kalman-consensus filter for two-target tracking sensor networks. *ISA Transactions* 71 (1): 103–111.
18. Trimpe, S., and M.C. Campi. 2015. On the choice of the event trigger in event-based estimation. In *2015 International Conference on Event-based Control, Communication, and Signal Processing*, pages 1–8, June 17–19 Krakow, Poland.
19. Trimpe, S., and R. D'Andrea. 2011. Reduced communication state estimation for control of an unstable networked control system. In *50th IEEE Conference on Decision and Control and European Control Conference*, pages 2361–2368, Dec 12–15 Orlando, FL, USA.
20. Wu, Junfeng, Qingshan Jia, Karl Henrik Johansson, and Ling Shi. 2013. Event-based sensor data scheduling: Trade-off between communication rate and estimation quality. *IEEE Transactions on Automatic Control* 58 (4): 1041–1046.

21. Yan, Liping, X. Rong Li, and Yuanqing Xia. 2013. Optimal sequential and distributed fusion for state estimation in cross-correlated noise. *Automatica* 49 (12): 3607–3612.
22. Zhang, Dan, Peng Shi, Qingguo Wang, and Li Yu. 2017. Distributed non-fragile filtering for t-s fuzzy systems with event-based communications. *Fuzzy Sets and Systems* 306 (2017): 137–152.
23. Zheng, Xiujuan, and Huajing Fang. 2016. Recursive state estimation for discretetime nonlinear systems with event-triggered data transmission, normbounded uncertainties and multiple missing measurements. *International Journal of Robust and Nonlinear Control* 26 (17): 3673–3695.

Part IV
Fusion Estimation for Systems with Heavy-Tailed Noises

Chapter 11
Distributed Fusion Estimation for Multisensor Systems with Heavy-Tailed Noises

As is known to all that for the linear dynamic systems with Gaussian noises, the classical Kalman filter can get the optimal estimates results in the sense of least mean square error (LMSE) [21, 23]. There are many Gaussian fusion algorithms under the framework of Kalman filter [4, 5, 22, 27, 28]. And they can generally be divided into two types, which are the centralized fusion and the distributed fusion [6]. If the fusion center can receive all measurements of every local sensor at once, the centralized fusion usually can get the optimal estimation in the sense of LMSE. However, due to the fact that in practical applications there may exist some bad impact caused by the limited bandwidth or delay of the transmission, to make the system be more robust in a poor environment, it is better to implement the distributed fusion strategy, which means that every local agent should use its local measurements to get the local estimates and send it to the fusion center [16, 32]. Then, the fusion center will generate a globally estimation with higher accuracy by using the local estimations. Briefly, the aim of the centralized fusion is to achieve the globally optimal estimates by using the measurements directly, while the kernel problem of the distributed fusion is how to process the measurements efficiently and fuse the local estimates to get the state estimation with higher accuracy and reliability [11].

In some real engineering applications, the outliers are produced inevitably by unreliable sensor or unknown disturbances, so the process and measurement noises are both non–Gaussian. In this part, we use two chapters to introduce the multisensor data fusion for state estimation of dynamic systems with non–Gaussian but heavy-tailed noises.

11.1 Introduction

For the state estimation of systems disturbed by non–Gaussian noises, there are already some adaptive algorithms on the base of modified Kalman filter [13, 14, 24]. Based on the mixture of Gaussians, a more general framework for filter of

L. Yan et al., *Multisensor Fusion Estimation Theory and Application*,
https://doi.org/10.1007/978-981-15-9426-7_11

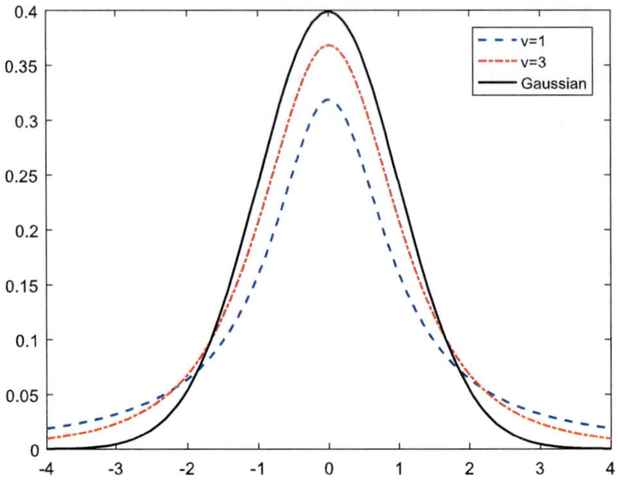

Fig. 11.1 Gaussian distribution and Student's t distribution with different dof

non-Gaussian systems is proposed [13]. In [14], based on mixture of Gaussian noises the statistical fusion approaches were divided into four types and under Bayesian framework a cost-effective fusion strategy has been presented. However, when there are outliers in the system, the Gaussian mixture cannot model the non–Gaussian noise well so that the results are unsatisfactory [20, 24].

Instead of Gaussian mixture, Student's t distribution can model the non–Gaussian noise more effectively [15, 18, 25]. As Fig. 11.1 shows, t distribution takes extreme value with higher possibility than the Gaussian distribution, so it has the heavy–tailed shape in both sides and is robust to outliers, where ν is the degrees of freedom (dof) of the t distribution. In recent years, the multivariate student t distribution is widely used to model the heavy–tailed noises by many researchers when they study the problem of filtering or smoothing for system with heavy–tailed noises since it has many advantages, such as, it can cope with the outliers, can be easily carry out with proper computing complexity, and does not have to tune the parameter [2, 3, 7, 8, 10, 15, 19].

For linear systems disturbed by heavy–tailed noises, the filter and the smoother are derived in reference [1, 30]. For filtering algorithm of nonlinear systems with heavy–tailed noises, Piché et al. proposed the recursive robust filter and smoother based on the Student's t distribution, which can cope with the outliers [15]. By use of the variational Bayesian algorithm, an improved Gaussian approximate filter for the nonlinear system with heavy–tailed noise is proposed [7]. The robust Student's t based nonlinear filter (RSTNF) and smoother (RSTNS) for dynamic nonlinear systems under heavy–tailed noises by use of the unscented transform (UT) were derived in [9]. Besides, based on cubature filter, a robust filtering method for systems with heavy–tailed noises is derived in [8]. Reference [3] analyzed the stability of the heavy–tailed dynamics. Literature [10] studied the modeling and stochastic linearization of heavy–tailed noises based systems. In addition, an unscented particle filter by using the multivariate t distribution is studied in [2].

However, the algorithms mentioned above only consider the state estimation by filter or smoother of heavy–tailed noises based single sensor systems. For fusion estimation of linear multisensor dynamic systems disturbed by heavy–tailed noises, the research results are few. References [20, 31] studied the centralized fusion of linear time–invariant and nonlinear systems with heavy–tailed noises, respectively.

From the above analysis, we can get that there are many open problems about filtering and fusion estimation of dynamic systems with heavy–tailed noises:

- The information filter based on the Student's t distribution for dynamic systems with heavy–tailed noise is not derived. Although literature [20] used the cubature information filter (CIF) to get the fusion estimation, the real information filter for systems with heavy–tailed noise or formulated by Student's t distribution are not derived yet.
- The distributed fusion for linear dynamic systems disturbed by heavy–tailed noises is not derived.

In this chapter, the information filter for linear time–variant dynamic systems disturbed by heavy–tailed noises is derived. Based on the information filter, the distributed fusion algorithm is proposed to solve the estimation problem of multisensor linear time–variant systems, and it is proved that the distributed fusion is equivalent to the derived centralized fusion in the sense of LMSE. To decrease computation complexity, a suboptimal distributed fusion is also given in this chapter.

The rest of this chapter is organized as follows. In Sect. 11.2, the problem is formulated. Section 11.3 derives the information filter for the linear time–variant dynamic system with heavy–tailed noises. Section 11.4 presents the centralized fusion and the distributed fusion algorithms in sequence. Section 11.5 shows the numerical simulation and Sect. 11.6 draws the conclusion.

11.2 Problem Formulation

A linear dynamic system where one target observed by N sensors can be formulated as [4, 17, 29]

$$x_{k+1} = F_k x_k + w_k, \quad k = 0, 1, \ldots \tag{11.1}$$

$$z_{i,k} = H_{i,k} x_k + v_{i,k}, \quad i = 1, 2, \ldots, N \tag{11.2}$$

where $i = 1, 2, \ldots, N$ denotes the ith sensor. $x_k \in R^q$ is the system state at the kth time instant, $F_k \in R^{q\times q}$ is the state transition matrix. $z_{i,k} \in R^{m_i}$ is the observation of sensor i at time k. $H_{i,k}$ is the observation matrix of the ith sensor. The system noise w_k and the measurement noise $v_{i,k}$ are heavy–tailed noises, which can be modeled by the multivariate t distribution as,

$$p(w_k) = \mathrm{St}(w_k; 0, Q_k, \nu_w) \tag{11.3}$$
$$p(v_{i,k}) = \mathrm{St}(v_{i,k}; 0, R_{i,k}, \nu_i) \tag{11.4}$$

where $\mathrm{St}(\cdot; \bar{x}, P, \nu)$ is the multivariate t distribution whose mean is $\bar{x}$, the scale matrix is P and degrees of freedom (dof) is ν. For a random vector x with the probability density function of $\mathrm{St}(\cdot; \bar{x}, P, \nu)$, $\frac{\nu}{\nu-2}P$ is the covariance matrix [12].

Similarly, it is assumed that the system state initial value x_0 is also heavy–tailed that meets the multivariate t distribution whose mean is $\hat{x}_{0|0}$, the scale matrix is $P_{0|0}$ and the dof is ν_0, i.e.,

$$p(x_0) = \mathrm{St}(x_0; \hat{x}_{0|0}, P_{0|0}, \nu_0) \tag{11.5}$$

It is assumed that x_0,w_k and $v_{i,k}$ are mutually independent.

The object of this chapter is to obtain the optimal estimation of x_l by using multisensor observations $Z_l = \{z_{i,t}, t = 1, 2, \ldots, l; i = 1, 2, \ldots, N\}$.

11.3 Linear Filter and Information Filter for Systems with Heavy-Tailed Noises

For system (11.1)–(11.2), when $N = 1$, it is the single sensor linear time–variant dynamic system. For simplicity, denote $z_k = z_{1,k}$, $H_k = H_{1,k}$, $v_k = v_{1,k}$, and $R_k = R_{1,k}$. Then, system (11.1)–(11.2) can be rewritten as

$$x_{k+1} = F_k x_k + w_k \tag{11.6}$$
$$z_k = H_k x_k + v_k \tag{11.7}$$

where,

$$\begin{cases} p(w_k) = \mathrm{St}(w_k; 0, Q_k, \nu_w) \\ p(v_k) = \mathrm{St}(v_k; 0, R_k, \nu_1) \\ p(x_0) = \mathrm{St}(x_0; \hat{x}_{0|0}, P_{0|0}, \nu_0) \end{cases} \tag{11.8}$$

In this section, the information filter for single sensor system (11.6)–(11.8) is derived. First, we will introduce three lemmas before that.

Lemma 11.1 *Random vector x meets the multivariate t distribution* $\mathrm{St}(x; \bar{x}, P, \nu)$ *if its probability density function (pdf) is:*

$$p(x) = \frac{\Gamma(\frac{\nu+2}{2})}{\Gamma(\frac{\nu}{2})} \frac{1}{(\nu\pi)^{\frac{d}{2}}} \frac{1}{\sqrt{\det(P)}} \left(1 + \frac{\Delta^2}{\nu}\right)^{-\frac{\nu+2}{2}} \tag{11.9}$$

where $\Delta^2 = (x - \bar{x})^{\mathrm{T}} P^{-1}(x - \bar{x})$. $\bar{x}$ *is the mean of* x, P *is the scale matrix, and* ν *is the degrees of freedom (dof).*

It has the following properties:

- *The covariance of* x *is* $\frac{\nu}{\nu-2}P$.
- *The distribution of* x *tends to Gaussian when* ν *increases to infinity.*
- *Let* $z = Ax + B$, *then* $p(z) = \mathrm{St}(z; A\bar{x} + B, APA^{\mathrm{T}}, \nu)$, *where* A *and* B *are matrix and vector of proper dimensions.*
- *If* $x_1 \in R^{d_1}$ *and* $x_2 \in R^{d_2}$ *are jointly t–distributed with pdf of*

$$p(x_1, x_2) = \mathrm{St}\left(\begin{bmatrix} x_1 \\ x_2 \end{bmatrix}; \begin{bmatrix} \bar{x}_1 \\ \bar{x}_2 \end{bmatrix}, \begin{bmatrix} P_{11} & P_{12} \\ P_{21} & P_{22} \end{bmatrix}, \nu\right) \tag{11.10}$$

where $P_{ii} \in R^{d_i \times d_i}$ *are the scale matrices of* x_i, $P_{ij} \in R^{d_i \times d_j}$ *are the joint scale matrices of* x_i *and* x_j, $i = 1, 2$, *then the marginal pdfs of* x_1 *and* x_2 *are*

$$\begin{cases} p(x_1) = \mathrm{St}(x_1; \bar{x}_1, P_{11}, \nu) \\ p(x_2) = \mathrm{St}(x_2; \bar{x}_2, P_{22}, \nu) \end{cases} \tag{11.11}$$

The conditional pdf $p(x_1|x_2)$ *is given by*

$$p(x_1|x_2) = \mathrm{St}(x_1; \bar{x}_{1|2}, P_{1|2}, \nu_{1|2}) \tag{11.12}$$

where

$$\nu_{1|2} = \nu + d_2 \tag{11.13}$$

$$\bar{x}_{1|2} = \bar{x}_1 + P_{12}P_{22}^{-1}(x_2 - \bar{x}_2) \tag{11.14}$$

$$P_{1|2} = \frac{\nu + \Delta_2^2}{\nu + d_2}(P_{11} - P_{12}P_{22}^{-1}P_{12}^{\mathrm{T}}) \tag{11.15}$$

and where $\Delta_2^2 = (x_2 - \bar{x}_2)^{\mathrm{T}} P_{22}^{-1}(x_2 - \bar{x}_2)$.

Lemma 11.2 (Matrix Inversion Lemma) *[21] Suppose* $M_1 \in R^{n\times n}$, $M_2 \in R^{n\times p}$, $M_3 \in R^{p\times n}$, $M_4 \in R^{p\times p}$, *where* M_1, $M_1 + M_2M_4^{-1}M_3$, $M_4 + M_3M_1^{-1}M_2$ *are full rank, then,*

$$(M_1 + M_2M_4^{-1}M_3)^{-1} = M_1^{-1} - M_1^{-1}M_2(M_4 + M_3M_1^{-1}M_2)^{-1}M_3M_1^{-1} \tag{11.16}$$

Lemma 11.3 (Filter for Linear Dynamic System with Heavy–tailed Noises) *The multivariate t distribution based filter for system (11.6)–(11.8) is given by [9]*

$$\begin{cases} \hat{x}_{k|k-1} = F_{k-1}\hat{x}_{k-1|k-1} \\ P_{k|k-1} = F_{k-1}P_{k-1|k-1}F_{k-1}^{\mathrm{T}} + \bar{Q}_{k-1} \\ \hat{z}_{k|k-1} = H_k\hat{x}_{k|k-1} \\ \hat{x}_{k|k} = \hat{x}_{k|k-1} + K_k(z_k - \hat{z}_{k|k-1}) \\ P_{k|k} = a_k(P_{k|k-1} - K_k P_{k|k-1}^{\tilde{z}\tilde{z}} K_k^{\mathrm{T}}) \\ K_k = P_{k|k-1}^{\tilde{x}\tilde{z}}(P_{k|k-1}^{\tilde{z}\tilde{z}})^{-1} \\ P_{k|k-1}^{\tilde{z}\tilde{z}} = H_k P_{k|k-1} H_k^{\mathrm{T}} + \bar{R}_k \\ P_{k|k-1}^{\tilde{x}\tilde{z}} = P_{k|k-1} H_k^{\mathrm{T}} \\ a_k = \frac{(\nu_0-2)(\nu_0+\Delta_i^2)}{\nu_0(\nu_0+m_1-2)} \\ \Delta_k = \sqrt{(z_k - \hat{z}_{k|k-1})^{\mathrm{T}}(P_{k|k-1}^{\tilde{z}\tilde{z}})^{-1}(z_k - \hat{z}_{k|k-1})} \\ \bar{Q}_k = \frac{\nu_w(\nu_0-2)}{(\nu_w-2)\nu_0} Q_k \\ \bar{R}_k = bR_k \\ b = \frac{(\nu_0-2)\nu_1}{\nu_0(\nu_1-2)} \end{cases} \tag{11.17}$$

where $\hat{x}_{k|k-1}$ and $\hat{x}_{k|k}$ are the state prediction and the state estimate, $P_{k|k-1}$ and $P_{k|k}$ are the scale matrices of the prediction error and the estimation error. m_1 is the dimension of measurement z_k. $\hat{z}_{k|k-1}$ is the measurement prediction, and $P_{k|k-1}^{\tilde{z}\tilde{z}}$ is the scale matrix of its prediction error. K_k is the innovation matrix.

Proof From Lemma 11.1 [9],

$$\begin{aligned} \hat{x}_{k|k-1} &= \int x_k p(x_k|Z_{k-1})\mathrm{d}x_k \\ &= F_{k-1}\hat{x}_{k-1|k-1} \end{aligned} \tag{11.18}$$

Denote $\tilde{z}_{k|k-1} = z_k - \hat{z}_{k|k-1}$, $\tilde{x}_{k|k-1} = x_k - \hat{x}_{k|k-1}$, then

$$\begin{aligned} P_{k|k-1} &= \frac{\nu_0 - 2}{\nu_0} E[\tilde{x}_{k|k-1}\tilde{x}_{k|k-1}^{\mathrm{T}}|Z_{k-1}] \\ &= \frac{\nu_0 - 2}{\nu_0} \int \tilde{x}_{k|k-1}\tilde{x}_{k|k-1}^{\mathrm{T}} p(x_k|Z_{k-1})\mathrm{d}x_k \\ &= \frac{\nu_0 - 2}{\nu_0} \int x_k x_k^{\mathrm{T}} p(x_k|Z_{k-1})\mathrm{d}x_k - \frac{\nu_0 - 2}{\nu_0}\hat{x}_{k|k-1}\hat{x}_{k|k-1}^{\mathrm{T}} \\ &= \frac{\nu_0 - 2}{\nu_0} \int x_k x_k^{\mathrm{T}} \{\int \mathrm{St}(x_k; F_{k-1}x_{k-1}, Q_{k-1}, \nu_w) \\ &\quad \times \mathrm{St}(x_{k-1}; \hat{x}_{k-1|k-1}, P_{k-1|k-1}, \nu_0)\mathrm{d}x_{k-1}\}\mathrm{d}x_k \\ &\quad - \frac{\nu_0 - 2}{\nu_0}\hat{x}_{k|k-1}\hat{x}_{k|k-1}^{\mathrm{T}} \\ &= \frac{\nu_0 - 2}{\nu_0} F_{k-1}(\hat{x}_{k-1|k-1}\hat{x}_{k-1|k-1}^{\mathrm{T}} + \frac{\nu_0}{\nu_0 - 2} P_{k-1|k-1})F_{k-1}^{\mathrm{T}} \end{aligned} \tag{11.19}$$

$$
\begin{aligned}
& -\frac{\nu_0-2}{\nu_0}\hat{x}_{k|k-1}\hat{x}_{k|k-1}^{\mathrm{T}}+\frac{\nu_w(\nu_0-2)}{(\nu_w-2)\nu_0}Q_{k-1} \\
& =F_{k-1}P_{k-1|k-1}F_{k-1}^{\mathrm{T}}+\frac{\nu_w(\nu_0-2)}{(\nu_w-2)\nu_0}Q_{k-1}
\end{aligned}
$$

$$
\begin{aligned}
\hat{z}_{k|k-1} &= E\{z_k|Z_{k-1}\}=\int z_k P(z_k|Z_{k-1})\mathrm{d}z_k \\
&= \int\{\int z_k \mathrm{St}(z_k;H_k x_k,R_k,\nu_1)\mathrm{d}z_k\}\times \mathrm{St}(x_k;\hat{x}_{k|k-1},P_{k|k-1},\nu_0)\mathrm{d}x_k \\
&= \int H_k x_k \mathrm{St}(x_k;\hat{x}_{k|k-1},P_{k|k-1},\nu_0)\mathrm{d}x_k \\
&= H_k\hat{x}_{k|k-1}
\end{aligned}
\tag{11.20}
$$

$$
\begin{aligned}
P_{k|k-1}^{\tilde{z}\tilde{z}} &= \frac{\nu_0-2}{\nu_0}E\{\tilde{z}_{k|k-1}\tilde{z}_{k|k-1}^{\mathrm{T}}|Z_{k-1}\} \\
&= \frac{\nu_0-2}{\nu_0}\int z_k z_k^{\mathrm{T}} p(z_k|Z_{k-1})\mathrm{d}z_k-\frac{\nu_0-2}{\nu_0}\hat{z}_{k|k-1}\hat{z}_{k|k-1}^{\mathrm{T}} \\
&= \frac{\nu_0-2}{\nu_0}\int\{\int z_k z_k^{\mathrm{T}}\mathrm{St}(z_k;H_k x_k,R_k,\nu_1)\mathrm{d}z_k\} \\
&\quad \times \mathrm{St}(x_k;\hat{x}_{k|k-1},P_{k|k-1},\nu_0)\mathrm{d}x_k-\frac{\nu_0-2}{\nu_0}\hat{z}_{k|k-1}\hat{z}_{k|k-1}^{\mathrm{T}} \\
&= \frac{\nu_0-2}{\nu_0}\int H_k(\hat{x}_{k|k-1}\hat{x}_{k|k-1}^{\mathrm{T}}+\frac{\nu_0}{\nu_0-2}P_{k|k-1})H_k^{\mathrm{T}}\mathrm{d}x_k \\
&\quad -\frac{\nu_0-2}{\nu_0}\hat{z}_{k|k-1}\hat{z}_{k|k-1}^{\mathrm{T}}+\frac{\nu_1(\nu_0-2)}{(\nu_1-2)\nu_0}R_k \\
&= H_k P_{k|k-1}H_k^{\mathrm{T}}+\frac{(\nu_0-2)\nu_1}{\nu_0(\nu_1-2)}R_k
\end{aligned}
\tag{11.21}
$$

$$
\begin{aligned}
P_{k|k-1}^{\tilde{x}\tilde{z}} &= \frac{\nu_0-2}{\nu_0}E\{\tilde{x}_{k|k-1}\tilde{z}_{k|k-1}^{\mathrm{T}}|Z_{k-1}\} \\
&= \frac{\nu_0-2}{\nu_0}\iint x_k z_k^{\mathrm{T}} p(x_k,z_k|Z_{k-1})\mathrm{d}x_k\mathrm{d}z_k-\frac{\nu_0-2}{\nu_0}\hat{x}_{k|k-1}\hat{z}_{k|k-1}^{\mathrm{T}} \\
&= \frac{\nu_0-2}{\nu_0}\int x_k\{\int z_k^{\mathrm{T}} p(z_k|x_k)\mathrm{d}z_k\}p(x_k|Z_{k-1})\mathrm{d}x_k-\frac{\nu_0-2}{\nu_0}\hat{x}_{k|k-1}\hat{z}_{k|k-1}^{\mathrm{T}} \\
&= \frac{\nu_0-2}{\nu_0}\int x_k\{\int z_k^{\mathrm{T}}\mathrm{St}(z_k;H_k x_k,R_k,\nu_1)\mathrm{d}z_k\} \\
&\quad \times \mathrm{St}(x_k;\hat{x}_{k|k-1},P_{k|k-1},\nu_0)\mathrm{d}x_k-\frac{\nu_0-2}{\nu_0}\hat{x}_{k|k-1}\hat{z}_{k|k-1}^{\mathrm{T}}
\end{aligned}
\tag{11.22}
$$

$$= \frac{\nu_0 - 2}{\nu_0}(\hat{x}_{k|k-1}\hat{z}^{\mathrm{T}}_{k|k-1} + \frac{\nu_0}{\nu_0 - 2}P_{k|k-1})H_k^{\mathrm{T}} - \frac{\nu_0 - 2}{\nu_0}\hat{x}_{k|k-1}\hat{z}^{\mathrm{T}}_{k|k-1}$$
$$= P_{k|k-1}H_k^{\mathrm{T}}$$

From Lemma 11.1,

$$p(x_k|Z_k) = \frac{p(x_k, z_k|Z_{k-1})}{p(z_k|Z_{k-1})} = \mathrm{St}(x_k; \hat{x}'_{k|k}, P'_{k|k}, \nu'_0) \tag{11.23}$$

where $\hat{x}'_{k|k}$, $P'_{k|k}$ and ν'_0 are given by

$$\nu'_0 = \nu_0 + m_1 \tag{11.24}$$
$$\hat{x}'_{k|k} = \hat{x}_{k|k-1} + K_k[z_k - \hat{z}_{k|k-1}] \tag{11.25}$$
$$P'_{k|k} = \frac{\nu_0 + \Delta_k^2}{\nu_0 + m_1}(P_{k|k-1} - K_k P^{\tilde{z}\tilde{z}}_{k|k-1} K_k^{\mathrm{T}}) \tag{11.26}$$
$$K_k = P^{\tilde{x}\tilde{z}}_{k|k-1}(P^{\tilde{z}\tilde{z}}_{k|k-1})^{-1}, \tag{11.27}$$
$$\Delta_k = \sqrt{(z_k - \hat{z}_{k|k-1})^{\mathrm{T}}(P^{\tilde{z}\tilde{z}}_{k|k-1})^{-1}(z_k - \hat{z}_{k|k-1})} \tag{11.28}$$

From [9, 19], the moment matching approach is used to keep the heavy–tailed property, and the final state estimation and the estimation error scale matrix are given by

$$\hat{x}_{k|k} = \hat{x}'_{k|k} = \hat{x}_{k|k-1} + K_k(z_k - \hat{z}_{k|k-1}) \tag{11.29}$$
$$P_{k|k} = \frac{(\nu_0 - 2)\nu'_0}{(\nu'_0 - 2)\nu_0}P'_{k|k} = \frac{(\nu_0 + \Delta_k^2)(\nu_0 - 2)}{\nu_0(\nu_0 + m_1 - 2)}(P_{k|k-1} - K_k P^{\tilde{z}\tilde{z}}_{k|k-1}K_k^{\mathrm{T}}) \tag{11.30}$$

Theorem 11.1 (Information Filter for Linear Dynamic Systems with Heavy–tailed Noises) *The multivariate t distribution based information filter for system (11.6)–(11.8) is computed by*

$$\begin{cases}
\hat{\xi}_{k|k} = \frac{1}{a_k}(\hat{\xi}_{k|k-1} + H_k^{\mathrm{T}}\bar{R}_k^{-1}z_k) \\
\hat{\xi}_{k|k-1} = [I - A_{k-1}(A_{k-1} + \bar{Q}_{k-1}^{-1})^{-1}]F_{k-1}^{-\mathrm{T}}\hat{\xi}_{k-1|k-1} \\
\Lambda_k = \frac{1}{a_k}(\Lambda_{k|k-1} + H_k^{\mathrm{T}}\bar{R}_k^{-1}H_k) \\
\Lambda_{k|k-1} = [I - A_{k-1}(A_{k-1} + \bar{Q}_{k-1}^{-1})^{-1}]A_{k-1} \\
a_k = \frac{(\nu_0-2)(\nu_0+\Delta_k^2)}{\nu_0(\nu_0+m_1-2)} \\
\bar{R}_k = \frac{\nu_1(\nu_0-2)}{(\nu_1-2)\nu_0}R_k = bR_k \\
A_{k-1} = F_{k-1}^{-\mathrm{T}}\Lambda_{k-1}F_{k-1}^{-1} \\
\bar{Q}_{k-1} = \frac{\nu_w(\nu_0-2)}{(\nu_w-2)\nu_0}Q_{k-1} \\
\Delta_k = \sqrt{(z_k - \hat{z}_{k|k-1})^{\mathrm{T}}(P^{\tilde{z}\tilde{z}}_{k|k-1})^{-1}(z_k - \hat{z}_{k|k-1})} \\
P^{\tilde{z}\tilde{z}}_{k|k-1} = H_k\Lambda_{k|k-1}^{-1}H_k^{\mathrm{T}} + \bar{R}_k \\
\hat{z}_{k|k-1} = H_k\Lambda_{k|k-1}^{-1}\hat{\xi}_{k|k-1}
\end{cases} \tag{11.31}$$

where m_1 is the dimension of z_k, and

$$\begin{cases} \Lambda_k = P_{k|k}^{-1}, & \hat{\xi}_{k|k} = P_{k|k}^{-1}\hat{x}_{k|k} = \Lambda_k\hat{x}_{k|k} \\ \Lambda_{k|k-1} = P_{k|k-1}^{-1}, & \hat{\xi}_{k|k-1} = P_{k|k-1}^{-1}\hat{x}_{k|k-1} = \Lambda_{k|k-1}\hat{x}_{k|k-1} \end{cases} \tag{11.32}$$

Proof From Lemmas 11.3 and 11.2, we have

$$\begin{aligned} P_{k|k}^{-1} &= \frac{1}{a_k}(P_{k|k-1} - K_k P_{k|k-1}^{\tilde{z}\tilde{z}} K_k^{\mathrm{T}})^{-1} \\ &= \frac{1}{a_k}(P_{k|k-1}^{-1} + H_k^{\mathrm{T}}\bar{R}_k^{-1}H_k) \end{aligned} \tag{11.33}$$

Let $\Lambda_k = P_{k|k}^{-1}$ and $\Lambda_{k|k-1} = P_{k|k-1}^{-1}$, we have

$$\Lambda_k = \frac{1}{a_k}(\Lambda_{k|k-1} + H_k^{\mathrm{T}}\bar{R}_k^{-1}H_k) \tag{11.34}$$

Let

$$A_{k-1} = F_{k-1}^{-T}P_{k-1|k-1}^{-1}F_{k-1}^{-1} = F_{k-1}^{-T}\Lambda_{k-1}F_{k-1}^{-1} \tag{11.35}$$

from Eq. (11.17), and by the use of Lemma 11.2, we have

$$\begin{aligned} \Lambda_{k|k-1} = P_{k|k-1}^{-1} &= [F_{k-1}P_{k-1|k-1}F_{k-1}^{\mathrm{T}} + \bar{Q}_{k-1}]^{-1} \\ &= A_{k-1} - A_{k-1}(I + \bar{Q}_{k-1}A_{k-1})^{-1}\bar{Q}_{k-1}A_{k-1} \\ &= [I - A_{k-1}(A_{k-1} + \bar{Q}_{k-1}^{-1})^{-1}]A_{k-1} \end{aligned} \tag{11.36}$$

Let $\hat{\xi}_{k|k-1} = \Lambda_{k|k-1}\hat{x}_{k|k-1}$, then from Eq. (11.17), we have

$$\hat{\xi}_{k|k-1} = \Lambda_{k|k-1}F_{k-1}\hat{x}_{k-1|k-1} \tag{11.37}$$

Substitute Eqs. (11.36) and (11.35) to Eq. (11.37), we have

$$\begin{aligned} \hat{\xi}_{k|k-1} &= \Lambda_{k|k-1}F_{k-1}\hat{x}_{k-1|k-1} \\ &= [I - A_{k-1}(A_{k-1} + \bar{Q}_{k-1}^{-1})^{-1}]A_{k-1}F_{k-1}\hat{x}_{k-1|k-1} \\ &= [I - A_{k-1}(A_{k-1} + \bar{Q}_{k-1}^{-1})^{-1}]F_{k-1}^{-T}\Lambda_{k-1}F_{k-1}^{-1}F_{k-1}\hat{x}_{k-1|k-1} \\ &= [I - A_{k-1}(A_{k-1} + \bar{Q}_{k-1}^{-1})^{-1}]F_{k-1}^{-T}\hat{\xi}_{k-1|k-1} \end{aligned} \tag{11.38}$$

Similar to Kalman filter, $P_{k|k}$ and K_k in Lemma 11.3 can be rewritten as

$$P_{k|k} = a_k[I - K_kH_k]P_{k|k-1} \tag{11.39}$$

$$K_k = \frac{1}{a_k}P_{k|k}H_k^{\mathrm{T}}\bar{R}_k^{-1} \tag{11.40}$$

Therefore, by the use of Eqs. (11.34), (11.39) and (11.40), we have

$$\begin{aligned}
\hat{\xi}_{k|k} &= \Lambda_k \hat{x}_{k|k} \\
&= \Lambda_k [\hat{x}_{k|k-1} + K_k (z_k - \hat{z}_{k|k-1})] \\
&= \Lambda_k [\hat{x}_{k|k-1} + \frac{1}{a_k} \Lambda_k^{-1} H_k^{\mathrm{T}} \bar{R}_k^{-1} (z_k - \hat{z}_{k|k-1})] \\
&= \frac{1}{a_k} (\Lambda_{k|k-1} + H_k^{\mathrm{T}} \bar{R}_k^{-1} H_k) \hat{x}_{k|k-1} + \frac{1}{a_k} H_k^{\mathrm{T}} \bar{R}_k^{-1} (z_k - H_k \hat{x}_{k|k-1}) \\
&= \frac{1}{a_k} [\hat{\xi}_{k|k-1} + H_k^{\mathrm{T}} \bar{R}_k^{-1} z_k]
\end{aligned} \tag{11.41}$$

The rest of Eq. (11.31) can be obtained from Eq. (11.17) directly.

11.4 The Information Fusion Algorithms

In this section, for simplicity, we consider the fusion algorithms under the assumption that $\nu_i = \nu_1$ for all $i = 2, 3, \ldots, N$. When the measurement noises meet t distributions with different dof, the fusion algorithms will be considered in Remark 11.3.

11.4.1 The Centralized Batch Fusion

Theorem 11.2 *For the linear dynamic system (11.1)–(11.5), when $\nu_i = \nu_1$ for all $i = 2, 3, \ldots, N$, the state estimation by centralized batch fusion of sensors 1 to N can be obtained by*

$$\begin{cases}
\hat{x}_{c,k|k-1} = F_{k-1} \hat{x}_{c,k-1|k-1} \\
P_{c,k|k-1} = F_{k-1} P_{c,k-1|k-1} F_{k-1}^{\mathrm{T}} + \bar{Q}_{k-1} \\
\hat{z}_{c,k|k-1} = H_k^a \hat{x}_{c,k|k-1} \\
\hat{x}_{c,k|k} = \hat{x}_{c,k|k-1} + K_{c,k} (z_k^a - \hat{z}_{c,k|k-1}) \\
P_{c,k|k} = a_{c,k} (P_{c,k|k-1} - K_{c,k} P_{c,k|k-1}^{\tilde{z}\tilde{z}} K_{c,k}^{\mathrm{T}}) \\
K_{c,k} = P_{c,k|k-1}^{\tilde{x}\tilde{z}} (P_{c,k|k-1}^{\tilde{z}\tilde{z}})^{-1} \\
P_{c,k|k-1}^{\tilde{z}\tilde{z}} = H_k^a P_{c,k|k-1} H_k^{a,\mathrm{T}} + \bar{R}_k^a \\
P_{c,k|k-1}^{\tilde{x}\tilde{z}} = P_{c,k|k-1} H_k^{a,\mathrm{T}} \\
a_{c,k} = \frac{(\nu_0 - 2)(\nu_0 + \Delta_{c,k}^2)}{\nu_0 (\nu_0 + m - 2)} \\
\Delta_{c,k} = \sqrt{(z_k^a - \hat{z}_{c,k|k-1})^{\mathrm{T}} (P_{c,k|k-1}^{\tilde{z}\tilde{z}})^{-1} (z_k^a - \hat{z}_{c,k|k-1})} \\
\bar{Q}_k = \frac{\nu_w (\nu_0 - 2)}{(\nu_w - 2) \nu_0} Q_k \\
\bar{R}_k^a = b R_k^a = \frac{\nu_1 (\nu_0 - 2)}{(\nu_1 - 2) \nu_0} R_k^a
\end{cases} \tag{11.42}$$

where the subscript c denotes the centralized fusion, $m = \sum_{i=1}^{N} m_i$, *and*

$$z_k^a = \begin{bmatrix} z_{1,k} \\ z_{2,k} \\ \vdots \\ z_{N,k} \end{bmatrix}, H_k^a = \begin{bmatrix} H_{1,k} \\ H_{2,k} \\ \vdots \\ H_{N,k} \end{bmatrix}, v_k^a = \begin{bmatrix} v_{1,k} \\ v_{2,k} \\ \vdots \\ v_{N,k} \end{bmatrix} \tag{11.43}$$

$$\begin{cases} p(v_k^a) = \mathrm{St}(v_k^a; 0, R_k^a, \nu_1) \\ R_k^a = \mathrm{diag}\{R_{1,k}, R_{2,k}, \ldots, R_{N,k}\} \end{cases} \tag{11.44}$$

Proof From Eq. (11.43), Eq. (11.2) can be rewritten as

$$z_k^a = H_k^a x_k + v_k^a \tag{11.45}$$

From the property of multivariate t distribution [12], we have Eq. (11.44). From Lemma 11.3, we have Eq. (11.42).

From Theorems 11.1 and 11.2, we have the following corollary.

Corollary 11.1 *The centralized fusion of linear dynamic system (11.1)–(11.5) by using the information filter introduced in Theorem 11.1 can be generated by*

$$\begin{cases} \hat{\xi}_{c,k|k} = \frac{1}{a_{c,k}}(\hat{\xi}_{c,k|k-1} + H_k^{a,\mathrm{T}} \bar{R}_k^{a,-1} z_k^a) \\ \hat{\xi}_{c,k|k-1} = [I - A_{c,k-1}(A_{c,k-1} + \bar{Q}_{k-1}^{-1})^{-1}] F_{k-1}^{-\mathrm{T}} \hat{\xi}_{c,k-1|k-1} \\ \Lambda_{c,k} = \frac{1}{a_{c,k}}(\Lambda_{c,k|k-1} + H_k^{a,\mathrm{T}} \bar{R}_k^{a,-1} H_k^a) \\ \Lambda_{c,k|k-1} = [I - A_{c,k-1}(A_{c,k-1} + \bar{Q}_{k-1}^{-1})^{-1}] A_{c,k-1} \\ A_{c,k-1} = F_{k-1}^{-\mathrm{T}} \Lambda_{c,k-1} F_{k-1}^{-1} \end{cases} \tag{11.46}$$

where

$$\begin{cases} \Lambda_{c,k} = P_{c,k|k}^{-1}, & \hat{\xi}_{c,k|k} = \Lambda_{c,k} \hat{x}_{c,k|k} \\ \Lambda_{c,k|k-1} = P_{c,k|k-1}^{-1}, & \hat{\xi}_{c,k|k-1} = \Lambda_{c,k|k-1} \hat{x}_{c,k|k-1} \end{cases} \tag{11.47}$$

and where H_k^a, $\bar{R}_k^a$, $a_{c,k}$, z_k^a *are computed by Eqs. (11.42)–(11.44).*

11.4.2 The Distributed Fusion Algorithms

Theorem 11.3 (The Optimal Distributed Fusion) *For the linear dynamic system (11.1)–(11.5), when* $\nu_i = \nu_1$ *for all* $i = 2, 3, \ldots, N$, *the optimal distributed fusion of sensors* 1 *to* N *for state estimation can be computed by*

$$\left\{\begin{array}{l}
\hat{x}_{d,k|k} = \Lambda_{d,k}^{-1}\hat{\xi}_{d,k|k} \\
P_{d,k|k} = \Lambda_{d,k}^{-1} \\
\Lambda_{d,k} = \frac{1}{a_{d,k}}[\Lambda_{d,k|k-1} + \frac{1}{b}\sum_{i=1}^{N}(a_{i,k}\Lambda_{i,k} - \Lambda_{d,k|k-1})] \\
\hat{\xi}_{d,k|k} = \frac{1}{a_{d,k}}(\hat{\xi}_{d,k|k-1} + \frac{1}{b}\sum_{i=1}^{N}[a_{i,k}\hat{\xi}_{i,k|k} - \hat{\xi}_{d,k|k-1}]) \\
\hat{\xi}_{d,k|k-1} = [I - A_{d,k-1}(A_{d,k-1} + \bar{Q}_{k-1}^{-1})^{-1}]F_{k-1}^{-\mathrm{T}}\hat{\xi}_{d,k-1|k-1} \\
\Lambda_{d,k|k-1} = [I - A_{d,k-1}(A_{d,k-1} + \bar{Q}_{k-1}^{-1})^{-1}]A_{d,k-1} \\
a_{d,k} = \frac{(\nu_0-2)(\nu_0+\Delta_{d,k}^2)}{\nu_0(\nu_0+m-2)} \\
A_{d,k-1} = F_{k-1}^{-\mathrm{T}}\Lambda_{d,k-1}F_{k-1}^{-1} \\
\Delta_{d,k}^2 = \left[\tilde{z}_{1,k}^{\mathrm{T}}\ \tilde{z}_{2,k}^{\mathrm{T}} \cdots \tilde{z}_{N,k}^{\mathrm{T}}\right](P_{d,k|k-1}^{\tilde{z}\tilde{z}})^{-1}\left[\tilde{z}_{1,k}^{\mathrm{T}}\ \tilde{z}_{2,k}^{\mathrm{T}} \cdots \tilde{z}_{N,k}^{\mathrm{T}}\right]^{\mathrm{T}} \\
P_{d,k|k-1}^{\tilde{z}\tilde{z}} = \begin{bmatrix} H_{1,k}P_{d,k|k-1}H_{1,k}^{\mathrm{T}} + bR_{1,k} & \cdots & H_{1,k}P_{d,k|k-1}H_{N,k}^{\mathrm{T}} \\ \vdots & \ddots & \vdots \\ H_{N,k}P_{d,k|k-1}H_{1,k}^{\mathrm{T}} & \cdots & H_{N,k}P_{d,k|k-1}H_{N,k}^{\mathrm{T}} + bR_{N,k} \end{bmatrix}
\end{array}\right. \tag{11.48}$$

where $m = \sum_{i=1}^{N} m_i$*, b and* $\bar{Q}_k$ *are the same as Lemma 11.3, the subscript d denotes the distributed fusion, the initial state meets Eq. (11.5). For* $i = 1, 2, \ldots, N$*, the local state estimation and the local state prediction are computed by*

$$\left\{\begin{array}{l}
\hat{x}_{i,k|k} = \Lambda_{i,k}^{-1}\hat{z}_{i,k|k} \\
P_{i,k|k} = \Lambda_{i,k}^{-1} \\
\hat{\xi}_{i,k|k} = \frac{1}{a_{i,k}}(\hat{\xi}_{d,k|k-1} + H_{i,k}^{\mathrm{T}}\bar{R}_{i,k}^{-1}z_{i,k}) \\
\Lambda_{i,k} = \frac{1}{a_{i,k}}(\Lambda_{d,k|k-1} + H_{i,k}^{\mathrm{T}}\bar{R}_{i,k}^{-1}H_{i,k}) \\
a_{i,k} = \frac{(\nu_0-2)(\nu_0+\Delta_{i,k}^2)}{\nu_0(\nu_0+m_i-2)} \\
\bar{R}_{i,k} = bR_{i,k} \\
\Delta_{i,k} = \sqrt{\tilde{z}_{i,k}^{\mathrm{T}}(P_{i,k|k-1}^{\tilde{z}\tilde{z}})^{-1}\tilde{z}_{i,k}} \\
P_{i,k|k-1}^{\tilde{z}\tilde{z}} = H_{i,k}\Lambda_{d,k|k-1}^{-1}H_{i,k}^{\mathrm{T}} + \bar{R}_{i,k} \\
\hat{z}_{i,k|k-1} = H_{i,k}\Lambda_{d,k|k-1}^{-1}\hat{\xi}_{d,k|k-1} \\
\tilde{z}_{i,k} = z_{i,k} - \hat{z}_{i,k|k-1}
\end{array}\right. \tag{11.49}$$

Proof From Theorem 11.1, based on the information filter to sensor i, the local state estimation can be computed by

$$\begin{cases} \hat{x}_{i,k|k} = \Lambda_{i,k}^{-1}\hat{\xi}_{i,k|k} \\ P_{i,k|k} = \Lambda_{i,k}^{-1} \\ \hat{\xi}_{i,k|k} = \frac{1}{a_{i,k}}(\hat{\xi}_{i,k|k-1} + H_{i,k}^{\mathrm{T}}\bar{R}_{i,k}^{-1}z_{i,k}) \\ \hat{\xi}_{i,k|k-1} = [I - A_{i,k-1}(A_{i,k-1} + \bar{Q}_{k-1}^{-1})^{-1}]F_{k-1}^{-\mathrm{T}}\hat{\xi}_{i,k-1|k-1} \\ \Lambda_{i,k} = \frac{1}{a_{i,k}}(\Lambda_{i,k|k-1} + H_{i,k}^{\mathrm{T}}\bar{R}_{i,k}^{-1}H_{i,k}) \\ \Lambda_{i,k|k-1} = [I - A_{i,k-1}(A_{i,k-1} + \bar{Q}_{k-1}^{-1})^{-1}]A_{i,k-1} \\ a_{i,k} = \frac{(\nu_0-2)(\nu_0+\Delta_{i,k}^2)}{\nu_0(\nu_0+m_i-2)} \\ \bar{R}_{i,k} = \frac{\nu_1(\nu_0-2)}{(\nu_1-2)\nu_0}R_{i,k} \\ A_{i,k-1} = F_{k-1}^{-\mathrm{T}}\Lambda_{i,k-1}F_{k-1}^{-1} \\ \bar{Q}_{k-1} = \frac{\nu_w(\nu_0-2)}{(\nu_w-2)\nu_0}Q_{k-1} \\ \Delta_{i,k} = \sqrt{\tilde{z}_{i,k}^{\mathrm{T}}(P_{i,k|k-1}^{\tilde{z}\tilde{z}})^{-1}\tilde{z}_{i,k}} \\ P_{i,k|k-1}^{\tilde{z}\tilde{z}} = H_{i,k}\Lambda_{i,k|k-1}^{-1}H_{i,k}^{\mathrm{T}} + \bar{R}_{i,k} \\ \hat{z}_{i,k|k-1} = H_{i,k}\Lambda_{i,k|k-1}^{-1}\hat{\xi}_{i,k|k-1} \\ \tilde{z}_{i,k} = z_{i,k} - \hat{z}_{i,k|k-1} \end{cases} \tag{11.50}$$

To improve the accuracy and the robustness of the local state estimations, the fused state estimation is transferred back to the local agent, i.e., let

$$\begin{cases} \hat{x}_{i,k|k-1} = F_{k-1}\hat{x}_{d,k-1|k-1} \\ P_{i,k|k-1} = F_{k-1}P_{d,k-1|k-1}F_{k-1}^{\mathrm{T}} + \bar{Q}_{k-1} \end{cases} \tag{11.51}$$

which imply

$$\begin{cases} \Lambda_{i,k|k-1} = P_{i,k|k-1}^{-1} = P_{d,k|k-1}^{-1} = \Lambda_{d,k|k-1} \\ \hat{z}_{i,k|k-1} = P_{i,k|k-1}^{-1}\hat{x}_{i,k|k-1} = P_{d,k|k-1}^{-1}\hat{x}_{d,k|k-1} = \hat{\xi}_{d,k|k-1} \end{cases} \tag{11.52}$$

Substitute Eq. (11.52) to Eq. (11.50), we have Eq. (11.49).

In the sequel, we will prove this theorem by the use of the deduction method.

Suppose $\hat{x}_{d,k-1|k-1} = \hat{x}_{c,k-1|k-1}$ and $P_{d,k-1|k-1} = P_{c,k-1|k-1}$, we will show that $\hat{x}_{d,k|k} = \hat{x}_{c,k|k}$ and $P_{d,k|k} = P_{c,k|k}$.

Since $\hat{x}_{d,k-1|k-1} = \hat{x}_{c,k-1|k-1}$ and $P_{d,k-1|k-1} = P_{c,k-1|k-1}$, we have

$$\Lambda_{d,k-1} = P_{d,k-1|k-1}^{-1} = P_{c,k-1|k-1}^{-1} = \Lambda_{c,k-1} \tag{11.53}$$

$$\hat{\xi}_{d,k-1|k-1} = \Lambda_{d,k-1}\hat{x}_{d,k-1|k-1} = \Lambda_{c,k-1}\hat{x}_{c,k-1|k-1} = \hat{\xi}_{c,k-1|k-1} \tag{11.54}$$

Thus

$$A_{d,k-1} = F_{k-1}^{-\mathrm{T}}\Lambda_{d,k-1}F_{k-1}^{-1} = A_{c,k-1} \tag{11.55}$$

$$\Lambda_{d,k|k-1} = [I - A_{d,k-1}(A_{d,k-1} + \bar{Q}_{k-1}^{-1})^{-1}]A_{d,k-1} = \Lambda_{c,k|k-1} \tag{11.56}$$

$$\hat{\xi}_{d,k|k-1} = [I - A_{d,k-1}(A_{d,k-1} + \bar{Q}_{k-1}^{-1})^{-1}]F_{k-1}^{-\mathrm{T}}\hat{\xi}_{d,k-1|k-1} = \hat{\xi}_{c,k|k-1} \tag{11.57}$$

From Corollary 11.1, and by the use of Eq. (11.49), we have

$$
\begin{aligned}
\hat{\xi}_{c,k|k} &= \frac{1}{a_{c,k}}(\hat{\xi}_{c,k|k-1} + H_k^{a,\mathrm{T}} \bar{R}_k^{a,-1} z_k^a) \\
&= \frac{1}{a_{c,k}}\left(\hat{\xi}_{d,k|k-1} + \frac{1}{b}\begin{bmatrix} H_{1,k} \\ H_{2,k} \\ \vdots \\ H_{N,k} \end{bmatrix}^{\mathrm{T}} \begin{bmatrix} R_{1,k} & 0 & \cdots & 0 \\ 0 & R_{2,k} & & \vdots \\ \vdots & & \ddots & \\ 0 & \cdots & & R_{N,k} \end{bmatrix}^{-1} \begin{bmatrix} z_{1,k} \\ z_{2,k} \\ \vdots \\ z_{N,k} \end{bmatrix}\right) \\
&= \frac{1}{a_{c,k}}(\hat{\xi}_{d,k|k-1} + \frac{1}{b}\sum_{i=1}^{N} H_{i,k}^{\mathrm{T}} R_{i,k}^{-1} z_{i,k}) \qquad (11.58) \\
&= \frac{1}{a_{c,k}}[\hat{\xi}_{d,k|k-1} + \frac{1}{b}\sum_{i=1}^{N}(a_{i,k}\hat{\xi}_{i,k|k} - \hat{\xi}_{d,k|k-1})]
\end{aligned}
$$

$$
\begin{aligned}
\Lambda_{c,k} &= \frac{1}{a_{c,k}}(\Lambda_{c,k|k-1} + H_k^{a,\mathrm{T}} \bar{R}_k^{a,-1} H_k^a) \\
&= \frac{1}{a_{c,k}}(\Lambda_{d,k|k-1} + \frac{1}{b}\sum_{i=1}^{N} H_{i,k}^{\mathrm{T}} \bar{R}_{i,k}^{-1} H_{i,k}) \qquad (11.59) \\
&= \frac{1}{a_{c,k}}[\Lambda_{d,k|k-1} + \frac{1}{b}\sum_{i=1}^{N}(a_{i,k}\Lambda_{i,k} - \Lambda_{d,k|k-1})]
\end{aligned}
$$

From Eq. (11.42),

$$
a_{c,k} = \frac{(\nu_0 - 2)(\nu_0 + \Delta_{c,k}^2)}{\nu_0(\nu_0 + m - 2)}, \qquad (11.60)
$$

$$
\Delta_{c,k} = \sqrt{(z_k^a - \hat{z}_{c,k|k-1})^{\mathrm{T}}(P_{c,k|k-1}^{\tilde{z}\tilde{z}})^{-1}(z_k^a - \hat{z}_{c,k|k-1})} \qquad (11.61)
$$

From Theorem 11.2, we have

$$
\begin{aligned}
P_{c,k|k-1}^{\tilde{z}\tilde{z}} &= H_k^a P_{c,k|k-1} H_k^{a,\mathrm{T}} + \bar{R}_k^a \\
&= \begin{bmatrix} H_{1,k} \\ H_{2,k} \\ \vdots \\ H_{N,k} \end{bmatrix} P_{d,k|k-1} \begin{bmatrix} H_{1,k} \\ H_{2,k} \\ \vdots \\ H_{N,k} \end{bmatrix}^{\mathrm{T}} + b \begin{bmatrix} R_{1,k} & 0 & \cdots & 0 \\ 0 & R_{2,k} & & \vdots \\ \vdots & & \ddots & \\ 0 & \cdots & & R_{N,k} \end{bmatrix}
\end{aligned}
$$

$$
= \begin{bmatrix} H_{1,k} P_{d,k|k-1} H_{1,k}^{\mathrm{T}} + bR_{1,k} & \cdots & H_{1,k} P_{d,k|k-1} H_{N,k}^{\mathrm{T}} \\ \vdots & \ddots & \vdots \\ H_{N,k} P_{d,k|k-1} H_{1,k}^{\mathrm{T}} & \cdots & H_{N,k} P_{d,k|k-1} H_{N,k}^{\mathrm{T}} + bR_{N,k} \end{bmatrix}
$$
$$
= P_{d,k|k-1}^{\tilde{z}\tilde{z}} \tag{11.62}
$$

Substitute Eq. (11.62) and $\hat{z}_{c,k|k-1} = H_k^a \hat{x}_{c,k|k-1}$, $\hat{x}_{d,k|k-1} = \hat{x}_{c,k|k-1}$, $P_{d,k|k-1} = P_{c,k|k-1}$, $\hat{x}_{i,k|k-1} = \hat{x}_{d,k|k-1} = \hat{x}_{c,k|k-1}$ to Eq. (11.61), we have

$$
\begin{aligned}
\Delta_{c,k}^2 &= (z_k^a - H_k^a \hat{x}_{c,k|k-1})^{\mathrm{T}} (P_{c,k|k-1}^{\tilde{z}\tilde{z}})^{-1} (z_k^a - H_k^a \hat{x}_{c,k|k-1}) \\
&= \left(\begin{bmatrix} z_{1,k} \\ z_{2,k} \\ \vdots \\ z_{N,k} \end{bmatrix} - \begin{bmatrix} H_{1,k}\hat{x}_{d,k|k-1} \\ H_{2,k}\hat{x}_{d,k|k-1} \\ \vdots \\ H_{N,k}\hat{x}_{d,k|k-1} \end{bmatrix} \right)^{\mathrm{T}} (P_{d,k|k-1}^{\tilde{z}\tilde{z}})^{-1} \\
&\quad \times \left(\begin{bmatrix} z_{1,k} \\ z_{2,k} \\ \vdots \\ z_{N,k} \end{bmatrix} - \begin{bmatrix} H_{1,k}\hat{x}_{d,k|k-1} \\ H_{2,k}\hat{x}_{d,k|k-1} \\ \vdots \\ H_{N,k}\hat{x}_{d,k|k-1} \end{bmatrix} \right) \\
&= \left[\tilde{z}_{1,k}^{\mathrm{T}} \cdots \tilde{z}_{N,k}^{\mathrm{T}} \right] (P_{d,k|k-1}^{\tilde{z}\tilde{z}})^{-1} \left[\tilde{z}_{1,k}^{\mathrm{T}} \cdots \tilde{z}_{N,k}^{\mathrm{T}} \right]^{\mathrm{T}} \\
&= \Delta_{d,k}^2
\end{aligned} \tag{11.63}
$$

$$
\Delta_{c,k}^2 = \Delta_{d,k}^2 \approx \sum_{i=1}^{N} \tilde{z}_{i,k}^{\mathrm{T}} (H_{i,k} P_{d,k|k-1} H_{i,k}^{\mathrm{T}} + bR_{i,k})^{-1} \tilde{z}_{i,k} \tag{11.64}
$$

Substitute Eq. (11.63) to Eq. (11.60), we have

$$
a_{c,k} = a_{d,k} \tag{11.65}
$$

Substitute Eq. (11.65) to Eqs. (11.58) and (11.59), we have

$$
\hat{\xi}_{c,k|k} = \frac{1}{a_{d,k}} [\hat{\xi}_{d,k|k-1} + \frac{1}{b} \sum_{i=1}^{N} (a_{i,k} \hat{\xi}_{i,k|k} - \hat{\xi}_{d,k|k-1})] = \hat{\xi}_{d,k|k} \tag{11.66}
$$

$$
\Lambda_{c,k} = \frac{1}{a_{d,k}} [\Lambda_{d,k|k-1} + \frac{1}{b} \sum_{i=1}^{N} (a_{i,k} \Lambda_{i,k} - \Lambda_{d,k|k-1})] = \Lambda_{d,k} \tag{11.67}
$$

The distributed fusion estimation is given by

$$\hat{x}_{d,k|k} = \Lambda_{d,k}^{-1}\hat{\xi}_{d,k|k} = \hat{x}_{c,k|k} \tag{11.68}$$

$$P_{d,k|k} = \Lambda_{d,k}^{-1} = P_{c,k|k} \tag{11.69}$$

The centralized fusion estimation and the distributed fusion estimation have the same initial state x_0 that meets Eq. (11.5), so, we have $\hat{x}_{d,0|0} = \hat{x}_{0|0} = \hat{x}_{c,0|0}$ and $P_{d,0|0} = P_{0|0} = P_{c,0|0}$. Similar to the above proof, it can be easily verified that $\hat{x}_{d,1|1} = \hat{x}_{c,1|1}$ and $P_{d,1|1} = P_{c,1|1}$. This completes the proof.

From the proof of Theorem 11.3, if $\Delta_{d,k}$ is simplified by using Eq. (11.64) to replace Eq. (11.63), a suboptimal distributed fusion algorithm is obtained. In the sequel, we will introduce another suboptimal distributed fusion algorithm, which is easier to carry out in real applications.

Corollary 11.2 (Suboptimal Distributed Fusion) *For the linear dynamic system (11.1)–(11.5), when $\nu_i = \nu_1$ for all $i = 2, 3, \ldots, N$, the state estimation by suboptimal distributed fusion of sensors* 1 *to* N *can be computed by*

$$\begin{cases} \hat{x}_{sd,k|k} = \Lambda_{sd,k}^{-1}\hat{\xi}_{sd,k|k} \\ P_{sd,k|k} = \Lambda_{sd,k}^{-1} \\ \Lambda_{sd,k} = \frac{1}{a_{sd,k}}[\Lambda_{sd,k|k-1} + \frac{1}{b}\sum_{i=1}^{N}(a_{i,k}\Lambda_{i,k} - \Lambda_{i,k|k-1})] \\ \hat{\xi}_{sd,k|k} = \frac{1}{a_{sd,k}}(\hat{\xi}_{sd,k|k-1} + \frac{1}{b}\sum_{i=1}^{N}[a_{i,k}\hat{\xi}_{i,k|k} - \hat{\xi}_{i,k|k-1}]) \\ \hat{\xi}_{sd,k|k-1} = [I - A_{sd,k-1}(A_{sd,k-1} + \bar{Q}_{k-1}^{-1})^{-1}]F_{k-1}^{-\mathrm{T}}\hat{\xi}_{sd,k-1|k-1} \\ \Lambda_{sd,k|k-1} = [I - A_{sd,k-1}(A_{sd,k-1} + \bar{Q}_{k-1}^{-1})^{-1}]A_{sd,k-1} \\ a_{sd,k} = \frac{(\nu_0-2)(\nu_0+\Delta_{sd,k}^2)}{\nu_0(\nu_0+m-2)} \\ A_{sd,k-1} = F_{k-1}^{-\mathrm{T}}\Lambda_{sd,k-1}F_{k-1}^{-1} \\ \Delta_{sd,k}^2 = \sum_{i=1}^{N}\tilde{z}_{i,k}^{\mathrm{T}}(H_{i,k}P_{i,k|k-1}H_{i,k}^{\mathrm{T}} + bR_{i,k})^{-1}\tilde{z}_{i,k} \end{cases} \tag{11.70}$$

where $m = \sum_{i=1}^{N} m_i$, b *and* $\bar{Q}_k$ *are the same as Lemma 11.3, the subscript* sd *denotes the suboptimal distributed fusion, the initial state meets Eq. (11.5). For* $i = 1, 2, \ldots, N$, *the local state estimation and the local state prediction are computed by information filter of sensor* i, *i.e., Eq. (11.50).*

Remark 11.1 From Lemma 11.3, one can easily get that in case of the dof of the initial state, the process noise and the measurement noise are equal, the only difference between the t distribution based filter and Gaussian based classical Kalman filter lies in the computation of scale matrix $P_{k|k}$ or the covariance of the state estimation error. It is similar to many adaptive filters, such as, the strong–tracking–filter (STF), which is given based on Gaussian assumption but forced the alteration of $P_{k|k}$ to improve the robustness of the classical Kalman filter under model inaccuracy [26]. The difference between this chapter and the STF or the related work is that the parameter that determines the alteration of $P_{k|k}$ is deduced under the formulation of t distribution.

Remark 11.2 From Lemma 11.3 and Theorem 11.1, it can be easily proven that when ν_0, ν_w and ν_1 tends to infinity, the filter given in Lemma 11.3 becomes to the classical Kalman filter, and Theorem 11.1 will reduce to the classical information filter. Meanwhile, the algorithms given in Theorems 11.2, 11.3, and Corollary 11.2 reduce to the centralized batch fusion, the optimal distributed fusion, and the suboptimal distributed fusion algorithms of Gaussian driven systems, respectively. Therefore, the algorithms derived in this chapter based on t distribution are the generalization of the traditional ones that based on Gaussian Kalman filter. Actually, it is known to all that the t distribution is very similar to Gaussian distribution when the dof is large enough. So, to describe the heavy–tailed noise better, the dof of the t distribution should not be too large. And the moment matching approach is used to keep the heavy–tailed property [9, 19].

Remark 11.3 In Eq. (11.4), if $\nu_i \neq \nu_1$ for $i = 2, 3, \ldots, N$, the moment matching approach is used to deal with the corresponding information fusion estimation problem. To best preserve the heavy–tailed property, let $\nu^* = \min\{\nu_i, i = 1, 2, \ldots, N\}$ [19]. By the use of the moment matching approach, i.e., approximate $p(v_{i,k}) = \mathrm{St}(v_{i,k}; 0, R_{i,k}, \nu_i)$ by $p(v^*_{i,k}) = \mathrm{St}(v^*_{i,k}; 0, R^*_{i,k}, \nu^*)$, where $R^*_{i,k} = \frac{(\nu^*-2)\nu_i}{(\nu_i-2)\nu^*} R_{i,k}$, $i = 1, 2, \ldots, N$, then $v_{i,k}$ and $v^*_{i,k}$ have the same mean and covariance. By the use of ν^* to replace ν_1, and $R^*_{i,k}$ to replace $R_{i,k}$ in Theorem 11.2, Corollary 11.1, Theorem 11.3, and Corollary 11.2, we have the t distribution filter based centralized fusion, information filter based centralized fusion, the optimal distributed fusion, and the suboptimal distributed fusion algorithms for systems disturbed by heavy–tailed noises with different dofs, respectively.

Remark 11.4 For the centralized fusion, the system parameters F_k, Q_k, $H_{i,k}$, $R_{i,k}$, ν_0 and the measurements $z_{i,k}$ for sensors $i = 1, 2, \ldots, N$ are transmitted to the fusion center. For the distributed fusion, both the measurements and the local estimations are transmitted to the fusion center. For the distributed fusion with feedback in Theorem 11.3, the fusion estimation is required to be sent to the local estimation too. Thus, the distributed fusion requires more communication cost for the systems with heavy–tailed noises. However, compared with the centralized fusion, the distributed fusion has the same estimation accuracy, but higher robustness and the reliability of the estimation result when the system lives in a hostile environment. Therefore, the distributed fusion is more practical and meaningful for real applications.

11.5 Simulation

An example is given to verify the feasibility and effectiveness of the proposed algorithms.

Considering a three-dimensional linear dynamic system with two sensors observing one target:

$$x_{k+1} = F_k x_k + w_k \tag{11.71}$$

$$z_k = H_{i,k} x_k + v_{i,k}, i = 1, 2, 3 \tag{11.72}$$

where

$$F_k = \begin{bmatrix} 1 & T_s & \frac{T_s^2}{2} \\ 0 & 1 & T_s \\ 0 & 0 & 1 \end{bmatrix}$$

where, T_s is the sampling interval which is taken value as 0.01s. State $x_k = [s_k\ \dot{s}_k\ \ddot{s}_k]^{\mathrm{T}}$, where s_k, $\dot{s}_k$ and $\ddot{s}_k$ denote position, velocity and acceleration, respectively. $H_1 = [1\ 0\ 0]$, $H_2 = [0\ 1\ 0]$, $H_3 = [0\ 0\ 1]$. The initial state is generated according to:

$$p(x_0) = \mathrm{St}(x_0; \hat{x}_{0|0}, P_{0|0}, \nu_0) \tag{11.73}$$

where $\hat{x}_{0|0} = [10\ 0\ 0]^{\mathrm{T}}$, $P_{0|0} = \mathrm{diag}\{1, 1, 1\}$ and $\nu_0 = 3$. The heavy–tailed noise w_k and $v_{i,k}$ are generated according to:

$$p(w_k) = \mathrm{St}(w_k; 0, Q, \nu_w) \tag{11.74}$$

$$p(v_{i,k}) = \mathrm{St}(v_{i,k}; 0, R_i, \nu_1), i = 1, 2, 3 \tag{11.75}$$

where $Q = 0.16 \times \mathrm{diag}[1\ 1\ 1]$, $R_1 = 0.36$, $R_2 = 0.25$, $R_3 = 0.20$ and $\nu_w = \nu_1 = 3$.

To evaluate performance of different algorithms, the root mean square error (RMSE) is employed as:

$$RMSE = \frac{1}{L}\sqrt{\sum_{i=1}^{L}(x_k^i - \hat{x}_{k|k}^i)^2} \tag{11.76}$$

where x_k^i and $\hat{x}_{k|k}^i$ denote the original signal and its estimate of the ith run, respectively. For each simulation, $L = 1000$ Monte Carlo simulations are run to get the RMSEs of the state estimates.

The RMSE of the position and velocity by using different algorithms are given in Figs. 11.2 and 11.3, where "G–CF" denotes the RMSE by use of the Gaussian information filter based centralized fusion that regard the heavy–tailed noises as Gaussian noises with covariances $\frac{\nu}{\nu-2}Q$ and $\frac{\nu}{\nu-2}R_i$, respectively. "H–S1" and "H–S2" denote the RMSE using the t distribution based information filter of Sensor 1 and Sensor 2, respectively. "H–CF", "H–DF" and "H–SDF" denote the RMSEs by using centralized fusion, distributed fusion and suboptimal distributed fusion based on the t distribution, respectively.

From Figs. 11.2 and 11.3, one can see that: (1) the proposed algorithms are more effective for estimation of both position and velocity than the Gaussian information filter based algorithms when the system noises are heavy–tailed; (2) the t distribution based fusion algorithms are effective since they has smaller RMSE than the single

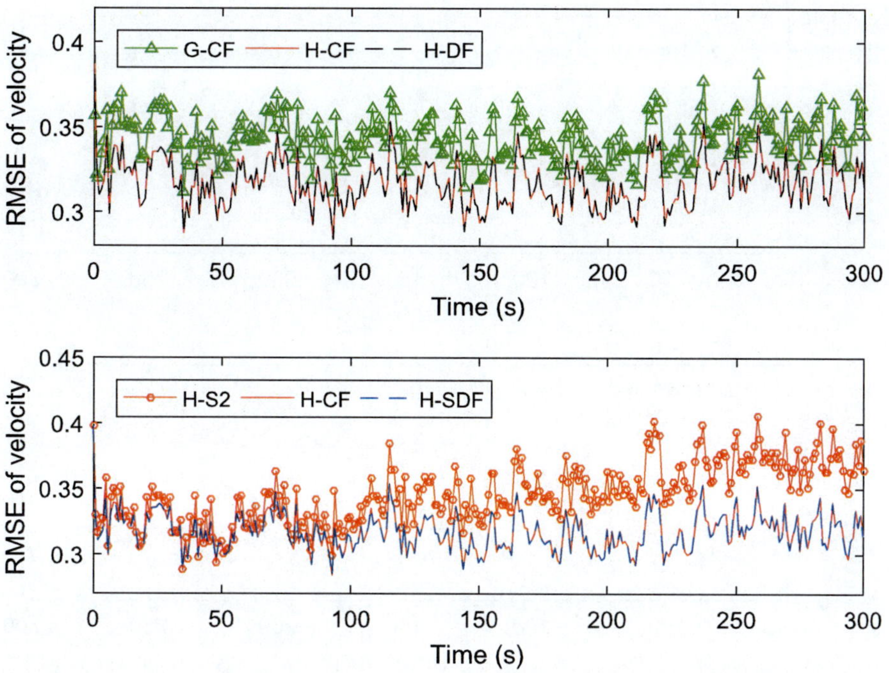

Fig. 11.2 RMSEs of the position

sensor; (3) the H–DF and the H–CF have equivalent estimation accuracy, which are a little bit superior to the H–SDF.

To better evaluate the proposed algorithms, Table 11.1 lists the averaged RMSE by different algorithms, where H–Si denote the results by using the proposed t distribution based information filter of Sensor i and G–Si denote the results by using the Gaussian information filter of Sensor i, $i = 1, 2, 3$, respectively. "G–CF" denotes the result by using the Gaussian Kalman filter based centralized fusion. "H–CF", "H–DF" and "H–SDF" denote the results by using the proposed t distribution based centralized fusion, distributed fusion, and the suboptimal distributed fusion, respectively. One can get that the t distribution based algorithms have smaller averaged RMSEs than the Gaussian Kalman filter based algorithms for both single sensor case and fusion case. Therefore, the effectiveness of the presented algorithms has been verified.

For further evaluation of the performance of different algorithms, Table 11.2 shows the CPU time of different algorithms. One can see that the Gaussian distribution based algorithms have shorter computation time than the corresponding t distribution based algorithms, and this is also consistent to our theory analysis that the computation of t distribution based algorithms is a little more complex than the Gaussian distribution based algorithms. For different t distribution based fusion algorithms, the H–SDF shows shortest CPU time and followed by the H–CF and then the H–DF. The H–CF

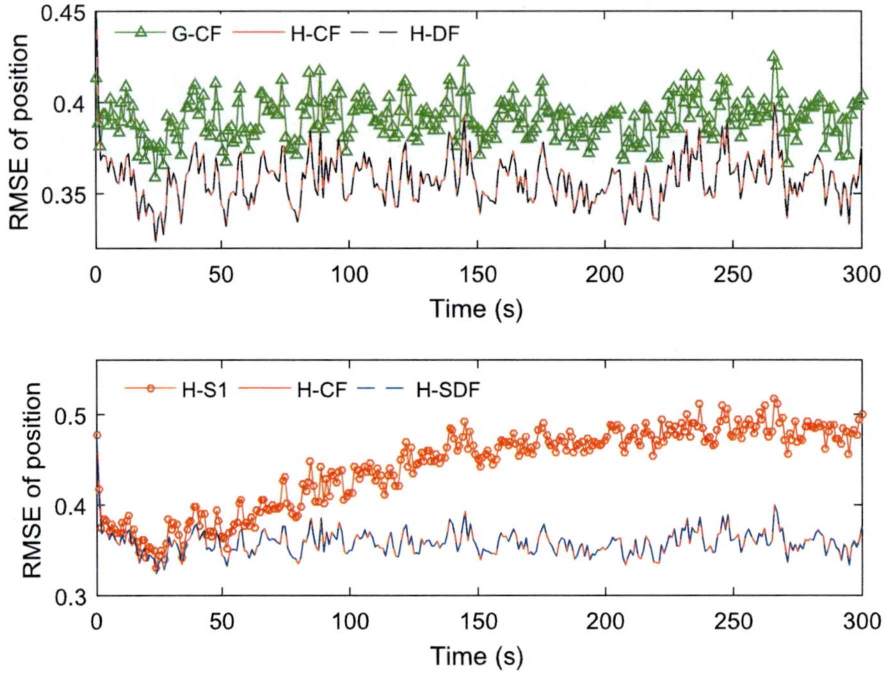

Fig. 11.3 RMSEs of the velocity

Table 11.1 Average RMSEs with different filtering algorithms by single sensor

Algorithms	Position	Velocity	Acceleration
G–S1	0.4752	0.3802	0.3101
H–S1	0.4414	0.3488	0.2934
G–CF	0.3908	0.3431	0.3038
H–DF/CF	0.3579	0.3170	0.2832
H–SDF	0.3665	0.3263	0.2912

seems more valuable than the H–DF since they have equivalent estimation accuracy but H–CF has less CPU time. However, in real applications, the observations of different sensors can hardly be received simultaneously, so the H–DF seems more efficient in practical applications. The H–SDF is promising in practice since it runs fastest among all the t distribution based fusion algorithms, and has competitive estimation accuracy.

According to the simulation results above, all the t distribution based algorithms have higher estimation accuracy compared with the Gaussian Kalman filter based algorithms when handle the estimation problem of multisensor systems with heavy–tailed noises. But the Gaussian Kalman filter based centralized fusion algorithm has

Table 11.2 Average CPU time per Monte Carlo run of different filtering algorithms

Algorithms	H–S1	H–S2	H–S3	G–S1	G–S2	G–S3
CPU time (ms)	18.19	18.15	18.62	13.25	13.39	13.49
Algorithms	H–CF	H–DF	H–SDF	G–CF		
CPU time (ms)	19.56	20.08	19.29	16.85		

shorter CPU time than the t distribution based fusion algorithms. Among three t distribution based fusion algorithms, the H–CF and the H–DF are shown to be equivalent and can obtain the most accurate estimations. The feedback is utilized in the H–DF, which leads to an increasement of communication cost, thus has longer CPU time than the H–CF. The H–SDF is a simplification of the H–DF, which decreased the computation time with little sacrifices of accuracy.

11.6 Conclusions

Based on multivariate t distribution, we derive the information filter, the centralized batch fusion and the distributed fusion algorithms for state estimation of linear dynamic systems with heavy–tailed noises. The theory analysis and simulations results draw the following conclusions: (1) the proposed t distribution based centralized batch fusion algorithm is effective, when the system noises are heavy–tailed, it is superior to the classical Kalman filter based centralized batch fusion; (2) the t distribution based optimal distributed fusion algorithm is effective, and it is proven that is equivalent to the centralized batch fusion in the sense of LMSE; (3) the t distribution based suboptimal distributed fusion algorithm is effective, which is a little worse in estimation accuracy than the optimal distributed fusion algorithm but has faster computation speed; (4) the traditional algorithms, including the classical Kalman filter, the Gaussian information filter, the Gaussian Kalman filter based centralized fusion and the optimal distributed fusion algorithms are the special case of the corresponding algorithms presented in this chapter when the dof of t distribution tends to infinity. Thus, the proposed algorithms have practical values in many real applications, such as target tracking, control system, and manufacturing industry.

There are still many open problems about the fusion estimation for the systems with heavy–tailed noises that can be further studied, such as the fusion estimation for asynchronous multirate multisensor systems with heavy–tailed noises, the filter and the fusion estimation for the dynamic systems with correlated heavy–tailed noises and the distributed fusion estimation of nonlinear systems with heavy–tailed noises. These are our future work.

References

1. Agamennoni, G., J.I. Nieto, and E.M. Nebot. 2012. Approximate inference in state-space models with heavy-tailed noise. *IEEE Transactions on Signal Processing* 60 (10): 5024–5037.
2. Amor, N., S. Kahlaoui, and S. Chebbi. 2018. Unscented particle filter using student–*t* distribution with non–Gaussian measurement noise. In *2018 International Conference on Advanced Systems and Electric Technologies*, pages 34–38.
3. Baccelli, Francois, Avhishek Chatterjee, and Sriram Vishwanath. 2017. Pairwise stochastic bounded confidence opinion dynamics: Heavy tails and stability. *IEEE Transactions on Automatic Control* 62 (11): 5678–5693.
4. Bar-Shalom, Y., X. Rong Li, and T Kirubarajan. 2001. *Estimation with Applications to Tracking and Navigation: Theory, Algorithms and Software*. New York: Wiley.
5. Elliott, Robert J., and John van der Hoek. 2006. Optimal linear estimation and data fusion. *IEEE Transactions on Automatic Control* 51 (4): 686–689.
6. Hu, Jun, Zidong Wang, Dongyan Chen, and Fuad E. Alsaadic. 2016. Estimation, filtering and fusion for networked systems with network-induced phenomena: New progress and prospects. *Information Fusion* 31: 65–75.
7. Huang, Yulong, Yonggang Zhang, Ning Li, and J. Chambers. 2016. A robust Gaussian approximate filter for nonlinear systems with heavy tailed measurement noises. In *IEEE International Conference on Acoustics, Speech and Signal Processing (ICASSP)*, pages 4209–4213.
8. Huang, Yulong, Yonggang Zhang, Ning Li, and J. Chambers. 2016. A robust student's t based cubature filter. In *19th International Conference on Information Fusion*, pages 9–16. Heidelberg, Germany.
9. Huang, Yulong, Yonggang Zhang, Ning Li, and J. Chambers. 2016. Robust student's *t* based nonlinear filter and smoother. *IEEE Transactions on Aerospace and Electronic Systems* 52 (5): 2586–2596.
10. Kashima, Kenji, Hiroki Aoyama, and Yoshito Ohta. 2019. Stable process approach to analysis of systems under heavy-tailed noise: Modeling and stochastic linearizatione. *IEEE Transactions on Automatic Control* 64 (4): 1344–1357.
11. Khaleghi, B., A. Khamis, F.O. Karray, et al. 2013. Multisensor data fusion: A review of the state-of-the-art. *Information Fusion* 14: 28–44.
12. Kotz, Samuel, and Saralees Nadarajah. 2004. *Multivariate t -Distributions and Their Applications*. Cambridge: Cambridge University Press.
13. Li, Wenling, and Yingmin Jia. 2012. Distributed consensus filtering for discrete-time nonlinear systems with non-Gaussian noise. *Signal Processing* 92: 2464–2470.
14. Niu, Wangqiang, Jin Zhu, Wei Gu, and Jianxin Chu. 2009. Four statistical approaches for multisensor data fusion under non–Gaussian noise. In *International Conference on Control, Automation and Systems Engineering*, pages 27–30.
15. Piché, R., S. Särkkä, and J. Hartikainen. 2012. Recursive outlier–robust filtering and smoothing for nonlinear systems using the multivariate student–*t* distribution. In *IEEE International Workshop on Machine Learning for Signal Processing*, pages 1–6. Santander, Spain, IEEE.
16. Qi, Hairong, S. Sitharama Iyengar, and Krishnendu Chakrabarty. 2001. Distributed sensor networks–a review of recent research. *Journal of the Franklin Institute* 338 (6): 655–668.
17. Ristic, Branko, Sanjeev Arulampalam, and Neil Gordon. 2004. *Beyond the Kalman Filter*. Artech House.
18. Roth, Michael. 2013. *On the Multivariate t Distribution*. Technical Report, Division of automatic control, Linköping university.
19. Roth, Michael, Emre Ożkan, and Fredrik Gustafsson. 2013. A student's filter for heavy tailed process and measurement noise. In *International Conference on Acoustics, Speech and Signal Processing (ICASSP)*, pages 5770–5774.
20. Shen, Chen, Jing Li, and Wei Huang. 2018. Robust centralized multi-sensor fusion using cubature information filter. In *The 30th Chinese Control and Decision Conference (2018 CCDC)*, pages 3297–3302.

21. Simon, D. 2006. *Optimal State Estimation: Kalman, H_∞, and Nonlinear Approaches*. New York: John Wiley and Sons, Inc., Publication.
22. Song, Enbin, Jie Xu, and Yunmin Zhu. 2014. Optimal distributed Kalman filtering fusion with singular covariances of filtering errors and measurement noises. *IEEE Transactions on Automatic Control* 59 (5): 1271–1282.
23. Sun, Shuli. 2013. Optimal linear filters for discrete-time systems with randomly delayed and lost measurements with/without time stamps. *IEEE Transactions on Automatic Control* 58 (6): 1551–1556.
24. Svensen, M., and C.M. Bishop. 2005. Robust bayesian mixture modeling. *Neurocomputing* 64: 235–252.
25. Wang, Zhu, and Xionglin Luo. 2017. A novel weight function-based robust iterative learning identification method for discrete box-jenkins models with student's t-distribution noises. *Journal of the Franklin Institute* 354 (18): 8645–8658.
26. Xie, Xiaoqing, Donghua Zhou, and Yihui Jin. 1999. Strong tracking filter based adaptive generic model control. *Journal of Process Control* 9 (4): 337–350.
27. Yan, Liping, X. Rong Li, and Yuanqing Xia. 2015. Modeling and estimation of asynchronous multirate multisensor system with unreliable measurements. *IEEE Transactions on Aerospace and Electronic Systems* 51 (3): 2012–2026.
28. Yan, Liping, X. Rong Li, Yuanqing Xia, and Mengyin Fu. 2013. Optimal sequential and distributed fusion for state estimation in cross-correlated noise. *Automatica* 49 (12): 3607–3612.
29. Yan, Liping, Yuanqing Xia, Baosheng Liu, and Mengyin Fu. 2015. *Multisensor Optimal Estimation Theory and its Application*. Beijing: The Science Publishing House.
30. Zhang, Le, Jian Lan, and X. Rong Li. 2018. A normal-Gamma filter for linear systems with heavy–tailed measurement noise. In *International Conference on Information Fusion (FUSION)*, pages 2552–2559. Xi'an, Shaanxi.
31. Zhu, Hao, Henry Leung, and Zhongshi He. 2013. A variational bayesian approach to robust sensor fusion based on student-*t* distribution. *Information Sciences* 221: 201–214.
32. Zhu, Junwei, Wen-An Zhang, Li Yu, and Dan Zhang. 2018. Robust distributed tracking control for linear multi-agent systems based on distributed intermediate estimator. *Journal of the Franklin Institute* 355 (1): 31–53.

Chapter 12
Sequential Fusion Estimation for Multisensor Systems with Heavy–Tailed Noises

12.1 Introduction

In this chapter, we intend to derive the sequential fusion estimation algorithm for linear multisensor dynamic systems disturbed by heavy–tailed process noise and measurement noise, which are obtained by using the multivariate t distributions. The performance of proposed sequential fusion algorithm will be evaluated by theoretical analysis and simulation examples as well. It will compare with the t filter based centralized fusion and the Gaussian Kalman filter based centralized fusion, besides the local estimators that barely using single sensor's information.

The rest of this chapter is organized as follows. Section 12.2 is the problem formulation. The sequential fusion algorithm for the multisensor linear time–variant dynamic systems with heavy–tailed noises is derived in Sect. 12.3. Section 12.4 shows the simulation and Sect. 12.5 draws the conclusions.

12.2 Problem Formulation

The linear multisensor dynamic system where there are N sensors observing one target can be formulated as follows [2, 8, 11]

$$x_{k+1} = F_k x_k + w_k, \quad k = 0, 1, \ldots \tag{12.1}$$

$$z_{i,k} = H_{i,k} x_k + v_{i,k}, \quad i = 1, 2, \ldots, N \tag{12.2}$$

where $i = 1, 2, \ldots, N$ denotes the ith sensor. $x_k \in R^n$ is the system state at the kth time instant, $F_k \in R^{n\times n}$ is the state transition matrix. $z_{i,k} \in R^{m_i}$ is the measurement of sensor i at time k. $H_{i,k}$ is the measurement matrix of sensor i. The process noise w_k and the measurement noise $v_{i,k}$ are heavy–tailed, which can be modeled by the multivariate t distribution as follows,

L. Yan et al., *Multisensor Fusion Estimation Theory and Application*,
https://doi.org/10.1007/978-981-15-9426-7_12

$$p(w_k) = \mathrm{St}(w_k; 0, Q_k, \nu_w) \tag{12.3}$$

$$p(v_{i,k}) = \mathrm{St}(v_{i,k}; 0, R_{i,k}, \nu_i), i = 1, 2, \ldots, N \tag{12.4}$$

where $\mathrm{St}(\cdot; \bar{x}, P, \nu)$ denotes a multivariate t distribution whose mean is $\bar{x}$, the scale matrix is P and degrees of freedom (dof) is ν.

Similarly, it is assumed that the system state initial value x_0 is also heavy–tailed that meets the multivariate t distribution with mean $\hat{x}_{0|0}$, the scale matrix $P_{0|0}$ and the dof ν_0, i.e.,

$$p(x_0) = \mathrm{St}(x_0; \hat{x}_{0|0}, P_{0|0}, \nu_0) \tag{12.5}$$

It is assumed that x_0, $v_{i,k}$ and w_k are mutually independent.

The aim of our work is to obtain the estimation of state x_k by sequentially use of the multisensor observations, i.e., to find

$$\hat{x}_{k|k} = E\{x_k|Z_k\} = \int x_k p(x_k|Z_k)\mathrm{d}x_k \tag{12.6}$$

$$P_{k|k} = \frac{\nu-2}{\nu}E[\tilde{x}_{k|k}\tilde{x}_{k|k}^T|Z_k] = \frac{\nu-2}{\nu}\int \tilde{x}_{k|k}\tilde{x}_{k|k}^T p(x_k|Z_k)\mathrm{d}x_k \tag{12.7}$$

where

$$p(x_k|Z_k) = \mathrm{St}(x_k; \hat{x}_{k|k}, P_{k|k}, \nu) \tag{12.8}$$

$$Z_k = \{z_{i,t}, t = 1, 2, \ldots, k; i = 1, 2, \ldots, N\} \tag{12.9}$$

12.3 The Sequential Fusion Algorithm

In this section, it is assumed that the dofs of the initial state, the process noise, and the measurement noises are equal, namely, $\nu_w = \nu_i = \nu$ for $i = 0, 1, 2, \ldots, N$. The unequal case will be presented in Remark 12.2.

Before presenting the sequential fusion algorithm, a Lemma will be introduced first. All the results of Lemma 12.1 can be found in [3, 6, 10], and the detailed proof of the last property of Lemma 12.1 can be found in [9].

Lemma 12.1 *x meets the multivariate t distribution* $\mathrm{St}(x; \bar{x}, P, \nu)$ *whose mean is $\bar{x}$, scale matrix is P, and dof is ν. It's probability density function (pdf) is:*

$$p(x) = \frac{\Gamma(\frac{\nu+2}{2})}{\Gamma(\frac{\nu}{2})}\frac{1}{(\nu\pi)^{\frac{d}{2}}}\frac{1}{\sqrt{\det(P)}}\left(1+\frac{\Delta^2}{\nu}\right)^{-\frac{\nu+2}{2}} \tag{12.10}$$

where $\Delta^2 = (x-\bar{x})^T P^{-1}(x-\bar{x})$.

It has the following properties:

- *The covariance of x is $\Sigma = \frac{\nu}{\nu-2}P$.*
- *When ν tends to infinity, the distribution of x reduces to Gaussian.*
- *Let $z = Ax + b$, then $p(z) = \mathrm{St}(z; A\bar{x} + b, APA^T, \nu)$, where A and b are matrix and vector of proper dimensions.*
- *If $x_1 \in R^{d_1}$ and $x_2 \in R^{d_2}$ meet joint t distribution whose pdf is:*

$$p(x_1, x_2) = \mathrm{St}\left(\begin{bmatrix} x_1 \\ x_2 \end{bmatrix}; \begin{bmatrix} \mu_1 \\ \mu_2 \end{bmatrix}, \begin{bmatrix} P_{11} & P_{12} \\ P_{21} & P_{22} \end{bmatrix}, \nu\right) \tag{12.11}$$

where $P_{ii} \in R^{d_i \times d_i}$, $i = 1, 2$, then the marginal pdf of x_1 and x_2 are

$$\begin{cases} p(x_1) = \mathrm{St}(x_1; \mu_1, P_{11}, \nu) \\ p(x_2) = \mathrm{St}(x_2; \mu_2, P_{22}, \nu) \end{cases} \tag{12.12}$$

The conditional pdf $p(x_1|x_2)$ is given by

$$p(x_1|x_2) = \mathrm{St}(x_1; \mu_{1|2}, P_{1|2}, \nu_{1|2}) \tag{12.13}$$

where

$$\nu_{1|2} = \nu + d_2 \tag{12.14}$$

$$\mu_{1|2} = \mu_1 + P_{12}P_{22}^{-1}(x_2 - \mu_2) \tag{12.15}$$

$$P_{1|2} = \frac{\nu + \Delta_2^2}{\nu + d_2}(P_{11} - P_{12}P_{22}^{-1}P_{12}^T) \tag{12.16}$$

and where $\Delta_2^2 = (x_2 - \mu_2)^T P_{22}^{-1}(x_2 - \mu_2)$.

Denote

$$z_k^a = \begin{bmatrix} z_{1,k} \\ z_{2,k} \\ \vdots \\ z_{N,k} \end{bmatrix}, H_k^a = \begin{bmatrix} H_{1,k} \\ H_{2,k} \\ \vdots \\ H_{N,k} \end{bmatrix}, v_k^a = \begin{bmatrix} v_{1,k} \\ v_{2,k} \\ \vdots \\ v_{N,k} \end{bmatrix} \tag{12.17}$$

From Eq. (12.17), Eq. (12.2) has a new form:

$$z_k^a = H_k^a x_k + v_k^a \tag{12.18}$$

From Lemma 12.1,

$$p(v_k^a) = \mathrm{St}(v_k^a; 0, R_k^a, \nu) \tag{12.19}$$

where

$$R_k^a = \text{diag}\{R_{1,k}, R_{2,k}, \ldots, R_{N,k}\} \tag{12.20}$$

For system (12.1) and (12.18), using the properties in Lemma 12.1, the state estimation by the centralized fusion of sensors 1 to N can be obtained by [1, 4, 10]:

$$\begin{cases} \hat{x}_{c,k|k-1} = F_{k-1}\hat{x}_{c,k-1|k-1} \\ P_{c,k|k-1} = F_{k-1}P_{c,k-1|k-1}F_{k-1}^{\mathrm{T}} + Q_{k-1} \\ \hat{x}_{c,k|k} = \hat{x}_{c,k|k-1} + K_{c,k}\tilde{z}_{c,k} \\ P_{c,k|k} = \frac{(\nu-2)(\nu+\Delta_{c,k}^2)}{\nu(\nu+m-2)}(I - K_{c,k}H_k^a)P_{c,k|k-1} \\ K_{c,k} = P_{c,k|k-1}^{\tilde{x}\tilde{z}}(P_{c,k|k-1}^{\tilde{z}\tilde{z}})^{-1} \\ P_{c,k|k-1}^{\tilde{z}\tilde{z}} = H_k^a P_{c,k|k-1}H_k^{a,T} + R_k^a \\ P_{c,k|k-1}^{\tilde{x}\tilde{z}} = P_{c,k|k-1}H_k^T \\ \Delta_{c,k}^2 = \tilde{z}_{c,k}^T(P_{c,k|k-1}^{\tilde{z}\tilde{z}})^{-1}\tilde{z}_{c,k} \\ \tilde{z}_{c,k} = z_k^a - \hat{z}_{c,k} \\ \hat{z}_{c,k} = H_k^a\hat{x}_{c,k|k-1} \end{cases} \tag{12.21}$$

where $m = \sum_{i=1}^{N} m_i$, and the subscript c denotes the centralized fusion.

In order to avoid augmentation of matrices and vectors, and to improve the efficiency of fusion estimation, the sequential fusion algorithm for linear multisensor dynamic systems with heavy–tailed noises will be deduced in the sequel.

Theorem 12.1 *For the linear dynamic system (12.1)–(12.5), the state estimation by sequential fusion of sensors* 1 *to* N *can be computed by*

$$\begin{cases} \hat{x}_{s,k|k} = \hat{x}_{s,k|k-1} + \sum_{i=1}^{N} K_{i,k}[z_{i,k} - H_{i,k}\hat{x}_{i-1,k|k}] \\ P_{s,k|k} = (\frac{\nu-2}{\nu})^N \prod_{1}^{i=N} (\frac{\nu+\Delta_{i,k}^2}{\nu+m_i-2})(I - K_{i,k}H_{i,k})P_{s,k|k-1} \end{cases} \tag{12.22}$$

where for $i = 1, 2, \ldots, N$,

$$\begin{cases} K_{i,k} = P_k^{\tilde{x}\tilde{z}_i}(P_k^{\tilde{z}_i\tilde{z}_i})^{-1} \\ P_k^{\tilde{x}\tilde{z}_i} = P_{i-1,k|k}H_{i,k}^T \\ P_k^{\tilde{z}_i\tilde{z}_i} = H_{i,k}P_{i-1,k|k}H_{i,k}^T + R_{i,k} \\ \Delta_{i,k}^2 = \tilde{z}_{i,k}^T(P_k^{\tilde{z}_i\tilde{z}_i})^{-1}\tilde{z}_{i,k} \\ \tilde{z}_{i,k} = z_{i,k} - \hat{z}_{i,k} \\ \hat{z}_{i,k} = H_{i,k}\hat{x}_{i-1,k|k} \\ \hat{x}_{0,k|k} = \hat{x}_{s,k|k-1} = F_{k-1}\hat{x}_{s,k-1|k-1} \\ P_{0,k|k} = P_{s,k|k-1} = F_{k-1}P_{s,k-1|k-1}F_{k-1}^{\mathrm{T}} + Q_{k-1} \end{cases} \tag{12.23}$$

and where $\prod_{1}^{i=N} D_i = D_N D_{N-1} \cdots D_1$ *is the product of N terms from the larger index N to the smaller index* 1. *Note that* $\prod_{1}^{i=N} D_i \neq \prod_{i=1}^{N} D_i = D_1 D_2 \cdots D_N$ *because matrix multiplication does not commute.*

Proof *Step 1: time update*

From Lemma 12.1 and Eqs. (12.1)–(12.5), we have

$$p(x_{k-1}, w_{k-1}|Z_{k-1}) = \mathrm{St}\left(\begin{bmatrix} x_{k-1} \\ w_{k-1} \end{bmatrix}; \begin{bmatrix} \hat{x}_{s,k-1|k-1} \\ 0 \end{bmatrix}, \begin{bmatrix} P_{s,k-1|k-1} & 0 \\ 0 & Q_{k-1} \end{bmatrix}, \nu\right) \tag{12.24}$$

From Eq. (12.1) and Lemma 12.1, we have

$$p(x_k|Z_{k-1}) = \mathrm{St}(x_k; \hat{x}_{s,k|k-1}, P_{s,k|k-1}, \nu) \tag{12.25}$$

where

$$\hat{x}_{s,k|k-1} = F_{k-1}\hat{x}_{s,k-1|k-1} \tag{12.26}$$

$$P_{s,k|k-1} = F_{k-1}P_{s,k-1|k-1}F_{k-1}^T + Q_{k-1} \tag{12.27}$$

Step 2: Measurement update–step by step

From

$$p(x_k, v_{1,k}|Z_{k-1}) = \mathrm{St}\left(\begin{bmatrix} x_k \\ v_{1,k} \end{bmatrix}; \begin{bmatrix} \hat{x}_{s,k|k-1} \\ 0 \end{bmatrix}, \begin{bmatrix} P_{s,k|k-1} & 0 \\ 0 & R_{1,k} \end{bmatrix}, \nu\right) \tag{12.28}$$

we have

$$p(x_k, z_{1,k}|Z_{k-1}) = \mathrm{St}\left(\begin{bmatrix} x_k \\ z_{1,k} \end{bmatrix}; \begin{bmatrix} \hat{x}_{s,k|k-1} \\ \hat{z}_{1,k} \end{bmatrix}, \begin{bmatrix} P_{s,k|k-1} & P_k^{\tilde{x}\tilde{z}_1} \\ P_k^{\tilde{x}\tilde{z}_1,T} & P_k^{\tilde{z}_1\tilde{z}_1} \end{bmatrix}, \nu\right) \tag{12.29}$$

where

$$\hat{z}_{1,k} = H_{1,k}\hat{x}_{s,k|k-1}, \tag{12.30}$$

$$P_k^{\tilde{x}\tilde{z}_1} = P_{s,k|k-1}H_{1,k}^T \tag{12.31}$$

$$P_k^{\tilde{z}_1\tilde{z}_1} = H_{1,k}P_{s,k|k-1}H_{1,k}^T + R_{1,k} \tag{12.32}$$

From the last property of Lemma 12.1, the conditional probability

$$p(x_k|Z_{k-1}, z_{1,k}) = \mathrm{St}(x_k; \hat{x}'_{1,k|k}, P'_{1,k|k}, \nu^{(1)}) \tag{12.33}$$

can be obtained by

$$\nu^{(1)} = \nu + m_1 \tag{12.34}$$
$$\hat{x}'_{1,k|k} = \hat{x}_{s,k|k-1} + K_{1,k}\tilde{z}_{1,k} \tag{12.35}$$
$$P'_{1,k|k} = \frac{\nu + \Delta^2_{1,k}}{\nu + m_1}[P_{s,k|k-1} - P_k^{\tilde{x}\tilde{z}_1}(P_k^{\tilde{z}_1\tilde{z}_1})^{-1}P_k^{\tilde{x}\tilde{z}_1,T}] \tag{12.36}$$
$$\Delta^2_{1,k} = \tilde{z}^T_{1,k}(P_k^{\tilde{z}_1\tilde{z}_1})^{-1}\tilde{z}_{1,k} \tag{12.37}$$
$$\tilde{z}_{1,k} = z_{1,k} - \hat{z}_{1,k} \tag{12.38}$$

Denote

$$K_{1,k} = P_k^{\tilde{x}\tilde{z}_1}(P_k^{\tilde{z}_1\tilde{z}_1})^{-1} \tag{12.39}$$

then, substitute Eqs. (12.39) and (12.31) to Eq. (12.36), we have

$$P'_{1,k|k} = \frac{\nu + \Delta^2_{1,k}}{\nu + m_1}(I - K_{1,k}H_{1,k})P_{s,k|l-1} \tag{12.40}$$

To preserve the heavy–tailed property, by using the moment matching approach, we have the approximate t distribution [1, 4, 10]

$$p(x_k|Z_{k-1}, z_{1,k}) \approx \mathrm{St}(x_k; \hat{x}_{1,k|k}, P_{1,k|k}, \nu) \tag{12.41}$$

where

$$\hat{x}_{1,k|k} = \hat{x}'_{1,k|k} = \hat{x}_{s,k|k-1} + K_{1,k}(z_{1,k} - H_{1,k}\hat{x}_{s,k|k-1}) \tag{12.42}$$
$$\begin{aligned} P_{1,k|k} &= \frac{\nu^{(1)}(\nu - 2)}{\nu(\nu^{(1)} - 2)}P'_{1,k|k} \\ &= \frac{(\nu - 2)(\nu + \Delta^2_{1,k})}{\nu(\nu + m_1 - 2)}(I - K_{1,k}H_{1,k})P_{s,k|k-1} \end{aligned} \tag{12.43}$$

Similarly, from

$$p(x_k, v_{2,k}|Z_{k-1}, z_{1,k}) = \mathrm{St}\left(\begin{bmatrix} x_k \\ v_{2,k} \end{bmatrix}; \begin{bmatrix} \hat{x}_{1,k|k} \\ 0 \end{bmatrix}, \begin{bmatrix} P_{1,k|k} & 0 \\ 0 & R_{2,k} \end{bmatrix}, \nu\right) \tag{12.44}$$

we have

$$p(x_k, z_{2,k}|Z_{k-1}, z_{1,k}) = \mathrm{St}\left(\begin{bmatrix} x_k \\ z_{2,k} \end{bmatrix}; \begin{bmatrix} \hat{x}_{1,k|k} \\ \hat{z}_{2,k} \end{bmatrix}, \begin{bmatrix} P_{1,k|k} & P_k^{\tilde{x}\tilde{z}_2} \\ P_k^{\tilde{x}\tilde{z}_2,T} & P_k^{\tilde{z}_2\tilde{z}_2} \end{bmatrix}, \nu\right) \tag{12.45}$$

where

$$\hat{z}_{2,k} = H_{2,k}\hat{x}_{1,k|k} \tag{12.46}$$

$$P_k^{\tilde{x}\tilde{z}_2} = P_{1,k|k}H_{2,k}^T \tag{12.47}$$

$$P_k^{\tilde{z}_2\tilde{z}_2} = H_{2,k}P_{1,k|k}H_{2,k}^T + R_{2,k} \tag{12.48}$$

The conditional probability

$$p(x_k|Z_{k-1}, z_{1,k}, z_{2,k}) = \mathrm{St}(x_k; \hat{x}'_{2,k|k}, P'_{2,k|k}, \nu^{(2)}) \tag{12.49}$$

can be obtained by

$$\nu^{(2)} = \nu + m_2 \tag{12.50}$$

$$\hat{x}'_{2,k|k} = \hat{x}_{1,k|k} + K_{2,k}\tilde{z}_{2,k} \tag{12.51}$$

$$P'_{2,k|k} = \frac{\nu + \Delta_{2,k}^2}{\nu + m_2}(I - K_{2,k}H_{2,k})P_{1,k|k} \tag{12.52}$$

$$K_{2,k} = P_k^{\tilde{x}\tilde{z}_2}(P_k^{\tilde{z}_2\tilde{z}_2})^{-1} \tag{12.53}$$

$$\Delta_{2,k}^2 = \tilde{z}_{2,k}^T(P_k^{\tilde{z}_2\tilde{z}_2})^{-1}\tilde{z}_{2,k} \tag{12.54}$$

$$\tilde{z}_{2,k} = z_{2,k} - \hat{z}_{2,k} \tag{12.55}$$

By using the moment matching approach, we have the approximate t distribution

$$p(x_k|Z_{k-1}, z_{1,k}, z_{2,k}) \approx \mathrm{St}(x_k; \hat{x}_{2,k|k}, P_{2,k|k}, \nu) \tag{12.56}$$

where

$$\hat{x}_{2,k|k} = \hat{x}'_{2,k|k} = \hat{x}_{1,k|k} + K_{2,k}(z_{2,k} - H_{2,k}\hat{x}_{1,k|k}) \tag{12.57}$$

$$\begin{aligned} P_{2,k|k} &= \frac{\nu^{(2)}(\nu - 2)}{\nu(\nu^{(2)} - 2)}P'_{2,k|k} \\ &= \frac{(\nu - 2)(\nu + \Delta_{2,k}^2)}{\nu(\nu + m_2 - 2)}(I - K_{2,k}H_{2,k})P_{1,k|k} \end{aligned} \tag{12.58}$$

Generally speaking, for $2 \leqslant i \leqslant N$, we have the approximate t distribution

$$p(x_k|Z_{k-1}, z_{1,k}, z_{2,k}, \ldots, z_{i,k}) \approx \mathrm{St}(x_k; \hat{x}_{i,k|k}, P_{i,k|k}, \nu) \tag{12.59}$$

where

$$\hat{x}_{i,k|k} = \hat{x}_{i-1,k|k} + K_{i,k}[z_{i,k} - H_{i,k}\hat{x}_{i-1,k|k}] \tag{12.60}$$

$$P_{i,k|k} = \frac{(\nu-2)(\nu+\Delta_{i,k}^2)}{\nu(\nu+m_i-2)}(I - K_{i,k}H_{i,k})P_{i-1,k|k} \tag{12.61}$$

and where

$$K_{i,k} = P_k^{\tilde{x}\tilde{z}_i}(P_k^{\tilde{z}_i\tilde{z}_i})^{-1} \tag{12.62}$$

$$\Delta_{i,k}^2 = \tilde{z}_{i,k}^T(P_k^{\tilde{z}_i\tilde{z}_i})^{-1}\tilde{z}_{i,k} \tag{12.63}$$

$$\tilde{z}_{i,k} = z_{i,k} - \hat{z}_{i,k} \tag{12.64}$$

$$\hat{z}_{i,k} = H_{i,k}\hat{x}_{i-1,k|k} \tag{12.65}$$

When $i = N$, the state estimation by sequentially fusing of Sensors 1 through N can be deduced,

$$\begin{aligned}\hat{x}_{s,k|k} &= \hat{x}_{N,k|k} = \hat{x}_{N-1,k|k} + K_{N,k}(z_{N,k} - H_{N,k}\hat{x}_{N-1,k|k}) \\ &= \hat{x}_{s,k|k-1} + \sum_{i=1}^{N} K_{i,k}(z_{i,k} - H_{i,k}\hat{x}_{i-1,k|k})\end{aligned} \tag{12.66}$$

$$\begin{aligned}P_{s,k|k} &= P_{N,k|k} = \frac{(\nu-2)(\nu+\Delta_{N,k}^2)}{\nu(\nu+m_N-2)}(I - K_{N,k}H_{N,k})P_{N-1,k|k} \\ &= \left(\frac{\nu-2}{\nu}\right)^N \prod_{1}^{i=N}\left(\frac{\nu+\Delta_{i,k}^2}{\nu+m_i-2}\right)(I - K_{i,k}H_{i,k})P_{s,k|k-1}\end{aligned} \tag{12.67}$$

where

$$\begin{cases}\hat{x}_{0,k|k} = \hat{x}_{s,k|k-1} \\ P_{0,k|k} = P_{s,k|k-1}\end{cases} \tag{12.68}$$

This completes the proof.

We all know that for the state estimation of Gaussian noise based linear dynamic systems, the Gaussian Kalman filter based centralized batch fusion and the optimal sequential fusion are equivalent in the sense of LMSE [2, 5, 7]. However, for the t distribution based filter, the moment matching approach is used to keep the heavy–tailed property when generating the state estimation, so it is actually an approximate filter. Thus, for the systems with heavy–tailed noises, the approximate t filter based centralized fusion estimation given by Eq. (12.21) and the sequential fusion estimation computed by Theorem 12.1 are not equivalent. Actually, we have the following theorem.

Theorem 12.2 *The sequential fusion estimation given in Theorem 12.1 is inequivalent to the centralized fusion estimation result given in Eq. (12.21) in the sense of least mean square error (LMSE), and any of them could be better, which is deter-*

mined by the dof of the noises, the dimension of the measurements, and the quantity of the residuals.

Proof Denote

$$a_{c,k} = \frac{(\nu-2)(\nu+\Delta_{c,k}^2)}{\nu(\nu+m-2)} \tag{12.69}$$

$$a_{s,k} = (\frac{\nu-2}{\nu})^N \prod_{1}^{i=N} \left(\frac{\nu+\Delta_{i,k}^2}{\nu+m_i-2} \right) \tag{12.70}$$

Similar to the proof of the equivalence property of Kalman filter based optimal centralized fusion and the optimal sequential fusion [2, 5, 7], it can be easily verified that

$$\frac{1}{a_{c,k}} P_{c,k|k} = \frac{1}{a_{s,k}} P_{s,k|k} \tag{12.71}$$

Therefore

$$P_{s,k|k} = \frac{a_{s,k}}{a_{c,k}} P_{c,k|k} \tag{12.72}$$

Denote

$$b_{sc,k} = \frac{a_{s,k}}{a_{c,k}} \tag{12.73}$$

Substitute Eqs. (12.69) and (12.70) to the above, we have

$$b_{sc,k} = (\frac{\upsilon-2}{\upsilon})^{N-1} \prod_{i=1}^{N} \left(\frac{\upsilon+\Delta_{i,k}^2}{\upsilon+m_i-2} \right) \left(\frac{\nu+m-2}{\nu+\Delta_{c,k}^2} \right) \tag{12.74}$$

From Eqs. (12.72)–(12.74), it can be easily seen that when $b_{sc,k} > 1$, $P_{s,k|k} > P_{c,k|k}$; otherwise, when $b_{sc,k} < 1$, $P_{s,k|k} < P_{c,k|k}$. Only when $b_{sc,k} = 1$, we have $P_{s,k|k} = P_{c,k|k}$. So, which of $P_{s,k|k}$ and $P_{c,k|k}$ is larger depends on the dof of the noises ν, m_i (i.e., the dimension of $z_{i,k}$), and the residual of the measurements (i.e., $\tilde{z}_{i,k}$ and $\tilde{z}_{c,k}$).

Remark 12.1 It can be seen that when ν increases to infinity, the algorithm given in Eq. (12.21) and Theorem 12.1 become to the classical Gaussian Kalman filter based optimal centralized fusion and optimal sequential fusion, respectively. Thus, the algorithms derived in this chapter are the generalization of the traditional ones that based on Gaussian distribution. Actually, the t distribution is quite similar to Gaussian distribution when the dof large enough. So, to model the heavy–tailed noise better, the dof of the t distribution should not be too large. That is one of the reasons why the moment matching approach is used when we deriving the approximate t filter and fusion algorithms for systems with heavy–tailed noises.

Remark 12.2 If the dof of the process noise, the measurement noises, and the initial state are different in the problem formulation, we may use the moment matching approach to get the centralized fusion and the sequential fusion algorithms. For example, under the following formulation:

$$\begin{cases} p(x_0) = \mathrm{St}(x_0; \hat{x}_{0|0}, P_{0|0}, \nu_0) \\ p(w_k) = \mathrm{St}(w_k; 0, Q_k, \nu_w) \\ p(v_{i,k}) = \mathrm{St}(v_{i,k}; 0, R_{i,k}, \nu_i), i = 1, 2, \ldots, N \end{cases} \tag{12.75}$$

To keep the heavy–tailed property as well as possible, let $\nu = \min\{\nu_i, i = w, 0, 1, 2, \ldots, N\}$ [10]. By the use of the moment matching approach, $p(x_0)$, $p(w_k)$ and $p(v_{i,k})$ can be approximated by

$$\begin{cases} p(x_0^\star) = \mathrm{St}(x_0; \hat{x}_{0|0}, P_{0|0}^\star, \nu) \\ p(w_k^\star) = \mathrm{St}(w_k^\star; 0, Q_k^\star, \nu) \\ p(v_{i,k}^\star) = \mathrm{St}(v_{i,k}^\star; 0, R_{i,k}^\star, \nu) \end{cases} \tag{12.76}$$

where

$$\begin{cases} P_{0|0}^\star = \frac{(\nu-2)\nu_0}{(\nu_0-2)\nu} P_{0|0} \\ Q_k^\star = \frac{(\nu-2)\nu_w}{(\nu_w-2)\nu} Q_k \\ R_{i,k}^\star = \frac{(\nu-2)\nu_i}{(\nu_i-2)\nu} R_{i,k} \end{cases} \tag{12.77}$$

Then, $w_k^\star$ and w_k, $v_{i,k}^\star$ and $v_{i,k}$, $x_0^\star$ and x_0 have the same mean and covariance, respectively. By using $Q_k^\star$, $P_{0|0}^\star$ and $R_{i,k}^\star$ to replace Q_k, $P_{0|0}$ and $R_{i,k}$, respectively, in Eq. (12.21) and Theorem 12.1, we obtain the t distribution filter based centralized fusion and sequential fusion algorithms, respectively, for systems with heavy–tailed noises of different dofs.

12.4 Numerical Example

In this section, we use an example to illustrate the effectiveness and robustness of the proposed algorithm.

Consider a two-dimensional linear target tracking system observed by three sensors [7]:

$$x_{k+1} = F x_k + w_k \tag{12.78}$$

$$z_{i,k} = H_i x_k + v_{i,k}, i = 1, 2, 3 \tag{12.79}$$

where

$$F = \begin{bmatrix} 0.95 & T \\ 0 & 0.95 \end{bmatrix} \tag{12.80}$$

$T = 1s$ is the sampling interval. State vector $x_k = [s_k \; \dot{s}_k]^{\mathrm{T}}$, where s_k and $\dot{s}_k$ denote position and velocity of the target, respectively. $H_1 = [1 \; 1]$, $H_2 = [0.9 \; 0.7]$, $H_3 = [0.8 \; 0.5]$. The initial state and the scale matrices are $\hat{x}_{0|0} = [10 \; 0]^{\mathrm{T}}$ and $P_{0|0} = \mathrm{diag}[2 \; 2]$. In this example, the heavy–tailed process noise w_k and measurement noise $v_{i,k}$ are generated according to

$$p(w_k) = \mathrm{St}(w_k; 0, Q, \nu) \tag{12.81}$$

$$p(v_{i,k}) = \mathrm{St}(v_{i,k}; 0, R_i, \nu), i = 1, 2, 3 \tag{12.82}$$

where $Q = \mathrm{diag}[1 \; 1]$, $R_1 = 8$, $R_2 = 16$, $R_3 = 20$ and $\nu = 3$. To analyze the filtering performance, we utilize the root mean square error (RMSE) of position and velocity:

$$\begin{cases} \mathrm{RMSE_p} = \frac{1}{L}\sqrt{\sum_{i=1}^{L}(s_k - \hat{s}_{k|k})^2} \\ \mathrm{RMSE_v} = \frac{1}{L}\sqrt{\sum_{i=1}^{L}(\dot{s}_k - \hat{\dot{s}}_{k|k})^2} \end{cases} \tag{12.83}$$

To get the RMSE of the state estimates, $L = 200$ Monte Carlo simulations are run.

Figure 12.1 presents the true values and the estimates of position and velocity by using different fusion algorithms, where the G–CF denotes the state estimates by using the Kalman filter based centralized batch fusion that regard the heavy–tailed noises as Gaussian noises with covariances $Q' = \frac{\nu}{\nu-2}Q$ and $R_i' = \frac{\nu}{\nu-2}R_i$, respectively. The CF denotes the estimates by using the heavy–tailed centralized fusion algorithm and the SF denotes the estimate by using the sequential fusion proposed in Sect. 12.3. Figure 12.1 shows that the estimates by the proposed sequential fusion (SF) best match the original signal in both position and velocity, followed by the CF, and then the G–CF, which shows the superiority of the presented sequential fusion among these fusion algorithms.

Figures 12.2 and 12.3 give the RMSEs of the position and the velocity by use of the single sensor and different fusion algorithms. The upper subgraphs of both Figs. 12.2 and 12.3 compare the RMSEs of position and velocity of different fusion algorithms, respectively, where one can see that the proposed sequential fusion can get the estimates with higher accuracy in both position and velocity compared with the G–CF, and the CF shows nearly the same performance as the SF. The second subgraphs show the RMSEs of the position and the velocity of single sensors and the SF algorithm. Three sensors observe the target with different accuracy. The SF has the smallest RMSEs in position and velocity than any single sensor and the effectiveness of the proposed t distribution based sequential fusion algorithm has been illustrated.

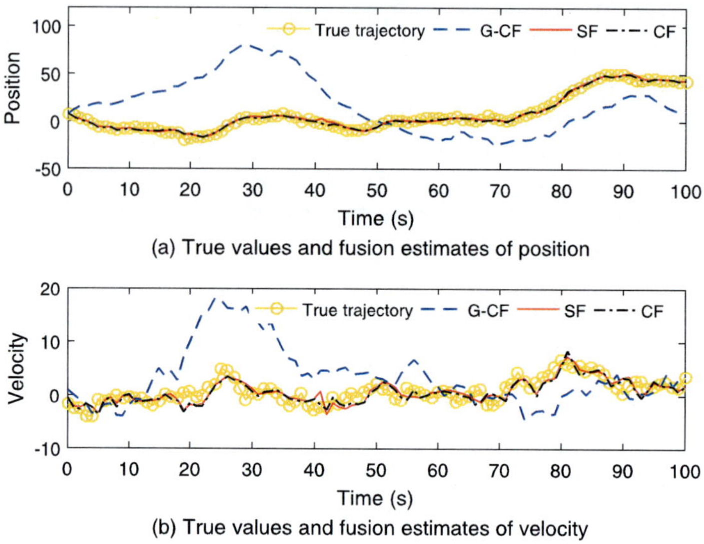

(a) True values and fusion estimates of position

(b) True values and fusion estimates of velocity

Fig. 12.1 True values and the fusion estimates

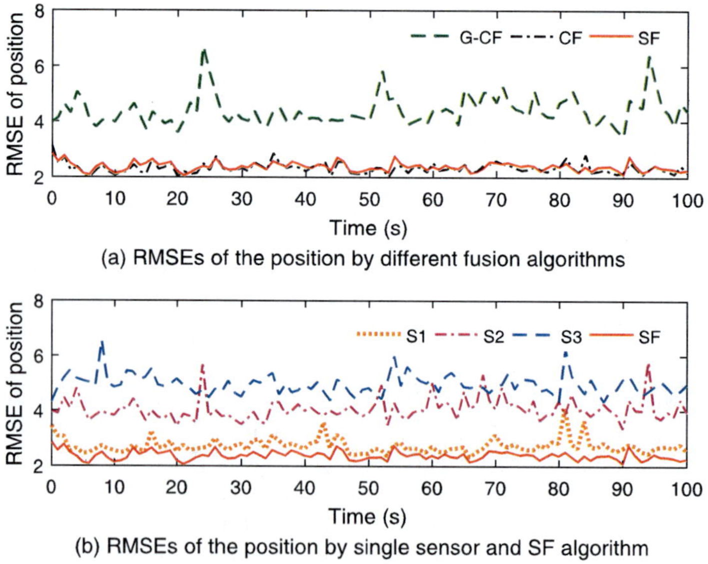

(a) RMSEs of the position by different fusion algorithms

(b) RMSEs of the position by single sensor and SF algorithm

Fig. 12.2 RMSEs of the position

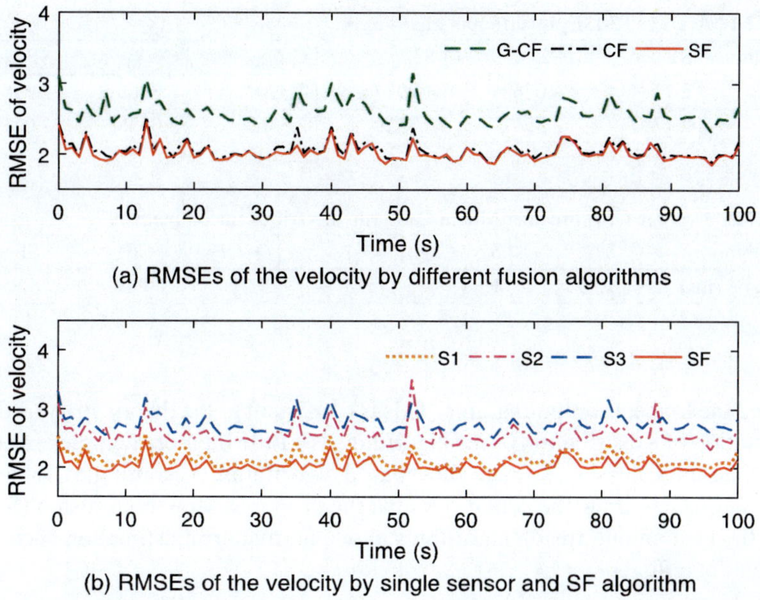

Fig. 12.3 RMSEs of the velocity

In order to evaluate the performance of different fusion algorithms better, Table 12.1 presents the time average RMSEs (RMSE_p for position and RMSE_v for velocity) by single sensor and by different fusion algorithms.

It can be seen from Table 12.1 that the proposed sequential fusion (SF) gets the smallest RMSE of the velocity, and the centralized fusion (CF) performs better in RMSE of the position. The Gaussian Kalman filter based centralized fusion (G–CF) gives perform worst among these fusion algorithms. Comparing the fourth column with the first, one can find that the average position RMSE by Gaussian centralized fusion (G–CF) is larger than that by single sensor S1. Due to the fact that the RMSEs by single sensor are obtained by using the approximate t filter, it is more effective than Gaussian Kalman filter for linear systems with heavy-tailed noises. Therefore, the sensor with the highest accuracy is possible to get better estimation results than the Gaussian Kalman filter based centralized fusion. The CF and the SF, which are the fusion algorithms based on t approximate filter have better performance than all single sensor's.

For further performance evaluation of different fusion algorithms, the CPU time of three fusion algorithms above and three single sensors' are listed in Table 12.2. Combining Tables 12.1 and 12.2, one can find that the Gaussian Kalman filter based centralized fusion (G–CF) has the shortest CPU time but the worst estimation accuracy among different fusion algorithms, followed by the proposed sequential fusion (SF) with a little bit more computing time and nearly the same accuracy as the centralized fusion (CF). The centralized fusion (CF) also can obtain good estimating results

Table 12.1 Average RMSE$_s$ by different algorithms

Algorithms	S1	S2	S3	G-CF	CF	SF
RMSE$_p$	2.7184	4.0762	4.9977	4.3904	2.3565	2.3585
RMSE$_v$	2.1397	2.5567	2.7529	2.5869	2.0628	2.0211

Table 12.2 Average CPU time per Monte Carlo run of various fusion methods

Algorithms	S1	S2	S3	G–CF	CF	SF
CPU time (ms)	1.89	1.88	1.92	5.21	9.49	5.85

but need the longest computing time. This is consistent to the theory analysis that the computation of t distribution based algorithms is a little bit more complex in computation than the Gaussian Kalman filter based algorithms. According to Tables 12.1 and 12.2, one can draw the conclusion that the proposed sequential fusion (SF) performs the best among fusion algorithms above no matter in estimation accuracy or computation efficiency.

12.5 Conclusion

The sequential fusion algorithm for linear multisensor dynamic systems disturbed by heavy–tailed process noise and measurement noises is derived. Theoretical analysis and simulation results draw following conclusions: for the linear multisensor dynamic systems with heavy–tailed noises, (1) the proposed sequential fusion algorithm is effective and is superior to the classical Gaussian Kalman filter based optimal centralized batch fusion; (2) the t distribution based sequential fusion algorithm and the centralized batch fusion algorithm are inequivalent, and any of them could get better estimation; (3) the traditional optimal sequential fusion algorithm that on the base of classical Kalman filter is a special case of the proposed algorithm. Thus, the proposed algorithm is valuable in many real applications, such as aerospace, detection, control systems, the surveillance and automation.

References

1. Agamennoni, G., J.I. Nieto, and E.M. Nebot. 2012. Approximate inference in state-space models with heavy-tailed noise. *IEEE Transactions on Signal Processing* 60 (10): 5024–5037.
2. Bar-Shalom, Yaakov, X. Rong Li, and Thiagalingam Kirubarajan. 2001. *Estimation with Applications to Tracking and Navigation: Theory Algorithm and Software*. Wiley-Interscience.
3. DeGroot, Morris H. 2006. *Optimal Statistical Decisions*. Beijing: Publishing House of Tsinghua University.

4. Huang, Yulong, Yonggang Zhang, Ning Li, and J. Chambers. 2016. Robust student's t based nonlinear filter and smoother. *IEEE Transactions on Aerospace and Electronic Systems* 52 (5): 2586–2596.
5. Kailath, T., A.H. Sayed, and B. Hassibi. 2000. *Linear Estimation*. Upper Saddle River, NJ, USA: Prentice-Hall.
6. Koop, Gary. 2003. *Bayesian Econometrics*. Wiley-Interscience.
7. Lin, Honglei, and Shuli Sun. 2018. Optimal sequential fusion estimation with stochastic parameter perturbations, fading measurements, and correlated noises. *IEEE Transactions on Signal Processing* 66 (13): 3571–3583.
8. Ristic, Branko, Sanjeev Arulampalam, and Neil Gordon. 2004. *Beyond the Kalman Filter*. Artech House.
9. Roth, Michael. 2013. *On the Multivariate t Distribution*. Technical Report, Division of automatic control, Linköping university.
10. Roth, Michael, Emre Ožkan, and Fredrik Gustafsson. 2013. A student's filter for heavy tailed process and measurement noise. In *International Conference on Acoustics, Speech and Signal Processing (ICASSP)*, pages 5770–5774.
11. Yan, Liping, Yuanqing Xia, Baosheng Liu, and Mengyin Fu. 2015. *Multisensor Optimal Estimation Theory and its Application*. Beijing: The Science Publishing House.